open sources 2.0

open sources 2.0

Edited by Chris DiBona • Danese Cooper • Mark Stone

THE CONTINUING
EVOLUTION

O'REILLY®

Beijing • Cambridge • Köln • London • Paris • Sebastopol • Taipei • Tokyo

Open Sources 2.0: The Continuing Evolution

Edited by Chris DiBona, Danese Cooper, and Mark Stone

Published by O'Reilly Media, Inc., 1005 Gravenstein Highway North, Sebastopol, CA 95472.

O'Reilly books may be purchased for educational, business, or sales promotional use. Online editions are also available for most titles (*safari.oreilly.com*). For more information, contact our corporate/institutional sales department: (800) 998-9938 or *corporate@oreilly.com*.

Executive Editor:	Mike Hendrickson
Production Editor:	Jamie Peppard
Cover Designer:	Mike Kohnke
Interior Designer:	Mike Kohnke

Printing History:

October 2005:	First Edition.

RepKover™ This book uses RepKover™, a durable and flexible lay-flat binding.

ISBN: 0-596-00802-3

[M]

Table of Contents

Foreword: Source Is Everything

The software industry has always been caught between two perspectives: one anchored in supply, the other in demand. To the market's supply side (the vendors), commodities and "commoditization" have always been threats. To the demand side (the customers), commodities have always been useful.

The latter view is winning, thanks to open source. And we're only beginning to discover how much larger the market will be, now that it's filling with useful open source commodities. These commodities, in most cases, have little or no sale value, but are useful for building countless other businesses. The combined revenues of those businesses will far exceed revenues of companies that make their money selling software.

Use value precedes sale value in every market category. Think about it. Agriculture started with gardening. Textiles started with weaving and knitting. Meat packing started with herding. Construction started with hut building.

What did the software industry start with? In a word, programming. As Eric S. Raymond said in the first chapter of the original *Open Sources*, "In the beginning were the Real Programmers." To Raymond, real programming was both legacy and destiny—a source that began with "guys in polyester shirts, writing in machine and assembler and FORTRAN," and ran through Unix programming and the free software movement, to arrive at "Linux development and mainstreaming of the Internet."

That latter phrase captured where the industry was when *Open Sources* was published in January 1999. The Internet, Raymond noted, "has even brought hacker culture to the beginnings of mainstream respectability and political clout."

A half-decade later, open source has grown far beyond the mainstream. It has become the bedrock over which the mainstream flows. Today it is hard to find a Fortune 500 company with an IT infrastructure that does not depend, in some fundamental way, on open source software.

The Internet mainstreamed programming by putting every programmer zero distance from every other programmer in the world. Supported by this extreme convenience the demand side began supplying itself in a global way. "Real Programmers" were back in power. This time, however, real programmers were a legion, not a mere handful, and the tools they needed could be found on any PC, not just on the infrastructure inside large organizations.

Today power comes from everybody who creates anything that's useful to anybody—and that any other programmer can improve.

Today there are hundreds of thousands of hackers, perhaps millions. Whatever the number, many more will soon arrive from Asia, South America, Africa, and other formerly Net-less places all over the world.

The technology sector's industrial age—the one in which manufacturers built platforms for silos in which customers and users were trapped—will, in retrospect, appear to be an early growing stage: a necessary but temporary step toward a healthy and mature marketplace.

Let's give credit where it's due: developing software and hardware in the early days of the computing industry was like settling Mars. Every company had to build its own airtight habitat, from the ground up. Making hardware and compatible software inside your own habitat was hard enough. Trying to become interoperable with anybody else's environment was nearly unthinkable. Even within large companies like IBM, whole systems were incompatible with whole other systems. Remember Systems 34, 36, and 38, which used twin-ax cabling while IBM mainframes used co-ax?

All these closed habitats naturally fell in to fighting. The press supported the battling-vendor view of the marketplace, which became a form of entertainment, dramatized via continuous coverage of the vendor wars.

The Net obviates the need to build closed habitats. You don't need to make your own bedrock anymore. Open source commodities provide all the base infrastructural building materials you need, and then some.

Naturally, the old supply side felt threatened by that. For a while.

Then the demand side (the programmers) *inside* those silos began using open source to build solutions to all kinds of problems. Today IBM, Novell, HP, Sun—nearly every big platform and silo company other than Microsoft—have shifted their product strategies to take advantage of abundant open source commodities in the marketplace. They also contribute, in most cases, to the development projects that continue to produce those commodities. While the decisions to "go open source" were made at the tops of those companies, in every case they also involved ratification of decisions already made by the companies' own engineers.

You can still build platforms today, of course, but for practical considerations, it only makes sense to build them on top of open source infrastructure. Amazon and Google are familiar platform businesses (one for retailing, the other for advertising), built on cheap or free open source building materials.

The portfolio of open source building materials now runs to 100,000 or more products, each the project of a hacker or community of hackers, each producing goods, the primary purpose of which is to be useful, not just to sell. Because that software is useful, and most of it doesn't have a proprietary agenda, whole market categories can be opened where once only proprietary platforms and silos grew.

Take the private branch exchange (PBX) telephone business. In the old days, which are now starting to end, companies had to choose corporate phone systems from Toshiba, Panasonic, NEC, Nortel, and other manufacturers of closed proprietary platforms and silos. Then a small device maker, Digium, released Asterisk, an open source PBX. In addition to a vigorous development community, Asterisk attracted countless varieties of businesses made possible by its wide-open use value. In the long run, far more money will surely be made *because* of Asterisk than could ever have been made *by selling* Asterisk—or even by selling the proprietary PBXes that Asterisk now obsoletes.

So, thanks to open source, the software market is finally growing up. It is becoming mature. Its healthy new ecosystem is made possible by countless commodities, growing more numerous every day.

There is an important difference, however, between open source commodities and those derived from raw materials (like wood or steel) that is harvested or mined. It's a difference that will make the new, mature, software marketplace incalculably large.

The difference is this: open source commodities are produced by creative and resourceful human minds. Not by geology, biology, and botany. This means there is neither a limit to the number of open source products, nor a limit to the number of improvements.

Yet every one of those open source projects is concerned mostly with the improvement of their own products. While they care about how those products interoperate with other products, they can't begin to account for all the combined possibilities where interoperation is required. That means there is room for businesses to test, certify, and support *combinations* of open source products.

That's what attracted me into the vast and growing new marketplace opened by a growing abundance of open source building materials. Like many industry veterans, I didn't see that opportunity until I moved my point of view from the supply side that felt threatened to the demand side that felt empowered.

And I'm hardly alone.

Some will say we're at the beginning of another boom—or worse, another bubble. Those views are both limited and misleading. Open source has changed the world of software into one in which raw materials are literally limitless. Every mature industry—such as construction, automobiles, computing hardware—has experienced hyper-accelerated growth resulting from commoditization of its core building blocks. But the impact on the software industry has the potential to be far more profound, because software is so malleable, so easily shared now, and so increasingly ubiquitous in everyday life.

So, while open source software, and the commoditization it brings to the software industry, may seem a threat today, it is actually ushering in a wave of enormous innovation and productivity—the impact of which has already reached far beyond this industry.

It comes down to one simple truth: we humans naturally desire to improve our own world through building useful tools. In sharing those tools, we've learned that the world around us gets better too. So the idea of open source is as old as civilization itself. And our very modern industry is finally realizing the power contained in this simple fact: the first source for everything we make is ourselves.

—KIM POLESE, CEO, SPIKESOURCE

Acknowledgments

Chris DiBona: Like its predecessor, the publication of *Open Sources 2.0* represents the work of dozens of people both inside and outside of O'Reilly. It was my pleasure and privilege to work with Mark Stone again, and Danese Cooper was invaluable to the creation of the book. The inspiration behind the international section was hers, and she should be called out for that. In the first *Open Sources* I noted that Mark had said that "a book could be written about how this book was written." And while this one's creation was hardly as dramatic, it was no less challenging.

Special thanks to Mike Hendrickson, our O'Reilly editor, who made it almost too easy. Thanks to Tim, Rael, and Nat for the books, conferences and knowledge that your company has crammed into my brain. Keep up the great work! Additionally, thanks to the folks at Google who allowed me the spare cycles to produce the book—so, thanks Bill Coughran! I'd also like to extend my love to Denise and Neil Kruse, my parents Bennie and Cynthia, and especially my sister Trish—we miss you.

Finally, in the last *Open Sources*, I dedicated it to my patient girlfriend Christine, and now I dedicate this book and all my life's works to my wife Christine and our daughter Frannie. I love you more than I can say.

Danese Cooper: Thanks to my family—Joey, Adi, Zoe & Marie—who have put up with lots of absences as I've traveled the world meeting open source people. Thanks also to my friends, especially Brian Behlendorf and Tim O'Reilly, and to all essay writers for agreeing that there should be an update on *Open Sources*. Thanks to my employers at Sun, and now Intel, for giving me space to work on the book, and to my colleagues at the Open Source Initiative for including me in the work. And finally, thanks to my co-editors, Chris and Mark, who had all the experience from creating the first book and generously shared it.

Mark Stone: This book is dedicated to my wife Karen and my son Alex; may your future always be open. Looking at the list of contributors here, I realize the three of us really do stand on the shoulders of giants and are privileged to facilitate what they have created. Several people at O'Reilly deserve special praise: Tim O'Reilly, for having the vision to recognize that the time was right for this book; Mike Hendrickson, who waited so patiently for the final manuscript; and Jamie Peppard, Marlowe Shaeffer, Audrey Doyle, and Rob Romano, who had the difficult production task of turning a wide range of formats and styles into a unified whole. My co-editors, Chris and Danese, have been invaluable and inspiring colleagues throughout this whole process. Finally, I'd like to thank xeno42, elcoronel, and beret for the example they set and the education they've given me; you guys live the ideal every day that the rest of us can only talk about.

List of Contributors

Danese Cooper has a 15-year history in the software industry, and has long been an advocate for transparent development methodologies. Danese worked for six years at Sun Microsystems, Inc. on the inception and growth of the various open source projects sponsored by Sun (including OpenOffice.org, java.net, and blogs.sun.com). She was Sun's chief open source evangelist and founded Sun's Open Source Programs Office. She has unique experience implementing open source projects from within a large proprietary company. She joined the Open Source Initiative (OSI) board in December 2001 and currently serves as secretary and treasurer. As of March 2005, Danese is with Intel to advise on open source projects, investment, and support. She speaks internationally on open source and licensing issues.

Chris DiBona is the open source programs manager for Mountain View, California-based Google, Inc. Before joining Google, Chris was an editor/author for the popular online web site, Slashdot. He is an internationally known advocate of open source software and related methodologies. Along with Mark Stone and Sam Ockman, he edited the original *Open Sources*. He writes for many publications and speaks internationally on software development and digital rights issues. His home page and blog can be found at *http://dibona.com*.

Mark Stone has made a career of studying collaborative communities. As a university professor with a Ph.D. in philosophy of science, he has studied and published on the disruptive community conditions that create scientific revolutions. More recent work has involved the open source community, as editor for Morgan Kaufmann Publishers covering operating systems and web technology, then as executive editor for open source topics at O'Reilly, and as the editor-in-chief of the *Journal of Linux Technology*. While at O'Reilly he co-edited, with Chris DiBona and Sam Ockman, the seminal *Open Sources* in 1999. For the last six years he has worked with various dot-coms on tools and practices for collaboration and online community building, including as part of the executive team managing top-tier technology sites such as Slashdot (3.5 million page views per day served), and SourceForge.net (1 million registered users). As director of product development for ManyOne Networks, he is currently working on the next evolution of online community, leveraging 3D environments and new tools for knowledge management. Mark holds a Ph.D. in philosophy of science from the University of Rochester, and earned his B.A. in philosophy from the University of Maryland. Mark can be reached at *mark.stone@gmail.com*.

Robert Adkins is cofounder of Technetra, a Silicon Valley software company which implements and deploys large-scale software projects specializing in open source solutions. Robert has more than 20 years of experience in the information technology industry, having led products and services groups at Apple Computer, IBM, BBN Communications, and Litton/PRC. He has an M.S. in computer science from Johns Hopkins University. He has published in many technology magazines and journals including *Linux Journal*, *LINUX For You*, the *Journal of the ACM,* and *Government Computer News* and speaks frequently at international technology events. Robert can be reached at *radkins@technetra.com*.

Jeremy Allison is one of the lead developers on the Samba Team, a group of programmers developing an open source Windows-compatible file and print server product for Unix systems. Developed over the Internet in a distributed manner similar to the Linux operating system, Samba is used by multinational corporations and educational establishments worldwide. Jeremy handles the coordination of Samba development efforts worldwide and acts as a corporate liaison to companies using the Samba code commercially. He works for Novell, which funds him to work full time on improving Samba and solving the problems of Windows and Linux interoperability.

Matt Asay has been involved with open source since 1999, and has made a fetish of understanding novel ways to monetize open source software. To this end, Matt founded the Open Source Business Conference as a place to aggregate and cluster people much more intelligent than he to figure out promising open source business strategies; cofounded Novell's Linux Business Office and helped to kick-start the company's growing Linux business; served as an entrepreneur-in-residence at Thomas Weisel Venture Partners, dedicated to finding and developing open source investment opportunities; and ran embedded Linux startup Lineo, a network and communications business, until its acquisition by Motorola in 2002. Matt speaks and publishes frequently on open source business strategy, and consults frequently for several open source startups and venture capital firms.

Matt is currently the general manager at Volantis Systems, where he manages the company's growing business with content providers (like eBay, Disney, and Yahoo!). He is applying the lessons of open source to the fragmented mobile world, hoping it will yield the same standardization and opportunity in mobile/embedded that open source did for the server world.

Matt holds a J.D. from Stanford, where he worked with Professor Larry Lessig on analyzing the GPL and other open source licenses.

Mitchell Baker has been the general manager of the Mozilla project (officially known as its Chief Lizard Wrangler) since 1999. The Mozilla project strives to create great software and maintain choice and innovation in key Internet client applications, such as its flagship Mozilla Firefox and Mozilla Thunderbird products. It is one of the largest open source software development projects in existence. The Mozilla project combines dedicated volunteers, a set of paid contributors, and its own flavor of engineering management.

With the formation of the Mozilla Foundation in 2003, Mitchell also took on the role of president of the Mozilla Foundation. Mitchell is also a board member of the Open Source Applications Foundation, which is developing a new-style personal information manager, known as Chandler.

Jeff Bates brings many years of strategic management and editorial leadership to the Open Source Technology Group (OSTG). As vice president of editorial operations and executive editor of Slashdot, Jeff is responsible for setting strategy and integration for the company's business development partnerships and for driving new site and product development, and for fun he helps manage strategic story editing and placement for the leading proprietary news site, Slashdot. While at Slashdot, Jeff has been responsible for the site winning several industry awards including a Webby People's Voice Award for Community, as well as Yahoo!'s "Top 100" Best of the Internet Award. Slashdot has also been cited by *The Washington Post, Brill's Content, TIME, USA Today, Rolling Stone*, and other industry-leading publications as one of the most innovative and important sites for the technical community.

Jeff has spoken at numerous academic institutions and industry-leading conferences and events, including MIT, LinuxWorld, Worcester Polytechnic Institute, Northern Michigan University, Sun Developers Group, the Asian Open Source Symposium, Conference of Australian Linux Users, O'Reilly's p2p Conference, and the University of Michigan. He's also a member of the Open Source Advisory Panel for the U.S. government. Jeff holds a bachelor's degree in history from Hope College.

Jesus M. Gonzalez-Barahona teaches and conducts research at the Universidad Rey Juan Carlos, Mostoles (Spain). He started to work in the promotion of libre software in 1991. Since then, he has carried on several activities in this area, including organizing seminars and courses and participating in working groups on libre software, in Spain and throughout the rest of Europe. Currently he collaborates with several libre software projects (including Debian) and associations, writes in several media about topics related to libre software, and consults for companies and public administrations on issues related to their strategy on these topics. His research interests include libre software engineering, and in particular, quantitative measures of libre software development and distributed tools for collaboration in libre software projects. In this area, he has published several papers, and is participating in some international research projects (visit *http://libresoft.urjc.es* for more information). He is also one of the promoters of the idea of a European masters program on libre software, and has specific interest in education in that area. On the personal side, he enjoys living, sleeping, and staying with his family (and not in that order).

Andrew Hessel is a biologist and programmer who has worked at the interface of industry and academia to facilitate scientific initiatives, usually in the area of genomics. He is fascinated by the functional similarities between electronic and biological systems, and the lessons that can be learned by comparing them. Andrew lives in Toronto, Canada, with his wife Stephanie, and works to advance collaborative breast cancer research and therapeutic development.

Pamela Jones is the founder of Groklaw (*http://www.groklaw.net*), an experiment in applying open source principles to the field of legal research. Groklaw is also an independent journalistic voice, covering legal news stories from the point of view of the Free and Open Source (FOSS) community. Groklaw is also an anti-FUD web site. It has focused heavily on the SCO litigation, because the community is, after all, while not a direct party to any of the lawsuits, directly interested in and affected by the outcome, since it is their code and their community that is under attack. For that reason, Pamela found it is both natural and appropriate that Groklaw try to contribute to a positive outcome.

Eugene Kim is the cofounder and principal of Blue Oxen Associates, a think tank and consultancy focused on improving collaboration. He has developed collaborative strategies for a number of organizations, focusing especially on interorganizational collaboration and collaborative learning. His research centers around identifying patterns of collaboration across different domains (with a special focus on open source communities) and on improving the interoperability of collaborative tools. Previously, Eugene worked closely with computer pioneer Doug Engelbart, who currently serves on the Blue Oxen Associates advisory board. He received his A.B. in history and science from Harvard University.

Ben Laurie is a founding director of the Apache Software Foundation, a founder and core team member of OpenSSL, the author of Apache-SSL, director of security for The Bunker Secure Hosting Ltd.1, coauthor of *Apache: The Definitive Guide,* and a frequent writer of articles and papers on security, cryptography, and anonymity. You can find his web page at *http://www.apache-ssl.org/ben.html.*

Louisa Liu is business development manager of the Channel Software Operation (CSO) in Intel China Ltd. She is responsible for strategic business development in China supporting the CSO. Louisa earned bachelor and master's degrees with honors in computer science from Fudan University and Tongji University.

Ian Murdock is cofounder, chairman, and chief strategist for Progeny. He is centrally involved in defining Progeny's technology and business strategies, and in establishing and maintaining key relationships with customers and partners. Ian has more than 10 years of experience in the software industry. He played an instrumental role in the transition of Linux from hobby project to mainstream technology by creating Debian, one of the first Linux-based operating systems, called distributions. Ian led Debian from its inception in 1993 to 1996, building it from an idea to a worldwide organization of more than 100 people in less than three years.

Today Debian is one of the most popular Linux platforms in the world, with millions of users worldwide. Debian is also widely considered one of the most successful and influential open source projects ever launched: more than 1,000 volunteers in all parts of the world are currently involved in Debian development, and the founding document of the open source movement itself was originally a Debian position statement.

An Indiana native, Ian holds a B.S. in computer science from Purdue University and was a founding director of Linux International and the Open Source Initiative.

Russ Nelson is a computer programmer and a founding board member of the Open Source Initiative. He is best known for his packet driver collection, begun while at Clarkson University in 1988. He started Crynwr Software to support his open source software, Freemacs (currently used by FreeDOS) and Painter's Apprentice (a MacPaint clone), and went full time with the packet driver collection in 1991. He has been making a living from open source support ever since then. His politics are both left and right of center, as he is a pacifist Quaker and a member of the Libertarian Party of the United States.

Michael Olson is president and chief executive officer of Sleepycat Software. Michael, one of the original authors of Berkeley DB, is a technology industry veteran with more than 20 years of experience in engineering, marketing, sales, and business management. He was named president and CEO of Sleepycat in 2001 after serving as vice president of sales and marketing. Prior to Sleepycat, he served in technical and business management positions at database vendors Britton Lee, Illustra, and Informix. He holds B.A. and M.A. degrees in computer science from the University of California at Berkeley.

Tim O'Reilly is founder and CEO of O'Reilly Media, thought by many to be the best computer-book publisher in the world. In addition to publishing pioneering books such as Ed Krol's *The Whole Internet User's Guide & Catalog* (selected by the New York

Public Library as one of the most significant books of the 20th century), O'Reilly Media has also been a pioneer in the popularization of the Internet. O'Reilly's Global Network Navigator site (GNN, which was sold to America Online in September 1995) was the first web portal and the first true commercial site on the World Wide Web.

O'Reilly Media continues to pioneer new content developments on the Web via its O'Reilly Network affiliate, which also manages sites such as Perl.com and XML.com. O'Reilly's conference arm hosts the popular Perl Conference, the Open Source Software Convention, and the O'Reilly Emerging Technology Conference.

Tim has been an activist for Internet standards and for open source software. He has led successful public relations campaigns on behalf of key Internet technologies, helping to block Microsoft's 1996 limits on TCP/IP in NT Workstation, organizing the "summit" of key free software leaders where the term *open source* was first widely agreed upon, and, more recently, organizing a series of protests against frivolous software patents. Tim received Infoworld's Industry Achievement Award in 1998 for his advocacy on behalf of the open source community.

Tim graduated from Harvard College in 1975 with a B.A. cum laude in classics. His honors thesis explored the tension between mysticism and logic in Plato's dialogs.

Gregorio Robles is a teaching assistant and a Ph.D. candidate at the Universidad Rey Juan Carlos in Madrid, Spain. His research work is centered in the empirical study of libre software development from a software engineering point of view. He has authored or coauthored many papers that were presented at both academic and community conferences, and has developed or collaborated in the design of programs to automate the analysis of libre software. He has also been involved in the seminal European Union FLOSS study and survey on libre software developers, the CALIBRE coordinated action to foster libre software development in Europe, and the FLOSS-World study which looks at libre software development worldwide, all of them financed by the European Commissions IST program.

Larry Sanger was the chief organizer/architect of the Wikipedia encyclopedia project in its first year, as well as of the now-moribund Nupedia encyclopedia project. Since 2000 he has thought and written about the best ways to develop a collaboratively built online encyclopedia. He is now working on that problem, among others, for the ambitious Digital Universe project as its director of distributed content programs. His Ph.D. (2000) from Ohio State University is in philosophy, with concentrations in epistemology and early modern philosophy, and his B.A. in philosophy is from Reed College in Portland, Oregon. He taught a wide range of philosophy courses off and on between 1992 and 2005 for Ohio State University and nearby institutions. He also plays Irish traditional music on the fiddle and has taught that too, off and on since 1997.

Sunil Saxena is senior principal architect in the Software and Solutions Group (SSG) at Intel Corporation. SSG is responsible for operating system enabling on Intel architecture products. Sunil received his Ph.D. in computer science from the University of Waterloo and received his B. Tech. in electrical engineering from Indian Institute of Technology, Delhi, in 1975.

Doc Searls is a writer and speaker on topics that arise where technology and business meet. He is the senior editor of *Linux Journal*, the premier Linux monthly and one of the world's leading technology magazines. He also runs the new Doc Searls IT Garage, an online journal published by *Linux Journal*'s parent company, SSC. He is coauthor of *The Cluetrain Manifesto: The End of Business as Usual*, a *New York Times, Wall Street Journal, Business Week*, Borders Books, and Amazon.com bestseller. (It was Amazon's #1 sales and marketing bestseller for 13 months and sells around the world in nine languages.) He also writes the Doc Searls weblog. J.D. Lasica of Annenberg's *Online Journalism Review* calls Doc "one of the deep thinkers in the blog movement." Doc's blog is consistently listed among the top few blogs, out of millions, by Technorati, Blogstreet, and others.

Wendy Seltzer is an attorney and special projects coordinator with the Electronic Frontier Foundation, where she specializes in intellectual property and free speech issues. In the fall of 2005, she will be at Brooklyn Law School as a visiting professor of law. As a fellow with Harvard's Berkman Center for Internet & Society, Wendy founded and leads the Chilling Effects Clearinghouse, helping Internet users to understand their rights in response to cease-and-desist threats. Prior to joining EFF, Wendy taught Internet law as an adjunct professor at St. John's University School of Law and practiced intellectual property and technology litigation with Kramer Levin in New York. Wendy speaks frequently on copyright, trademark, open source, and the public interest online. She has an A.B. from Harvard College and a J.D. from Harvard Law School, and occasionally takes a break from legal code to program (Perl).

Sonali K. Shah is an assistant professor at the University of Illinois at Urbana-Champaign. Her research focuses on the creation and maintenance of novel organizing innovation communities that support innovation development and diffusion. She studies innovation communities in fields as diverse as open source software, sports equipment, and medical products. A second stream of work examines the processes underlying the formation of new industries and product markets. Previously she worked at Morgan Stanley & Co. and McKinsey & Co. She holds degrees in biomedical engineering, finance, and management. She is a graduate of the University of Pennsylvania and the Massachusetts Institute of Technology.

Alolita Sharma is cofounder and CEO of Technetra, a Silicon Valley software company which implements and deploys large-scale software projects specializing in open source solutions. Alolita has more than 14 years of experience in the information technology industry, having engineered and led services groups at IBM, MCI

Worldcom, Intelsat, and SWIFT. While pursuing the Ph.D. program in computer science at George Washington University (GWU), she concentrated on networking, security, and parallel computing. She received an M.S. in computer science also from GWU. She speaks at technology forums and has published in many technology magazines and journals including *Linux Journal*, *Linux Gazette*, and is a monthly columnist for India's only open source magazine, *LINUX For You*. She is a proponent of Linux and open source software in India. Alolita can be reached at *as@technetra.com*.

Bruno Souza is a senior consultant at Summa Technologies. He helps large companies to successfully use and develop open source products and projects. Bruno is president of SouJava, Brazil's largest Java User Group, where he has led the group's Javali Project, an ambitious umbrella project that hosts 10 large open source projects. Javali, which includes a project to create an open source Java runtime, is targeted to bring software development into Brazil's open source discussions. Bruno also co-authored SouJava's Open Source Manifest, which discusses open source and open standards as the way to correctly apply and succeed with open source in Brazil. The document positively influenced the adoption of open source in Brazil. Bruno is a member of the Management Board of java.net, one of the largest open source hosting sites for Java developers, where he leads the World Wide Java User Group Community, is an activist for the creation of open source–compatible implementation of Java standards, and is an active participant in several Java open source projects.

Stephen R. Walli has worked in the IT industry since 1980 as both customer and vendor. He is presently the vice president of Open Source Development Strategy for Optaros. Stephen is responsible for architecting and managing Optaros's relationships with the open source community. Most recently, Stephen was a business development manager at Microsoft on the Windows Platform team, where he operated in the space between community development, standards, and intellectual property concerns. While at Microsoft, he also worked on the Rotor project (Shared Source CLR), and started as the product unit manager for Interix in Services for Unix.

Prior to Microsoft, Stephen was the vice president of R&D and a founder at Softway Systems, Inc., a venture-backed startup that developed the Interix environment to re-host Unix applications on Windows NT. Stephen has also worked as an independent consultant for X/Open, SunSoft, UNISYS, and the Canadian government. He was once a development manager at Mortice Kern Systems, and a systems analyst at Electronic Data Systems.

Stephen was a longtime participant and officer at the IEEE and ISO POSIX standards groups, representing both USENIX and EurOpen (E.U.U.G.), and has been a regular speaker and writer on open systems standards since 1991.

He blogs at *http://stephesblog.blogs.com*, and occasionally podcasts from *http://stephenrwalli.users.blogmatrix.com/podcasts*.

Steven Weber, a specialist in international relations, is an associate with the Berkeley Roundtable on the International Economy (BRIE) and the International Computer Science Institute, and affiliated faculty of the Energy and Resources Group. His areas of special interest include international politics, and the political economy of knowledge-intensive industries.

Steven went to medical school at Stanford and then earned his Ph.D. in the political science department at Stanford. In 1992, he served as special consultant to the president of the European Bank for Reconstruction and Development in London. He has held academic fellowships with the Council on Foreign Relations and the Center for Advanced Study in the Behavioral Sciences. He is a member of the Global Business Network in Emeryville, California, and actively consults with government agencies on foreign policy issues, risk analysis, strategy, and forecasting.

Boon-Lock Yeo is currently director of the ICSC (Intel China Software Center) of the SSG (Software and Solutions Group) in Intel China Ltd. He received his Ph.D. in electrical engineering from Princeton University and a BSEE from Purdue University. He received an IEEE Transactions Best Paper Award in 1996, has published more than 40 technical papers, and holds 25 U.S. patents.

Chris DiBona, Danese Cooper,
and Mark Stone

Introduction

Midnight comes to the Nevada desert, and nothing is visible but a line of taillights ahead and a line of headlights behind.

"You'll see the lights from Gerlach first, and then Black Rock City," the driver says to the passenger. They drive another 20 minutes in silence before the highway crests a ridge line. On the horizon, the blackness is broken by a band of multicolored light.

"Is that Gerlach?" the passenger asks.

"No." The driver points to a small, dim cluster of yellow lights in the middle distance. "That's Gerlach. Way out there, those lights are Black Rock City. We're still about an hour away."

The passenger ponders this for a moment, then asks, "An hour? How big is it?"

"For one week each year, Black Rock City is the fourth largest city in Nevada. Population 30,000, give or take."

Gerlach rolls by, with its one bar, one gas station, and one motel. The caravan of cars bunches up after Gerlach. Then they turn off the highway, rumbling over the packed mud playa of the Black Rock Desert. Lights, neon, and thousands of RVs spread out before them. Music and drums—especially drums—can be heard in the distance. At the gate they're approached by someone who looks like a transplant from Mardi Gras: face paint, bright-colored suit, and a carnival hat. He checks their tickets and flashes them a big grin.

"Welcome to Burning Man!"

* * *

When the original *Open Sources* was published in 1999, it served mainly as an affirmation that open source existed. The book brought together the leading voices in open source, demonstrating that we were a community, that we were indeed a movement to be taken seriously. To put that time in context:

- Microsoft had, only a year earlier, leaked the "Halloween Memo," its first semi-public acknowledgment that open source was a competitive threat.

- IBM had provided some initial backing for Apache, but had yet to announce its $1 billion Linux initiative.

- Linux was in only the 2.2 stage of kernel development.

- SourceForge.net was a relatively new site with only a few hundred projects hosted.

The mainstream press could not separate rising interest in open source from dot-com bubble hype. The media took the surface ideas of Eric Raymond's *The Cathedral & the Bazaar* and created a caricature of legions of hobbyist programmers distributed across the globe, competing against the technology Goliaths of the day. That picture bore no more resemblance to reality than the tale of King Arthur does to the historical Middle Ages. Yet like any good mythology, it serves as a useful point of departure for understanding the real history from which it arose.

Today open source is an accepted fact of business life, with many companies engaged in core open source business models (Sleepycat, MySQL) or significant hybrid models blending open source and proprietary software (IBM, Novell, Red Hat). Many companies have striven to incorporate open source development into their range of software development practices—even Microsoft has projects hosted on SourceForge.net. How much do we really understand about the dynamics of open source software development or the communities that stand behind those projects?

The essays presented in this volume take a major step forward in our understanding since 1999, when the original *Open Sources* was published.

* * *

Burning Man is approaching its 20th anniversary. Conceived and inspired by Larry Harvey, it began in 1986 as a gathering of dozens of participants at Baker Beach in San Francisco. The centerpiece of the event was then, as it is now, the construction of a wooden effigy of a man, which is burned in celebration.

Celebration of what? The answer to that question may differ for every participant.

The event was originally timed with the summer solstice—it's now held the week leading up to Labor Day—and has always had a pagan, tribal feel to it. Participants began bringing their own art projects, many of which were also burned in celebration at the climax of the event. By 1990 the gathering numbered in the hundreds, and even the unusually tolerant San Francisco police made it clear that the event needed to find another venue.

Burning Man then moved to its current location—the Black Rock Desert—an empty stretch of Nevada desert on federal Bureau of Land Management land, roughly two hours north of Reno. The extreme remoteness and the harsh environment have become an indelible part of the event. To be there, you have to really want to be there.

* * *

Mitchell Baker makes clear in her essay that part of the strength of the Firefox community is its size. Thousands of people have contributed to Firefox, a community of contributors larger than the core project leaders can really envision. Firefox seems very much like one of those mythical "legion of programmers" projects that comes to mind when people think of the metaphor suggested by Eric Raymond's *The Cathedral & the Bazaar*.

Yet the open source development model remains the most enigmatic aspect of the open source community. One striking and unexpected outcome of the years since the original *Open Sources* is how little technology companies have been able to leverage the open source development process. Projects such as Linux and Apache have had world-changing success, yet no commercial software company has been able to replicate this development process for its own products or its own success. AOL, for example, has never figured out how to integrate the Mozilla/Firefox developer community into its product development process. Sun has struggled to open up both Java and Solaris, and the jury is very much out on the success of those projects.

Read the essays here by Chris DiBona and Jeremy Allison and you will see how little proprietary software development differs from open source software development. The differences their essays suggest are subtle: an emphasis on knowledge reuse, not just code reuse; a recognition that open standards matter; and that architecture needs to be created with openness in mind. Why, though, are these and other open source lessons so hard for commercial companies to use?

The real paradox is as old as Fred Brooks' classic, *The Mythical Man-Month*. In this work, Brooks formulates what has become known as Brooks' Law: that while the amount of programming work completed increases linearly as the number of programmers increases, the complexity of a project increases as the square of the number of programmers. The result is that large programming teams fail to reduce the time to project completion. The rationale is that the number of communication interfaces, which is roughly equivalent to the amount of coordination effort the project requires, increases geometrically as more people are added to the project.

Brooks' Law appears to set a fundamental limit on the optimal size of programming teams—and a rather small limit at that. Empirical evidence supports Brooks' Law. For example, since its inception SourceForge.net has maintained very close to a 10:1 ratio of registered users to registered projects, suggesting that open source development projects seldom have more than 10 active developers.

What are we to make, then, of the thousands of Firefox contributors? The key is to recognize that they are not a homogeneous mass of contributors, and Firefox is not a monolithic piece of software. In fact, the design is highly modular, enabling small teams to work on separate components of the code without interfering with each other. By far the most common characteristic open source projects have is a highly modular design. That design architecture is more a choice of social engineering than technical engineering, however. In the original *Open Sources,* Linus Torvalds commented that he started with a monolithic kernel design for Linux because he knew it would provide higher performance. Only later was Linux's distinctive system of loadable kernel modules developed, and it was developed as much out of a project management need as anything else. Could Apache provide higher performance as a purely monolithic piece of code? Probably. But the development process would become unmanageable. The original release of Mozilla suffered terribly because it was a monolithic piece of code. Firefox is the result of years of rearchitecting to achieve a modular, and thus manageable, design architecture.

Once it's clear that programming teams must necessarily be small, and that modularity is driven by communication and management needs more than by engineering needs, the structure of the open source community makes a lot more sense. The "bazaar" looks less like a bustling, homogenous mass and more like a structured community. More than anything, it resembles a tribe.

In fact, there is very little of Brooks' Law that is unique to software development. Any creative, collaborative knowledge enterprise faces the same constraints, and as a result, many collaborative communities adopt the same tribal structure.

* * *

Traditionally Black Rock City is laid out in a half-circle, with a grid of "streets," some of which form the spokes of the wheel, and others of which form concentric rings spreading from the center. At the very center is the Man, 40 feet of elegantly assembled wood, awaiting his night of conflagration at the end of the weeklong event.

A "burner" approaches a small tent encampment at the corner of 7 O'Clock and Justice. Out of the back of a pickup truck, two men from the encampment haul bags of sand and large rocks painted in black-light colors. They place the rocks inside a wooden sandbox and pour sand into it. Above them, a hand-painted sign reads "Reflections in Sand."

"So, what's your project?" asks the passerby.

One of the men looks up, squinting under the relentless desert sun. "We're making a Zen garden," he replies. He gestures toward a black light and generator lying nearby next to a couple of small, handmade wooden rakes. "Only this will be one you can enjoy at night."

"Where are you guys from?"

"San Francisco."

"You drove 350 miles to bring sand to the desert? Cool."

Burning Man is a participatory event. According to its mission statement (*http:// www.burningman.com/whatisburningman/about_burningman/mission.html*):

> Our intention is to generate society that connects each individual to his or her creative powers, to participation in community, to the larger realm of civic life, and to the even greater world of nature that exists beyond society.

While not everyone who comes brings an art project, the expectation is that projects will be interactive in nature and that no one is there simply to observe. All are there to participate. Burning Man has no "audience." While the event is not a political gathering, many projects have a definite message connecting to that "larger realm of civic life," and to encourage this, each year Burning Man has an overall theme. Recent themes have included "The Vault of Heaven," "Beyond Belief," and "The Floating World."

Several other ideas underpin the spirit of Burning Man. Black Rock City LLC has had a delicate relationship with the Bureau of Land Management over the years, striving to show that the event is a positive part of the area. A key part of this is Burning Man's "leave no trace" philosophy. Art projects are not permanent, but are designed to be temporary, something can be removed or destroyed at the end of the event without leaving trash behind. Volunteers spend weeks after each event restoring the Black Rock Desert to its pre-event conditions. The "leave no trace" policy is not just a practical matter, though: it is also a good philosophical fit with the civic aspect of the event and with the fragile sense of the temporary that has been at the heart of the event since the first Burn.

Burning Man is also basically a nocturnal event. At roughly 6,000 feet elevation, the Black Rock Desert air is thin and dry. Daytime temperatures can hover around 100 degrees, so dehydration is the most common condition treated at the infirmary. As the sun goes down each day, people pause at whatever they are doing, look toward the craggy western hills, and cheer. Drums begin to beat. Music begins to play. Black Rock City comes alive.

Temporary, participatory, and nocturnal: those are the key elements of a Burning Man art project.

* * *

As puzzling as *how* open source projects organize themselves is *why*. To the casual outside observer, it appears that open source developers spend enormous amounts of time developing software that, in the end, they are simply going to give away without the prospect of compensation in return. While open source certainly has an altruistic side, altruism is neither the only nor the most important motivation.

First, one must realize that many open source projects started out of a developer's desire to solve an immediate problem. Linux was born of Linus Torvalds' desire to have a development platform on his PC at home. Apache came together from a group of people who had been relying on the NCSC web server and wanted to continue its development after NCSC stopped maintaining it.

Read Sonali Shah's essay, and you'll understand that the pattern here extends far beyond software development. Many consumer communities have come together collaboratively to innovate the products they consume, often when producers fail to produce innovation on their own. Software companies created just such a stagnant environment, out of which Linux and the rest of open source software was born. In this sense, the initial catalyst was quite selfish: developers wanted the software that companies were unwilling or unable to produce, so open source developers created it to "scratch their own itches."

What started from largely selfish motivations has evolved into something quite complex. In Steve Weber's essay, we get a clear analysis of just how complex the governance structures and processes of this community have become, as well as an intriguing view of where these enabling governance structures might foster collaborative communities in other endeavors. Andrew Hessel provides in his essay a very tangible example of where open source ideas are taking hold in an entirely new realm.

Lurking in the background, though, is still a question of motivation. If the inspiration for Linux was a selfish one, why make the choice to give the result away, and furthermore to do so under open source terms? How does selfishness become altruism?

The apparent paradox rests on the assumption that acts of charity necessarily conflict with acts of self-interest. From the point of view of a modern market economy, it often appears that charity and self-interest do conflict. What drives the open source developer, however, is clearly self-interest—even if it is based on an older notion of self-interest not easily captured by modern market economics.

The answer to the paradox lies in the reputation game played within the open source community. After all, monetary compensation is only a means to an end; it is a means of providing survival resources. Yet monetary compensation is not the only means to that end. While open source luminaries such as Linus Torvalds and Brian Behlendorf may not have the personal wealth of fortunate dot-commers, neither will ever lack for gainful employment. They have sufficient reputations based on their open source accomplishments to always be able to earn a living from their expertise.

Consider another fact: the largest age group among open source hackers is college students and graduate students: those under 25 for whom gainful employment is not an immediate issue, but one that certainly looms in their thoughts and plans. Because they are students, we can assume that they have the immediate survival resources needed for one to become a student in the first place. However, we can also assume that those survival resources are finite. Securing future survival resources is very much a part of the agenda of a student, indeed one of the main reasons for becoming a student in the first place. While a degree may provide a measure of that future security, a degree is not the exclusive means to that end: reputation as a creator of good code may provide that future security as well as or better than a degree.

Once this separation is made between monetary compensation and survival resources, we can see that there are historical precedents for this kind of behavior, and that there is a social model that loosely fits that of open source hacker culture. In Western civilization, we can look to medieval Europe, when nomadic groups like the Franks and the Vikings had settled into the agrarian-based feudal system. The mature feudal system of the 13th and 14th centuries has some interesting and instructive social structures; we'll focus on the concept of chivalry.

The good knight adhered to a code of behavior that transcended the laws of any particular kingdom and encouraged an attitude with some similarities to the attitude of today's hacker: a knight should be humble and should regard himself at the service of others, yet he would be judged by his prowess at his trade and would succeed to the extent that he could spread the reputation of his prowess.

Behind shield and visor, and upon adopting a particular set of heraldic emblems, a knight took on a kind of persona, creating a public identity that might be quite different from his private identity. There were regular events for testing and publicizing one's prowess at arms, such as the tournament or the hunt. There were orders of knighthood—again, often transcending the boundaries of kingdoms—that would certify one's prowess. To belong to such an order was an honor; to be of sufficient repute to be able to found an order and have other knights flock to it, was perhaps the greatest measure of success in the chivalric reputation game.

While a knight was part of the nobility, knighthood was a terrible burden financially. A suit of armor cost, relative to the medieval standard of living, the equivalent of a brand-new Mercedes today. Horses were expensive, and a knight was expected to have several. In addition, a knight had to maintain an entourage of squires, pages, and men-at-arms. He also officially owed 40 days of service to his feudal lord each campaign season, 40 days that in reality could often drag into several months. A landed knight with a small manor could easily spend any excess capital just to maintain his position; a landless knight, or knight errant, would likely live in perpetual debt.

What motivation, then, would a young man in the Middle Ages have to aspire to knighthood? Did one live according to the code of chivalry out of pure selflessness, and a desire to serve others? Or was there some more pragmatic force at work? To a knight of repute, money was not really important. His lord, or anyone else interested in retaining his services, would see that his needs were met, that he had survival resources. To flourish, a skilled squire aspiring to knighthood need only hone his skills and act to establish and further his reputation.

This attitude and this kind of behavior seem quite similar to that of the young student who is an aspiring hacker. While academia, and academic computer science, is a reputation game of its own, what is fascinating about computer science is that there is a large body of practitioners that refuse to play this particular reputation game. To this latter group, an education really is just a means to an end, and the end is to develop the skills necessary to create good code. Some may go far enough to pick up a university degree, but many do not, and the degree is clearly secondary to the ability to code. In other words, one can build a reputation without having to acquire the academic pedigree. It is not simply a distaste for academia that fosters this kind of behavior. The hacker who is more interested in picking up skills than in picking up a degree is often the same hacker who is unwilling to be tied down to a steady job, preferring to move from assignment to assignment as a freelancer and consultant. These are the knights errant of the open source movement.

In reality, the medieval knight errant was essentially a mercenary, hardly the noble figure portrayed by Malory or Chaucer. The fact that these soldiers were mercenaries made chivalry no less important. A mercenary captain had to be trustworthy; his word of honor alone was a binding contract. Otherwise, he simply was not employable. These men lived by the chivalric code. What they found was that that code alone assured them survival resources. A skilled and honorable mercenary captain was never without employment and never lacked for resources.

Chivalry and pragmatism are not conflicting goals, but that pragmatism can indeed be served by chivalry. The mercenary captain lacked money, land, and all other tangible resources. He had only one form of collateral: his reputation. That reputation could be maintained only if his behavior was seen to be genuinely honorable. Chivalry, then, was a necessity: it was an essential ingredient in building the only available collateral that could be parlayed into survival resources.

From this analogy, we can learn several lessons about today's hackers. First, the open source gift culture need not be seen as strictly, or even predominantly, altruistic. Pragmatism and altruism are not mutually exclusive. Today's hackers, like the knights errant and mercenaries of old, can and do trade in the coin of reputation as a means of achieving survival resources.

Second, as a culture matures, the pragmatism becomes more apparent. This was true in the Middle Ages. In the high Middle Ages, the era of the crusades, the knight errant made a flamboyant pretext of making a gift of his skills and services; one has to look below the surface to see a rational exchange of skills for resources in such gift acts. By the late Middle Ages, though honor and reputation were still essential, when a nobleman retained the services of a mercenary captain, the transaction was explicitly and without apology for the mutual benefit of nobleman and captain.

We see this same trend among today's hackers. While reputation alone has always provided survival resources for some, the trend to switch the meme about their activity from "Free Software" to "Open Source" reflected a maturing shift from altruistic pretext to honest pragmatism.

Today we see a symbiotic balance between the chivalric open source hackers and the companies that employ them. In fact, this development is foreshadowed in the events Eugene Kim describes in his essay, which concerns the development of the first commercial compilers. Even in the 1950s it was possible for a company such as IBM to see the advantages of transforming a competitive relationship into a collaborative one.

Today a number of prominent open source developers are employed at major technology teams. Novell transformed itself in a matter of months into a major player in the enterprise open source space through the acquisition of Ximian and SuSE. Of contributors to this volume, Jeremy Allison works for HP, and Chris DiBona works for Google. Neither works at an open source company per se, but both have an understanding that it is in their employer's interest that they be allowed time and resources to continue working on open source projects. Other contributors here, such as Ian Murdock (the "ian" of "debian"), Michael Olson, and Stephen Walli, are involved in more purely open source business models.

* * *

Burning Man is, in some sense, a commercial operation. There is a significant admission charge for the event (more than $200), and the event is run by a limited liability corporation. The corporation's main purpose, however, is to sustainably manage the event. There is a permit to obtain from the Bureau of Land Management every year. There is insurance for the event. There is preparation before and cleanup after the event, as well as basic infrastructure, such as sanitation services, that must be provided every year. Finally, there is a small paid staff responsible for everything from event promotion and organization to informal lobbying efforts with the Department of the Interior, Washoe County, and the state of Nevada.

Once inside the gates, however, participants are forbidden from engaging in monetary commerce. The primary form of commerce is barter. In the spirit of the event, barter is as much a pretext for participation as an exchange of goods. It may take the form of a scavenger hunt, where admission to an art project requires a ticket stamped by several other art projects. It may take the form of a raid by the "Viking Longship" art car, which "pillages" camps but always leaves some small gift behind. Or it may be in the form of a quiet bar on a back street of Black Rock City that asks only some small trinket from the day's events as the price of a drink.

Unfettered from monetary exchange, however, most denizens of Burning Man gravitate toward a gift economy. Acts of giving range from the mundane to the extravagant: the accordion player who serenades those in the porta-potty line with his renditions of AC/DC; the massage therapist volunteering her services; the water-gun brigade, spraying people down for a moment of cool relief from the midday sun; or the man who brings along a week's supply of dry ice so he can serve cold ice cream every day.

* * *

One of the most ironic developments since the publication of the original *Open Sources* has been the rapid adoption of open source business models by technology companies.

In 1999, the consensus view of the business community was that giving away intellectual property for free was a poor basis for doing business. At that point, Michael Olson and Sleepycat Software had been quietly pursuing their dual licensing model for Berkeley DB for three years. Now several open source database companies are pursuing similar models.

Sleepycat's approach is an example of the more general business dynamics at work behind open source. One of the key effects is commoditization, discussed in different aspects in essays by Matt Asay, Ian Murdock, and Stephen Walli. Commoditizing a complement to one's core business serves to enhance that business. It brings down the cost of entry for customers, thus expanding the potential market size for the core business. The key is to have a complementary core that can be monetized. Sleepycat achieves this through dual licensing, charging for a proprietary license for customers who are unwilling or unable to open their own source. Novell achieves this through a hybrid business model, with a service business and a proprietary software business further up the "application stack" from its commoditized Linux business.

Commoditization is not the only benefit. Open source business models lower the cost of both sales and marketing. The common fear with any free product is that "you get what you pay for." With open source, however, the source code is entirely open to inspection so that there are no hidden surprises. Further, the source code can be freely redistributed. Several market effects result from this. First, those most likely to avail themselves of open source are those with the greatest understanding of its benefits—namely other software developers who actually have the skill and desire to examine source code. Second, the distribution model creates a user community of like-minded enthusiasts without intervention or incurred marketing costs by the originating company. Finally, that user base will, at some point, approach the originating company with a request for additional features, services, or complementary software. In other words, by its very nature, open source has very low marketing costs that create an inbound sales channel of prequalified leads.

Open source software companies that exploit this dynamic can thus maintain lower overall operating costs, consequently passing on lower prices while still maintaining healthy profit margins. All of this accelerates the commoditization process, making a well-established open source software product quite difficult to compete against.

Yet these very business advantages inherent in open source bring us to another aspect of the same paradox: why is it so difficult for companies to leverage open source as a development model, rather than as a business or marketing model? Consider Sleepycat's dual licensing scheme. The model works only if Sleepycat holds full copyright to all of the software in Berkeley DB. Otherwise, it is not permitted to offer the second, proprietary license in addition to the open source license. If it must have all rights to the software, though, that means that the software must essentially be developed in-house. Sleepycat does its own development instead of leveraging outside, open source development.

Perhaps Russ Nelson offers a purer example of an open source business model, one where he both develops open source software and leverages the open source developments of others. The complimentary values Russ Nelson offers are his reputation and his expertise, both carefully maintained over the years. The resulting business may not be a large one, but it is one where he alone is the master of his own destiny.

* * *

Burning Man certainly has the feel of an organic, grass-roots movement. Certainly that grass-roots element is part of the dynamic that makes the Burning Man community what it is. But simply thinking of "grass roots" makes it too easy to overlook what a complex community structure Burning Man has and requires.

First, there is the structure of Black Rock City itself. Maintaining order in Black Rock City is primarily the responsibility of the Black Rock Rangers. They describe themselves as a "non-confrontational mediating agency" (see *http://www.rangers.org*). They are all volunteers, and they have no official authority. They pay admission like everyone else; rangering is not a way around the admission price. Further, each ranger is required to attend training prior to the event, and each ranger must enter the ranger program with the sponsorship of another ranger as mentor. While the rangers occasionally call in actual law enforcement, for the most part the rangers are accorded tremendous respect.

Black Rock City has many of the elements of any other city of 30,000. There is a radio station, some years several. There is a DMV (Department of Mutant Vehicles); you cannot drive around within Black Rock City or the playa beyond without registering with the DMV. There is an airport; dozens of attendees routinely fly in for

Burning Man. And of course, there is a newspaper, the *Black Rock Gazette*, published six times during each Burning Man event.

Residential areas of Black Rock City have structure as well. Look at a map of Black Rock City when you arrive at Burning Man, and you'll see a number of areas along the inner circle marked as Theme Camps or Villages. These are areas that are both residential and interactive, involving a large number of people working on a common art project where the residential area itself is the art project.

The inner circle faces toward the Man, and beyond on the playa is the Burning Man "gallery" of art installations. Here is where the larger projects are constructed and the larger group events are played out. Typically, there will be an opera and several other stage performances. Weddings are common.

For several years, there was a project called Solaria. It was a scale model of the solar system, where not only the distances between objects were proportional, but also the size of those objects relative to distance was proportional. Each object was a light source, with the sun represented by a small lamp about the size of a bowling ball. On that scale, Pluto could be reached only by a three-mile bike ride across the playa. Not even the Smithsonian can put on an exhibit of that scale.

* * *

No one has grasped the power of commoditization as quickly as developing nations. This is an international arena with general concern over globalization, anxiety about domination by American corporations, and fear of Microsoft's monopoly in particular. Open source has given developing nations a bargaining chip to pressure technology companies, especially Microsoft, on price. The natural response has indeed been a lowering of prices in countries ranging from Brazil to India.

Yet the significance of open source goes far beyond commoditization and price pressure. Read the essays here from Jesus M. Gonzalez-Barahona and Gregorio Robles, Alolita Sharma and Robert Adkins, and Boon-Lock Yeo, Louisa Liu, and Sunil Saxena, and you see that open source is really about controlling one's technology destiny. Outside the United States, people find it odd that we use the same word, *free*, to mean two very different things: *with no cost* or *liberated*. The open source community has adopted the slogans "free as in beer" versus "free as in speech" to draw attention to the difference. Commoditization is all about "free as in beer;" what developing nations care about the most is "free as in speech."

Open source provides greater intellectual property control than proprietary software that one does not own. Developing nations want control over the intellectual property on which their technology infrastructure depends. What emerges is a different sense of ownership from the traditional market economy sense of ownership, one that speaks not just to the motivations of policymakers in developing nations, but to the motivations of open source developers as well. Think again of chivalry and of our feudal heritage.

In a feudal system, a farmer could not own land, nor the harvest from that land. Serfs were indentured to the land and were entitled to only a portion of their harvest after paying their taxes to the feudal overlord and landowner.

Technology workers today face an analogous form of servitude. It is almost universal practice at technology companies to confront new employees with a hiring agreement that says, among other things, that any and all code and inventions created by the employee while in the employ of the company belong to the company; all copyrights and patents resulting from these creations must be transferred to the company. Technology workers may reap the fruits of their creative labor only under terms dictated by the company. Our modern notions of intellectual property and ownership of it are based on this relationship: that it is fundamentally companies, not individuals, that own intellectual property, and that individuals create new intellectual property primarily in the service of companies.

If open source hackers have one common attitude that ties them together into a community, it is the rejection of this notion of intellectual property. The conventional outsiders' view is to say that open source software is not owned. It is fear of the lack of accountability associated with this perceived lack of ownership that makes many companies reluctant to deploy open source software for mission-critical functions.

In fact, this conventional view is deeply mistaken. To understand why, we must make a distinction between "ownership" and "stewardship." Ownership is something that is fully transferable from one owner to another without loss of value. Money, and many (though not all) material objects, are examples of entities that can be simply owned. Stewardship, on the other hand, applies when something undergoes change, when it evolves, or when it has some kind of life cycle. In this case, something has value only if it is cared for in such a way as to sustain the life cycle. In an agrarian society, animals are a prime example of something requiring stewardship. Skills are required, and effort must be put forth, to maintain a herd. Transferring the herd to someone who lacks those skills or is unable to put forth the effort diminishes the value of the herd. In other words, only a good steward can realize the full value of that which is stewarded.

Historically, land has been, and continues to be, at the center of contention between these two notions of ownership. For example, Native Americans considered themselves stewards of the land, and thus fell victim to the European notion of landownership. Today the battle between environmentalists and certain corporations is over exactly these two conflicting senses of ownership. Andrew Hessel's essay points to a brewing conflict in these senses of ownership with the biotech industry.

In the technology sector, open source developers believe that software requires stewardship. The standard employment contract and the proprietary software it engenders preclude stewardship. Open source software, however, by its very nature encourages stewardship. Again, the motivation here is not altruism or charity. To an open source developer, stewarding software is the best way to see that the software evolves and improves, and hence it's just pragmatism to take a stance toward intellectual property that assures that software will be stewarded.

The proof is in the longevity of open source software projects and the stewards who tend them. Linus Torvalds is still at the head of the Linux kernel "tribe" more than a decade after the first public release of Linux. Eric Allman has guided Sendmail for more than 20 years. Larry Wall is still the guiding vision behind Perl, again after more than 20 years. In these and many more cases, a common core group stood behind the software for far longer than most proprietary software enjoys the benefits of a common development team. It is this—the dynamics of stewardship—far more than the "legions of programmers" that accounts for the success of open source software.

Further, it is this dynamic of stewardship that fosters the social network around open source software that is based on the reputation game. Having committed themselves to what they regard as the most pragmatic approach to intellectual property, open source hackers have then adopted the professional and social structure needed to support that approach to intellectual property. It isn't altruism. It's chivalry, a far subtler and more pragmatic thing.

* * *

The Burning Man event lasts for one week each year. Contained within that event is the remarkable community of Black Rock City. Yet it would be a mistake to assume that the community exists for only one week a year. The complex structure and intricate hierarchy that is the Burning Man community could not adapt, evolve, and sustain itself if it were not a year-round phenomenon.

That is the deeper truth not obvious to the casual observer: Burning Man is a permanent worldwide community whose members are connected to and engaged with each other continuously. Art projects that are on exhibit at Burning Man will often be shown at smaller gatherings in places such as San Francisco. Burners gather for regular social events throughout the year to talk about past events and plan for next year. When Washoe County or the Bureau of Land Management considers a change that may affect the permit for Burning Man, word travels through the community like wildfire, and burners show up in force to make their cases and state their views.

It's no accident that the growth of Burning Man parallels the growth of the Internet. The Burning Man web site is an impressive knowledge archive about the event and the community, providing a wealth of information and resources for anyone trying to understand Burning Man and learn how to get involved. The "Jack Rabbit Speaks" mailing list provides an announcement forum that goes out to the whole of the Burning Man community, but there are dozens of other mailing lists that tie together smaller communities within. Some of these are organized by geographical proximity, but many more are organized by common interest. A theme camp will often have a mailing list for its members. The Black Rock Rangers have their own web site, and their own mailing lists.

What the Internet has done is to free us from the constraints of geography in terms of whom we connect to, whom we share common interests with, and whom we form community with. The power of intentional community is abundantly clear in the open source movement, where developers from around the world can collaborate on software of common interest. Yet the larger lesson is that the power of collaboration and the power of community exhibited in open source have relatively little to do with software development.

We begin to see the lasting significance of open source only when we see that it is one instance of a general pattern of online, collaborative community. Even a very physical, tangible event like Burning Man is crucially dependent on this larger sense of community. If we look closely, we see that this pattern of collaboration is beginning to manifest itself in many other places beyond open source.

We see this strikingly in Eugene Kim's essay, contrasting an early example of software collaboration with the grass-roots collaboration that emerged around the Ground Zero cleanup. We see this in the power of consumer-driven innovation that Sonali Shah explores. We see it in the sense of community behind Slashdot, described by Jeff Bates and Mark Stone, and the spontaneous movement that became Groklaw, described by Pamela Jones. And we see a compelling attempt to distill the most general patterns of these communities in Steve Weber's essay.

The simplest elements are these:

- Recognizing that one has common cause with people who might otherwise have competing or divergent interests

- Acknowledging that small teams working on a component of a problem are the only scalable way to tackle large problems

- Improving solutions iteratively through a sense of stewardship, ecosystem, and evolution, rather than a sense of property and ownership

Taken together, these principles suggest an organizational structure that is at once novel and familiar. These intentional communities form hierarchically, but it is a hierarchy based on achievement and reputation rather than power, money, or authority. Communication flows easily up and down the hierarchy, but decision-making flows from the top down. "Everyone gets a voice, but not everyone gets a vote" (see Bates and Stone). The resulting organization is more tribal than democratic.

* * *

Each night when the two burners return to camp in the hours before dawn, "Reflections in Sand" has changed shape. Some new pattern has been carefully raked into the black-lit sand, and though they never see the visitors, the evidence of their passage and participation is there.

The last evening is the night of the Burn. All day the Man has been laid down flat as he is prepped. At sunset he is raised up again. While all week his arms have been down at his side, now they are raised high above his head. This is the signal for the ceremony to begin.

A ring of lights surrounds the Man, and the rangers walk the perimeter to ensure everyone keeps their distance. For hours, within the ring of lights, drummers, firedancers, and musicians perform. Pagan rituals from 1,000 years ago must not have looked so different. When at last the Man ignites, flames shooting 50 feet or more into the night sky, there is awed silence. Pieces begin to fall off, and he begins to tremble, as only guy wires hold him in place. The trembling increases, and at last the whole Man collapses into a burning mound. At that moment the crowd rushes the center in a wild, swirling dance that brings them as close to the flames as heat will permit.

Now when the two 'burners return to their camp, they find the sand has been raked again, into one last new pattern. And something more; there, in the very center of the little Zen garden, someone has left a bottle of water. One of them reaches in to retrieve the bottle and pulls off the lid.

"You going to drink that?" asks the other.

"Of course. Out here, water is the most precious gift of all."

An hour later their camp is packed. The black-light-painted rocks go back with them, the sand has been scattered, and the wooden frame and rakes have been heaped onto their neighborhood burn pile. They drive toward the gate, and the attendant waves them down.

"Heading out?" asks the attendant.

"Yeah, back to San Francisco," replies the driver.

"This your first time here?"

The passenger answers, "It is for me."

The attendant gives a knowing smile and waves them through. "See you next year. Welcome to the tribe!"

SECTION 1

Open Source: Competition and Evolution

In Section 1, we present essays tied directly to the history and development of open source software. These essays can be loosely grouped into three categories:

- Essays on the software development process (Baker, DiBona, Allison, and Laurie)
- Essays on business competition and open source (Olson, Murdock, Asay, Walli, and Nelson)
- Essays on policy issues related to open source (Seltzer; Gonzalez-Barahona; Sharma and Adkins; Yeo, Liu, and Saxena; and Souza)

The essays on the development process provide a natural extension from the original *Open Sources*. These essays explore the community and process that open source developers comprise, and explore the subtle similarities and differences between open source and proprietary development.

With the original publication of *Open Sources* in 1999, the idea of an open source business model was something of a novelty. Today, we see in these essays, that open source, both in its licensing structure and in the commoditizing effect of its distribution model, has become a powerful tool in the hands of businesses large and small.

One critical aspect of the business dynamics behind open source is the desire to avoid vendor lock-in through proprietary software, and to control one's own technology destiny. While these issues matter to businesses, they have become fundamental policy issues in Europe and developing nations. Control of technology resources in the coming decades will likely matter as much as control of natural resources has in the last century. Avoiding monopoly by a single company, or hegemony by a single nation, has become a paramount policy objective. Increasingly, open source is becoming the means of achieving that objective.

CHAPTER 1

Mitchell Baker

The Mozilla Project: Past and Future

The Mozilla project was launched on March 31, 1998. On this date, the source code for the Netscape Communicator product was made publicly available under an open source license, the "Mozilla Organization" was founded to guide the project, and development of the codebase began to move from a proprietary model into an open model coupled with commercial involvement and management practices.

Of these three elements, the release of the source code is discussed in *Open Sources*. In summary, the source code was prepared for public release by removing all code that Netscape didn't have the right to license under an open source license, and then replacing those pieces necessary for the code to compile and run. At the same time, a new open source license—the Mozilla Public License—was written, reviewed, and accepted by the open source community, including the Open Source Initiative (*http://www.opensource.org*). The other two topics—the story of mozilla.org and the development of the Mozilla project—are the subject of this essay. The creation of the Mozilla Public License is generally an untold story, but it occurred during the time covered by the original *Open Sources* book and isn't discussed in detail here.

Each of these three activities was a step into the unknown. Basic development principals of the open source model ("running code speaks," peer review, leadership based on technical merit) were known. But the combination of open source techniques with an active, focused commercial management structure was uncharted territory. The shift of authority from a commercial management structure to a separate organization was new, and presented many management challenges. The development of project management techniques and tools that could be shared by multiple commercial development teams and a volunteer community was new. Development of a large, complex end-user application in the open source space was new.

Of course, the Mozilla project was not the first open source project with commercial involvement. Cygnus, many of the Linux distributors, and Sendmail were all companies involved with open source development, and the Apache project was developing experience in coordinating open source development where some of the contributors were paid by their employers. But none of these projects provided more than a rough set of guidelines for how the Mozilla project might operate. The Mozilla project was unusual, and at the time perhaps unique, in the way project leadership interacted closely with both commercial teams (project managers, people managers, and engineers) and individual contributors.

Not all open source projects are interested in commercial project management and people management issues, but for us it was always a given. Today other projects are thinking about these issues as the development and use of open source software increase. Given our history, size, and scope, the Mozilla project remains a trendsetter in this arena.

Founding of the Mozilla Organization: Obvious for Developers, a Bold Step for Management

The Mozilla project originally grew out of Netscape Communications Corporation and its Netscape Communicator product. In early 1998, the Netscape management team made the decision to continue development of Netscape's flagship product, Netscape Communicator, through an open source development model. At the time, Netscape Communicator and Microsoft's Internet Explorer browser were locked in a fierce competitive battle often referred to as the "browser wars." Netscape's goal was to seed a broad-based development effort within the software development community to produce future browser products as a shared resource.

At its inception, the Mozilla project faced some paradoxes. First, the only people familiar enough with the code to participate actively in its development were Netscape employees. Those employees were still expected to work within the management system and practices that Netscape had developed in its proprietary days. There was no volunteer community. And yet, even at that early time, it was clear that the long-term success of the project required a broad constituency of people and companies working jointly on the project. It was not enough to have open source *code* (code available under an open source license). The project needed an open development *process*, and this required authority over the code's development to be based on technical merit and distributed outside Netscape. The question was how to get there from here.

One thing was clear: the success of the project depended on it being a real open source project. In other words, the project needed to have technical legitimacy and development decisions would need to be guided by technical considerations. This was intuitively clear to the group of Netscape employees who were familiar with open source, eager to help move the Mozilla code into the open source world and who ultimately became the founding members of the Mozilla Organization. This group made the need clear to Netscape management, which was receptive to trying to do the right thing.

When the Mozilla project was officially launched, Netscape executive management therefore took some bold steps. First, they officially anointed "mozilla.org" as the steward of the codebase and leader of the project. I say *officially* because it's quite possible that a group like mozilla.org would have developed even if Netscape hadn't officially helped to create one. But this step was important, as it allowed mozilla.org to focus on building the project rather than on proving the necessity of its role.

The creation of mozilla.org was a significant step that set the tenor of the project's development. It meant that the development of the Mozilla codebase was to be guided by something other than Netscape's own product and revenue plans, and also *that Netscape management would need to give up control.* This may seem like an obvious statement in an open source world, but it is one of the most difficult problems in moving from a proprietary to an open system. It is particularly difficult when the commercial vendor is actively trying to ship a product and the code has not yet reached a good, solid state that can serve as the basis of that product.

Some have said that the Mozilla project was not a true open source project during this time because Netscape employees contributed so much to the project and Netscape management was so involved for so long. It's possible this is true. But I believe that Netscape management lived in an intensely uncomfortable setting as control of the project moved from their hands into those of mozilla.org. And since I personally was the fulcrum for stresses between the project leadership and the Netscape management team, I'll warrant that Netscape management felt it was living in a real open source project. In 1999, Netscape Communications Corp. was acquired by America Online (AOL). This resulted in many changes, but the relationship between mozilla.org and the Netscape browser development group continued as before. For quite a while, we used the term *Netscape/AOL* to describe the Netscape browser development group after the AOL acquisition, and I'll use that phrase for the rest of this chapter.

The members of the Mozilla Organization are known as "mozilla.org staff." The original members of the 1998 launch were Netscape employees who had a vision for an open source Mozilla project and a determination to see it succeed. The most media-genic of these founders was Jamie Zawinski, who left the project after a year. But the most consistent and long-term contributor has been Brendan Eich, who was a founding member of mozilla.org and remains the technical and philosophical leader of the project to this day. Over time, the percentage of mozilla.org staff employed by Netscape decreased steadily. Today, the mozilla.org staff does not have any Netscape/AOL employees.

After the Mozilla project was launched, mozilla.org staff members began the process of changing development styles from a proprietary to an open source model. The early steps were logistical: establish public communications channels such as mailing lists and newsgroups; establish a public system for viewing and tracking bugs. A harder task was changing habits. For example, the existence of public communications channels was not enough. Old habits die hard, and there was a tendency for people to use the methods they had always used. This was complicated by the fact that Netscape as a company still had confidential data about itself and its business partners that couldn't go into public forums. So, it was not possible to eliminate all private channels. Eventually we changed the names of any remaining private mailing lists to something long and awkward that required conscious thought to use. This gave Netscape/AOL employees a way to disseminate confidential data when necessary, but made public disclosure the easiest path.

Even as basic a step as public communications in an open source project can be difficult for some management teams to accept. In a system that is public by default, everyone needs to learn what information must remain confidential, and to remember this while working. At first it's a big effort to work in public and some people see it as overhead. Then as the project progresses and the public interaction provides increasing value, the need to keep something private is seen as a burden. This is certainly the case for the Mozilla project today, where the only private development information we solicit are bugs which could have an impact on the security features of our products. We've set up a system for treating these bugs privately, and the system has overhead. We bear it in the security context because security is critical, but we avoid it in other contexts.

We also set up a public bug and issue-tracking system. This is commonplace today, but was innovative at the time. We made the bug-tracking system an open source project under the Mozilla umbrella, and today Bugzilla is a successful project in its own right (*http://www.bugzilla.org*). We also set up a continuous build system and web frontend (*http://tinderbox.mozilla.org/showbuilds.cgi*). This means that we have an automated process that builds and rebuilds the software continuously on multiple platforms to see if and when a new piece of code causes the software not to build. Then came a period of learning to "work in the fishbowl." Some people adapt easily to having all their work visible, and others struggle. Many simply walk down the hallway to talk with a buddy, and then forget to tell everyone else. This period takes some time and effort.

Updating the Codebase

About six months into the project, it became clear that the codebase was in need of updating. By late 1998, the inherited code was several generations old, had been patched over and over, and actually hindered ongoing innovation. Old and fragile, it looked backward toward the beginning of the Web, rather than forward to the new technologies a modern browser would need to support. So, in late 1998, a painful decision was made to rewrite the layout engine, a critical and complex core component.

This decision changed the scope of the project dramatically. The initial project was an incremental upgrade from the Netscape Communicator 4.*x* product line to a proposed 5.*x* product line. Moving to the new layout engine (known as Gecko) meant that the incremental model was gone; the Mozilla project would need to develop a complex new layout engine and then build a new browser application on top of it. And things got harder from there. As the new layout engine began to mature, it became clear that other significant parts of the codebase would also need rewriting. The development process turned out to be a lot like a remodeling project where fixing one problem leads to another. Then came the long, slow grind to producing something useful (Mozilla 1.0 in June 2002) and finally something great (Mozilla Firefox in November 2004).

During this time, many proclaimed us dead, a failure. What those people didn't see was the passion and commitment of the contributors to the project, including the Netscape employees. The contributors knew that they were developing good technology. They knew they had a shot at building a great browser and mail client. And they knew it mattered. The Web matters. Browsers matter. Much of the world decided that the days of browser innovation were over. Some mourned the loss of choice, and many didn't realize the dangers of accessing the World Wide Web only through a single access point. But the contributors to the Mozilla project realized both the danger and the potential for something innovative. They persevered. The prominence of Netscape often obscured the efforts and dedication of the individuals themselves. Yes, many contributors were paid by Netscape. Of these, many contributed far beyond the requirements of a job, doing extra work to make the product "theirs" and to make sure they were proud of it. Meanwhile, the individual volunteers provided critical expertise and contributions across the codebase.

During this period, almost all the code in the Mozilla browser and email client was rewritten. The focus was a modern layout engine, and a set of technologies designed to make the promise of cross-platform development a reality for the Web. We created a cross-platform component model (XPCOM),a cross-platform XML-based UI language known as XUL (pronounced *zool*), a new toolkit using XUL, and a set of cross-platform applications themselves. Developing these was a long process, but we felt that it was important to have technology that would help us move forward. The power of these technologies has been demonstrated through our new products: Mozilla Firefox and Mozilla Thunderbird, in which we were able to build award-winning cross-platform applications quickly on top of mature, preexisting infrastructure.

A Disciplined Methodology

Along the way, the Mozilla projects developed a highly disciplined method of distributed software development. Many people think that open source development is necessarily chaotic. Or they wonder about the quality of the code because anyone can create a patch and offer it for inclusion in the source base. Open source need not be chaotic, and the Mozilla project is not. For every piece of code checked into the Mozilla products, we track:

- Who checked it in
- When it was checked in (to the minute)
- What problem it was trying to address
- The complete history of the issue (bug) the code was trying to address
- Who did the code review (often two levels of review)
- Whether the next build of the software was broken on any of our main platforms
- Whether the code affected our performance metrics, by platform
- Build and optional log comments
- A comparison with the previous version of the code.

This information is available at any time; it does not require an expert to find or assemble the data. It is available online, in real time, through a web interface; all one needs is a web browser (*http://tinderbox.mozilla.org/showbuilds.cgi*). We do this so that many contributors can work on the same codebase simultaneously and know what's going on. We do this so that the source code "tree" stays healthy, and we address problems before more new code makes them worse. We also have policies determining who gets access to the CVS tree, what's required before code can be checked in, what to do when the tree doesn't compile and run, how authority is delegated to those with expertise, and so on.

Building an Open Source Project

The process of building software proceeded simultaneously with building a vibrant open source project. The creation of mozilla.org had gone hand in hand with the recognition that Netscape management would need to give up a great deal of control over the development process. Now it was time to figure out how to make the transfer. I describe in this section two of the most significant topics we addressed—control over the source code repository and control of the designated releases—in some detail, for I believe this shows how the Mozilla project came of age.

Implementing a transfer of control from Netscape management to mozilla.org caused a number of strains. Mozilla.org staff could have proceeded in opposition to Netscape. Indeed, we thought about it many times. However, Netscape was a large and valued contributor, whose involvement and work product remained very important for the project. So, we spent a great deal of energy figuring out techniques that addressed the concerns a commercial entity like Netscape brings to a project and simultaneously building a strong open source community.

Having control of the source code repository may seem like an obvious requirement for an open source project, but there were many sensitivities in our case. For example, the Mozilla project has processes to make sure that code is of good quality before it is checked in. This is often the case in open source projects, and many contributors understand that their code needs to meet project standards. However, this is not always the case in commercial settings. Often the employment decision is what matters. If someone is an employee, his code goes into the project he was hired to work on. Instituting code review for *everyone*, even people employed to contribute code, can be a surprise for new employees and for their managers. Suddenly the decision of what code goes into the source repository is not made by managers for a particular person through the employment process. Instead, the decision is made by engineers who review the code itself rather than the person's credibility or employment status. And the code review is not optional or auxiliary; code cannot be checked into the tree until an appropriate reviewer has given formal approval.

We instituted code review as a prerequisite to check-in relatively early in the process without too much controversy. Sometime later we implemented a second layer of review which we ended up calling "super-review." Super-review is an "integration review" or "plumbing review." Does the code conform to coding guidelines? Does it use the somewhat tricky XPCOM in appropriate places and not elsewhere? Does it conform to the overall architectural goals? Does it avoid needlessly diminishing modularity?

We instituted the super-review requirement because we felt we had to. The Mozilla codebase is large and complex, and we were worried about overall code quality. We needed to increase our confidence that the code we were accepting into the tree would solve the immediate problem, and still leave us enough flexibility for future development. Implementing the super-review was very painful for everyone involved. It was painful for the identified super-reviewers. Being a super-reviewer is actually a lousy job—who wants to review yet more code instead of writing it oneself? And of course, the super-reviewers were well regarded and had plenty of work to do themselves. They took on the super-review job not because they wanted to, but because they believed it was important to making Mozilla the project we wanted it to be. Super-review was also painful for the contributing engineers. Adding another layer of review to the pre-check-in requirements was seen as "yet more bureaucracy" by some.

There were many requests for absolute, guaranteed turnaround times for super-review. Engineers and managers complained they could not schedule work accurately due to the unknown lags caused by waiting for super-review. And yet, the super-reviewers had to balance review of code with writing their own. And since they were well regarded, their code was acutely needed. I was unwilling to agree to rigid turnaround times for super-review, believing that doing so would put the work of key contributors at risk. We eventually agreed on a timeframe for an initial response from the super-reviewer, which would contain some estimated time for full review.

Another issue was determining who has "write-access" to the source code repository. This was extremely sensitive. In a commercial setting, it is often the case that when people are hired they are given access to the part of the source code repository to which they are expected to contribute. In open source settings, one typically earns access by making valuable contributions. Moving the open source standard of "earn the right to touch the tree" to employees who need to contribute code to do their job can be difficult. In particular, the rationale for why this is necessary can be hard to explain in a commercial setting when all the code an employee writes must be reviewed and super-reviewed before it can be checked in. The question comes up: assume Employee X is hired. She writes code, and it passes review and super-review. Why on earth can that employee not go through the mechanical task of actually checking the code in? I was never able to provide a complete answer to this question. I know that open source projects regularly vet people before allowing access to the source code repository. And I know it would be very odd for a management chain in a company to make the decision about CVS access. So it's clear that this "is just not done." But I was not able to explain clearly how someone could do damage by having CVS access if all of her code was reviewed and approved before check-in, even though the technical leadership of the project felt very strongly about this. The idea of automatic access for employees had an emotional response because open source projects rely on peer review and technical leadership, and I shared this. But the key engineers were adamant that the quality of the code would suffer through automatic access, even though our pre-check-in code review requirements are quite stringent.

It was an awkward setting to institute a policy for which I couldn't give a crisp reply to the various management teams affected. (I've been responsible for policy for mozilla.org since 1999, so it was my job to write the policy, describe it to management, and address the concerns and complaints that might come up.) And the possibility existed that management groups would be surprised, distressed, or outraged that such a policy would be instituted without a clear answer as to why the code quality wasn't adequately protected. Nevertheless, we instituted this policy in 2001.

A second area where shifting control to mozilla.org was highly sensitive concerned control of the milestone releases. By "owning the releases," I mean several things: first, defining and implementing a planned milestone schedule; second, defining and implementing a process for getting a release into shipping condition; and finally, identifying and shipping the release.

When the Mozilla project was launched, the planning and release schedule was determined and implemented by Netscape employees contributing to the project, but not directly by mozilla.org staff. We worked on changing this for quite some time before we had proven ourselves trustworthy enough for Netscape to give up control. This may seem laughable—how can a so-called "open source" project not control its own releases? But it's important to remember that Netscape was under enormous pressure to release a product, and giving up control of the process by which releases are made is extremely uncomfortable. Mozilla.org was new and unproven. And of course, managing a software project with hundreds of people working on it is not easy in any setting.

Control of these aspects moved formally from the Netscape management to the mozilla.org staff following the release of the Netscape 6 browser. This shift had been discussed for some time but still involved a leap of faith for those who had previously exercised decision-making power.

This was a tempestuous time; mozilla.org staff had influence, but not control. We thought many times about whether we needed to create a fork in order to affect such a shift in control. Each time, the mozilla.org staff decided that Netscape's contributions were far too important to the project and outweighed the desire for open source purity or credibility. I suspect that the Netscape management team must have had similar discussions, weighing a fork and the ability to manage their releases as they felt best with the value gained from the open source project. In the end, we all hung in and mozilla.org staff became the official keeper of our releases after Netscape 6.

Owning our releases did not mean that we ignored Netscape. Netscape remained the largest single contributor, and its pool of talented and dedicated engineers was boggling. Netscape was also the largest single distributor of Mozilla-based products. Managing the project without taking Netscape's needs into account would have been stupid. By this time, Netscape didn't exercise control as it once had, but its leadership role in the project was greater than some might have liked. Mozilla.org staff continued to hear from the community that we were Netscape stooges and there wasn't a "real" open source project. However, to my mind, Netscape's role was now determined by classic open source principles: leadership and influence through the quality of one's contribution. Of course, this involves an acceptance of the role of a corporate entity in an open source project, which made some uncomfortable.

Mozilla.org was able to own our releases well, in part because we had developed an active, effective quality assurance (QA) community. Over the years, I've learned how few people intuitively grasp the importance of the QA effort (and I suspect, how important QA will be to other projects of similar scope). Both of our major efforts—web browsing and email—live in an extremely complex world. The Web is very diverse, and people use our software in a boggling array of environments and in wildly different ways. Hiring a QA team as full-time testers is part of a solution, but it is not the complete answer. I'm becoming more and more convinced that it may well be impossible to hire a QA team big enough and diverse enough to do thorough testing of a product like a web browser.

By 1999, it had become apparent that an active community of people was interested in contributing to our testing and QA effort. Christine Beagle joined us to lead an effort at mozilla.org to make this group cohesive, figure out ways to give the group some authority, and encourage them to step forward. The response to a bit of attention and appreciation directed at this nascent community was astounding. One mark of success is that shortly thereafter, we hired one of the more active and organized of these folks as our community QA lead. This person was Asa Dotzler, who has been a key figure in the project ever since, still coordinates QA activities, and is extremely active in managing our release process. With Asa's coordination, we began to see a set of people doing more organized testing of our products. This provided enormous value. The testers also did things such as look at all the bugs assigned to a particular engineer, verify that the bugs were legitimate, check to see if the bug existed on all platforms, create test cases, and then verify fixes across our main platforms. These efforts saved enormous amounts of time for the engineers trying to write the code to fix the bugs. I've learned that this type of work often gets little respect, but we value it highly and find these contributors to be extremely important to our project. Many of these efforts were coordinated through mozillaZine, an independent webzine dedicated to the Mozilla project that was founded by Chris Nelson in September 1998 (*http://www.mozillazine.org*).

Massive community testing remains important today. We provide Release Candidates for our major product releases precisely so we can get 50,000–100,000 downloads from our key community and get a good reach on quality.

Young Adulthood—the Mozilla Foundation

The idea of an independent legal organization to guide the Mozilla project had been discussed when the project was first launched in 1998. However, it was decided that the time was not quite right. At the time, there were no models for setting up such an organization and figuring out how it would be governed, who would participate, and so on. There was enough unknown and far too much work in getting the code ready, the project launched, and a browser developed to take on things we didn't absolutely have to do. Eventually we decided that the right time to create an independent Mozilla Foundation would be when a critical mass of people was interested in supporting a foundation. That critical mass would need to include a significant set of volunteers and a set of companies interested enough to fund browser developer and distribute Mozilla-based technology.

That critical mass began to develop with the release of Mozilla 1.0. Mozilla 1.0 showed that we could produce a good product, that the Mozilla releases where determined by Mozilla rather than by Netscape, and that the project had a positive future. At least one critical corporate participant came to us and told us that 1.0 proved our viability and that they were very interested in helping form and support an independent Mozilla Foundation.

Following the release of Mozilla 1.0, I spent a fair amount of time thinking about what an independent Mozilla Foundation would look like, how we might put it together, how many employees we would need, which companies would likely provide support, and how to finance employees in the early years. I had help from a set of mozilla.org staff members. In addition, I had the good fortune of hooking up with Mitch Kapor, who had recently joined the open source world with the launch of the Open Source Applications Foundation (*http://www.osafoundation.org*). Mitch was an immense help in thinking through various possible structures for the Mozilla Foundation and is an unsung hero in getting the Mozilla Foundation launched.

In the spring of 2003, the stars aligned. Mozilla.org staff was ready, the project had developed a critical mass, and we had some corporate support. In addtion, AOL decided it was ready to help spin out the Mozilla project. This was an important element for mozilla.org staff. Of course, we could have launched a project without AOL's support—that's the nature of open source—but the mozilla.org staff felt that AOL's support was important to the launch of an independent Mozilla project. We hoped that the use of the Mozilla trademarks would be transferred to a new organization, along with a set of machines. We wanted to be able to hire a group of people, some of whom were current AOL employees, without bad feelings. We felt it was very important to the project's stability to have a smooth transition from AOL to a successor. We also knew we needed to hire people to keep the project running well, and that it would take us time to find ongoing funding sources. So, the seed funding that AOL provided was another critical factor. Through July, I worked to reach agreement with AOL on how the Mozilla Foundation would be launched. Once again, Mitch Kapor provided invaluable assistance in helping to get the arrangements with AOL worked out.

On July 14, 2003, the Mozilla Foundation was launched as an independent non-profit organization. AOL contributed $2 million in seed funding for the Mozilla trademarks, the Mozilla Public License, the machines we were using to host the web site and other infrastructure, and the efforts of a transition team to help create a smooth handoff. We knew we had some additional funding from IBM and Sun, and Mitch Kapor donated $150,000 for each of the first two years. Based on this, Brendan Eich and I decided, with the help of Chris Hoffman and Mitch, to aim for an initial group of 10 employees.

The initial group was divided among (i) those focused on the projectwide resources (technical leadership, infrastructure, tools, web site management, builds, releases, QA), (ii) those focused on the codebase itself (Firefox, Thunderbird, Gecko, the DOM, and JavaScript) and (iii) a couple of people focused on all the other things the project and the Mozilla Foundation needed to be successful, including relationships with commercial contributors and other organizations, legal structure, trademarks, finances, and so on. Mitch Kapor offered to extend his organization for providing back-office services—payroll, benefits, accounting, donation processing, and human resources—to the Mozilla Foundation on very gracious terms, which has been a great boon. Securing high-quality services in these areas for the Mozilla Foundation had always been of concern to me and this has been a phenomenal solution for us.

This resulted in a group that was small for the scope of the project, but still big for a nonprofit open source project to support. We chose this route because we believed that the project was unlikely to reach its potential without a core group of at least this size. We felt this was the minimum size for critical mass for several reasons, including these:

The World Wide Web isn't finished.
It' changes all the time. New content types develop, new technologies develop, and new possibilities emerge. If the browser doesn't continue to develop, the consumer's ability to enjoy these enhancements stagnates.

Browsers and email clients aren't done yet.
There's a whole range of innovative ideas that interact with browsers and email clients. For example, RSS readers can be nicely integrated with both browsers and email. In addition, the underlying components on which the actual end-user applications are built require constant development.

Speed matters.
We need good Internet clients now. Having a core set of people able to devote full-time attention to this makes a big difference in accomplishing things quickly.

The size and scope of the project requires it.
Just keeping track of what's going on in the Mozilla project takes time. About 80 people actively check into the CVS repository each month, and of course, many more active participants don't have CVS access. We also have a high level of involvement with commercial entities and with Mozilla development teams at commercial entities. Providing the technical leadership and coordination for this large a group is a big job, even with a set of full-time employees. Doing so without a set of people available full time (or more than full time) would be beyond daunting.

The founding in July was followed by a hectic startup period through the fall. We assembled the team of employees. We found office space at an affordable rate, thanks again to friends of the Mozilla project who extended a helping hand. We moved our equipment from AOL to our co-location facility and our offices. We knew it was important that enterprises and other institutions got a good picture of the Mozilla Foundation and grew confident that we are not a naive, flaky group, so we spent a chunk of time talking with these groups.

We decided that a serious focus on the end user needed to be added to our traditional focus on developers. Product development continued at a fast clip through this period.

Firefox and Thunderbird

As if forming the Foundation, moving employees, and establishing and supporting ourselves wasn't enough, we also began a determined transition from the application known as Mozilla or the Mozilla Application Suite, or by its codename, Seamonkey, to our new products: Mozilla Firefox and Mozilla Thunderbird.

We knew that our future lay with the new applications. The integrated Mozilla Application Suite is a fine product that many love. But the integration caused difficulties, the UI had been built by accretion and had been added to over the years, and we knew we wanted updated, standalone browsing and mail applications. Given our limited resources, we had to place a bet, and we did.

The Mozilla Foundation hired the lead Firefox and Thunderbird developers, Ben Goodger and Scott McGregor. We talked with Ben and Scott about providing assistance to the community of people working on the Mozilla Application Suite. We continued with our releases of the Mozilla Application Suite, including improvements to the core components, performance, stability, and security, and coordinating feature work done by our community. But we did not hire people focused on the Mozilla Application Suite.

Both Firefox and Thunderbird were in the early stages of development when we made this decision. Indeed, Thunderbird had not even seen its 0.1 release when the Foundation was launched. Despite this, we knew that the then-current state of Thunderbird could probably have supported an 0.1 or 0.2, or maybe even an 0.3 label. This was borne out when we were contacted over the summer by a Fortune 100 company wanting information about Thunderbird. The company had already done a significant amount of due diligence and had realized that Thunderbird was the best option. They wondered if the Foundation would be interested in speeding development of certain enterprise features if we had some additional funding to do so. As a result, Thunderbird has had a rich set of enterprise features from its early days. It lacks an integrated calendar, but the Mozilla calendar project was reinvigorated and an integrated calendar project launched in the fall of 2004.

Firefox was further along the development path, but still quite young. It wasn't even called Firefox at that time; it was called Firebird, the second of two early names which we abandoned due to trademark issues. The application-eventually-known-as-Firefox was at the 0.5 stage, quite usable, but not a polished end-user application. The development goal had always been a strict focus on the end-user experience above all else. This continued, and Bart Decrem drove the end-user focus throughout all aspects of the Foundation's operations. Firefox began to be noticed in 2004 with the 0.6 release. It quickly began to capture the interest of much of our developer community. There were still millions of contented users of the Mozilla Application Suite, but the momentum had clearly begun shifting to Firefox.

In February 2004, we found a public posting by a visual designer named Steve Garrity, describing what Firefox needed for its icons, logos, and visual identity in general. The content of the post was excellent. Better yet, Steve seemed to have both knowledge of and an interest in tackling these problems, instead of simply complaining or pointing out problems. We asked him if he'd like to take the lead for a bit and show us what could be done. He said yes, and the Visual Identity Team was created. Both Firefox and Thunderbird took a giant step toward becoming sophisticated, polished end-user applications. By the 0.8 release of Firefox in June 2004, the momentum for Firefox was growing dramatically. Firefox was already an impressive product, offering features new to most users.

In addition, the Internet experience had become extremely painful. Malicious actors were everywhere. The Web was infested with viruses and security exploits, seemingly uncontrollable pop-up windows appeared almost everywhere, and distracting, bandwidth-chewing ads appeared long before desired web content. The browser, a piece of software many had come to take for granted, suddenly mattered. The browser is the mechanism through which one's computer—one's hard drive with its critical and private data—connects to the wild, wild world of the Web. A modern, high-quality browser is necessary to keep this connection from being increasingly painful and even dangerous. The Mozilla Foundation had a great browser in Mozilla Firefox, and people began to notice. By mid-2004 we began to see that the types of people who were interested in Mozilla Firefox were expanding. We began getting messages from people who clearly were neither early adopters nor even particularly savvy. So, we knew we were making a difference. And we knew that the difference was important enough for more people than ever before to pay attention. People who tried Firefox loved it. Around the 0.9 timeframe (June 2004), a groundswell began building.

The summer of 2004 was an even more painful time on the Web. A series of viruses and security issues caused enormous inconvenience and concern. Internet Explorer was a vector for many attacks. These problems caused consumers to pay more attention to their browser. They helped people realize why an alternative browser is so important to the health of the Internet and one's ability to interact comfortably with the Web. Security is a very difficult problem. A browser must be open to the content of the Web— that's the whole point. At the same time, it can't be *too* open. A browser needs to have a series of defenses to help filter out bad content. No browser can be perfect, and that includes Firefox. We know that we will be making security changes and improvements in our products on a continual basis, and we hope that others do as well.

We saw significant adoption of Firefox 0.9 through the summer and fall of 2004— almost 8 million people came to get a product that hadn't reached its 1.0 status yet. Mozilla Thunderbird adoption was also proceeding well, though not at the same fantastic rates. On the marketing side, Spread Firefox was launched in September 2004. This was a community marketing effort, perhaps the first of its kind. We knew that the great strength of the Mozilla project is the community of people dedicated to making it successful. We also knew that we would not have a traditional "marketing" or "PR" effort, spending large amounts of money on media events. And the mail we were receiving made it clear that people were excited about Firefox and wanted to help their friends and family switch.

The result was *www.spreadfirefox.com*, the home of a fervent evangelism community focused on increasing Firefox adoption. The most famous Spread Firefox campaign to date has been the *New York Times* campaign, which was proposed and initiated by a community member. This started out as a 10-day campaign to get 2,500 people to contribute funds to buy a full page ad in the New York Times supporting Firefox. Ten days was the wrong timeframe—2,500 people signed up in the first *two* days. We kept the campaign open for 10 days anyway and ended up with 10,000 choosing to participate. We had promised that contributors' names would be in the ad and would be legible, so we enlarged the ad and made it a full two-page ad. It ran on December 16, 2004. A while later I came to work to find two young men standing outside our door. The door is glass and we had taped the *NYT* ad to the door so that it was visible from the outside. The two young men looked lost, but one wore a Firefox T-shirt. So, as I reached the door to go in, I asked, "Can I help you find something?" The men were rather shy, looking at their feet and mumbling, "We just wanted to see the Mozilla Foundation. We're only in town for a few days and had to see it." Then one of them straightened up, looked me in the eye, jabbed his finger at the *New York Times* ad, and said proudly, "And there's my name, right there!" Sometimes I think people believe I'm exaggerating when I describe how passionate consumers are after they've tried Firefox, but it's actually hard to overstate the excitement that Firefox has generated.

Getting Firefox and Thunderbird to a 1.0 status and shipped was a very intense period. We knew we had great products in the works, but we had to finish them. We also had to get a set of related activities completed. These included revamping our web site, developing our communications plan, working with the Spread Firefox community, improving our localization process and working with the various localization communities, figuring out our search relationships, working with our affiliates—Mozilla Europe and Mozilla Japan—on the international aspects of the launch, and so on. The ferocious dedication of everyone involved was required. I cannot stress enough the commitment of the Mozilla community. On Sunday, November 7, I logged onto IRC at about 8:00 A.M.., which is early for me and for most of the Mountain View–based staff. I was bombarded with questions from our localization communities in Europe and Asia. Some were up early, many were up very, very late, and all were trying to figure out how to manage their schedules over the next 48 hours to be available whenever needed to get their localized versions finished, approved and shipped as part of the 1.0 release.

Mozilla Firefox 1.0 and Mozilla Thunderbird 1.0 were released on November 9, 2004. To say they have been well received is an understatement. Firefox 1.0 was downloaded from our mirror site about 2 million times in the first two days alone, and has plunged on at an average rate of almost 250,000 downloads per day since then. As of mid-April 2005, the number of downloads that we can track is very close to 50 million. On the usage side, Firefox has gained worldwide market share at a rate of nearly 1% per month from November to April. As of April 2005, surveys are beginning to show Mozilla browsers at or above 10% market share. Among technically focused sites, the market share of Mozilla products ranges up to much higher numbers.

It's extremely difficult to gain this sort of market share on the desktop in the face of a competitive product that people get when they buy a computer. The fact that Mozilla Firefox has done so is a reflection of a great product, a huge need, a fervent community, and the power of the Internet.

Many people have wondered whether open source development can produce great end-user applications. One school of thought says that open source developers can produce infrastructure and products that other developers like, but not applications aimed at the general end user. Mozilla Firefox and Thunderbird demonstrate that open source software can indeed produce great end-user products. I believe that we are only at the beginning, and we will see a range of innovative end-user products come from the open source world in the coming years.

The Future

The mission of the Mozilla project is to promote choice and innovation on the Web by creating great end-user offerings. We focus on innovation because the Web is still young—we've seen only the beginnings of its potential. That potential can be stifled if we don't have innovative work done on the client side.

We focus on choice because this allows people to have greater control over their Internet experience. This control over our life on the Web increases in importance each year, as more and more critical functions such as banking, health care, insurance, and commerce are done over the Web. A monoculture is rarely a healthy ecology. A single effective choice in browsers and email clients is dangerous, both to consumers and to the health of the Web itself.

Firefox in particular has shown that consumers will pay attention to a product that provides an alternative, and that the Mozilla project can create such a product. We have a number of challenges ahead of us. We need to continue to release products that people love. We have a set of responsibilities that come with the user base, adoption rate, and increased visibility of the project. Conditions will change, and we will need to adapt. These are challenges, but certainly no greater than those we have faced to date. These are the challenges that result from the project's achievements.

We have great talent, a powerful and creative community, a well-earned place in the Internet ecosystem, a growing user base, and, at long last, a legal home for the Mozilla project in the Mozilla Foundation.

As we go forward, there is no change in the mission of the project. Our basic approach of combining open source DNA with involvement by commercial entities will continue. The Mozilla Foundation has grown some and may grow some more, and we expect to continue working closely with a set of companies that are interested in developing and distributing Mozilla technology. The increasing acceptance of open source software by the commercial world opens up greater possibilities for collaboration. The emergence of web-based services provided through the browser also encourages business models for the service provider other than charging for each copy of software provided. This allows more entities to contribute to our project. Our focus on distributed development, technical excellence, and welcoming new participants will continue. The need for a vibrant, creative community of people focused on the Web will not change.

I expect the Mozilla project will continue to be a trendsetter in a number of arenas: development of open source end-user products, combining volunteer and commercial activity in an open source project, maintaining a critical mass of people as employees of the Mozilla Foundation, and funding that set of employees plus community marketing and adoption programs. We aren't the only ones doing these things, and we continue to learn and benefit enormously from the open source projects. We hope to contribute ever more in return.

Chris DiBona

CHAPTER 2

Open Source and Proprietary Software Development

In this chapter, I present a perspective on the similarities, differences, and interactions between open source and proprietary software development.

Proprietary Versus Open Source?

Before you go any further, throw off any notion that the proprietary developer is somehow a different person from the open source developer. It is uncommon for a member of the open source developer community to do only open source for a living. Only the most prominent, or loaded, members of the open source community come close to having this kind of freedom. It is indeed rare to find a developer who develops only with proprietary tools and libraries. Even Visual C++ and C# developers benefit from a great variety of code and libraries that are free for use in their programs.[1]

My career has focused on open source development for the last 10 years, and I'm constantly pleasantly surprised by how open source development and proprietary resemble each other. I believe this is because proprietary developers are educated by the adventures of their slightly crazy open source cousins, but I also know that open source developers have learned just as much from proprietary developers.

---

1   Traditionally, one difference between open source and proprietary development teams has been that open source teams are, in general, geographically quite dispersed. However, in this age of outsourced, offshored, and distributed development, even proprietary development has become highly dispersed geographically.

Don't read this as an attempt to muddy the difference between proprietary and open source *programs*. They are different, sometimes very much so. However, they come from the same people, and they're using a lot of the same methods and tools. It is the licenses and the ideals behind open source programs that make them remarkable, different, and revolutionary.

The Example Culture

A lot of people, when talking about open source software development, say that open source developers enjoy a great productivity gain from code reuse. This is true, but in my experience *all* developers, not just open source developers, benefit from the existence of free-of-charge standard libraries and code snippets. For decades, proprietary developers have had a great variety of prepackaged libraries to choose from, but these proprietary libraries haven't taken root in the same way that freely usable, open libraries have.[2]

Code reuse? Knowledge reuse!

In Linus Torvalds' essay from the first *Open Sources*, he talked about how the rise of open code was delivering on the promise of reuse touted by proponents of the Java™ programming language specifically and object-oriented programming in general.

That said, it has been my experience that there is a point at which software developers will go out of their way to avoid reusing code from other projects. In some shops, they call it "not invented here" (NIH) syndrome, and some companies are famous for it. But even those shops use standard kernels, libraries, and compilers. The real difficulty here is in figuring out where the NIH line lies. Although the answer is different for every single programmer and team, all still can (and still do) learn from the open code out there, which is a unique advantage of open code. While both open and proprietary code can be reused in a wide range of circumstances, open code enables something further: knowledge reuse. By examining the code itself, the developer can learn how a particular problem is solved, and often how that solution is an instance of a general solution type. It is this kind of reuse that Linus applauded and that the NIH developer misses.

Then why not simply use other people's code? There are a number of factors to consider before code is incorporated, and these must be understood before one can understand the role that Free Software has had in development.

---

2   This will likely inspire many to cite their favorite commercial library. A full survey of libraries, both commercial and open source, would be required to validate this statement properly. This is an educated assumption on my part, as when commercial libraries manage to gain any sort of prominence, open source developers tend to fill the gap, thus overshadowing the commercial project.

Speed of development

There are very real barriers to using other people's code. You have to examine how to interface with said code, and you need to review the code to make sure it meets your standards for security, license, style, and correctness. You also need to integrate it into your version control and build system.

None of these problems is insurmountable, but they have to be worth surmounting. To wit: if all I need is a routine to do something simple, such as iterate through an array of numbers and perform some simple operation on them, using someone else's software would be a waste of time.

When developing, I like to use large libraries only when I either don't want to deal with a technology, or I don't fully understand it and don't feel qualified to implement it. For a recent project, I was pulling newsfeeds from weblogs and performing a kind of natural English language processing on it. I thought that using a tool called a "stemmer" to normalize the data would make my later analysis more accurate.

Implementing the routines to download and process feeds could have taken a month or two, and this is exactly the kind of development I don't like to do. To properly implement a stemmer, I'd likely have to get my graduate degree and then write it— which would impact my deadline a bit—so I downloaded programmer-friendly libraries that did each of these tasks. The stemmer was available under the Berkeley Software License, and the feed parser was available under the Python Software License, both of which are very easy to deal with and do not require any onerous post-incorporation duties. I was thus able to save time and have better code.

That said, some things I'm very interested in developing myself. Since I was doing this project as an excuse to learn a natural-language processing algorithm, which was interesting to me, I wanted to write that part of the program myself. I was (and am) also fascinated with a problem I think I'll have in storing the results such that I can quickly retrieve them from a database. I haven't solved that problem as of this writing, but I don't necessarily want to use other people's code for that. I have read some code and examples in textbooks and online that will help me with the former, but the storage problem is mine, for now.

This gives you an idea where the line was for me in this particular project, but others have the same reticence for other, subtler reasons.

A particularly difficult codebase

What makes software difficult to add to your code? Sometimes the code is simply in the wrong language. Maybe you are using Perl and want to tie some code into a C or Python module. That's not always so easy. Maybe the code was really developed on only one platform—say, an Intel machine—and you want it to work on your iBook, which runs on a PowerPC processor.

The problems with using other people's code can be legion. Maybe their routines were implemented assuming a machine with a lot of memory or processor cache, making it perform poorly or, worse, unpredictably,[3] on your target platform. Maybe the software was developed for an earlier version of your programming language, so a lot of features you would have implemented with a standard library call are instead implemented from scratch, thus reducing future maintainability.

Problems arise with canned libraries as they get older. For instance, the aforementioned feed parser library is useful because its author, Mark Pilgrim, is very good at keeping it up to date with the 13 "standards" that lie behind that "xml" button on your favorite blog or web site. If the library were to fall into disuse, or Mark were to stop working on it and no one else picked up the work, I'd likely change to a different library or choose to maintain it myself.

There is another reason to not use someone else's code, and it will look amazingly petty to all but the programmers reading this.

Technically speaking, this:

```
int myfunction(int a)
{
    printf("My Function %d\n",a);
}
```

is the same as this:

```
int myfunction(int a) {
    printf("My Function %d\n",a);
}
```

which is the same as this:

```
int myfunction(int a){    printf("My Function %d\n",a);}
```

and this:

```
int myfunction(int a){
  printf("My Function %d\n",a);
}
```

They compile to the same result on any given compiler.

I could go on, but I won't. The point is that, depending on the programmer or dictated company style, each of these is wrong, evil, bad, or awful, or perhaps one is acceptable. Not all programmers and companies care about style, but many (one might argue the smartest) do. The ones that do care actively dislike the ones that don't and do not want to use their code. Should they have to touch the offending

---

3   This might seem strange, but programmers are OK with the odd performance hit sometimes. Unpredictable results lead to crashed programs, however. This is not good, no matter what you've been told.

library, they will inevitably have to make it "readable." Whether you call this refactoring or prettifying or whatever, it can drive a programmer away from a hunk of code, unless it really brings something fantastic along with it.

"My Goodness," you might consider asking, "are programmers delicate, petty creatures?" No, there are some very good reasons to have consistent code style. It aids in debugging. Some say it reduces bugs (I'd agree). It makes code navigation much faster and makes it easier for people to write tools to generate and manipulate code than they might otherwise. There are other reasons too, but I don't want to get too arcane. Some languages, such as Python, have very rigid appearance rules, as appearance can dictate how a variable can be used. Style may appear to be a trivial concern, but it isn't.

Comfort

Maybe you just want to do it yourself. Businesspeople in the industry who have grown up around open source often comment that duplication of effort, or "reinventing the wheel," is not time well spent. I rarely hear this from programmers. When people hear about KDE and GNOME, or Linux and BSD, or even more esoteric arguments about which window manager to use, inevitably someone will chime in, "Obviously, they had a lot of time on their hands. Otherwise, why would they have started from scratch?"

The implication is that the programmers have somehow wasted time. When I choose to reimplement some technology or program, I know what I'm doing, and even if it is a "waste" of time or duplication of effort, I think of it as practice. And when I can enjoy the luxury of implementing from scratch, I really like the results, because they're all mine and what I've developed works exactly the way I want it to.

But Why So Many of the Same Things?

Business, of course, is interested in productive developers, and productive developers don't rewrite things, right? No, not necessarily. People rewrite code all the time. The more-informed companies recognize that this type of thing is often inevitable, and the best and most resourceful encourage this kind of mental knife sharpening, because it leads to better developers and better code. Given the time, programmers often prefer to learn from other people's code without actually using the code, and if open source ends up as one big repository of example code, I call that a success.

Also, computers change. Computers, languages, compilers, and operating systems change so quickly that a periodic rewrite of some code becomes vital, from a performance perspective. To take advantage of the newest processors, architectures, and other advancements, a recompile will certainly be required and will likely expose issues with your code (architecture changes lead to this directly).[4]

---

4 For example, you write a program on your handy laptop, you compile it, and it runs great. Later, you run it on your fabulous dual Opteron server. It crashes because you assumed an integer was 32 bits and the Opteron (running a 64-bit OS like Linux, of course) has 64-bit integers. This is a basic error that comes up in a lot of different ways during 32–64 bit transitions.

But people are using libraries, code, and examples from open source code, copying them into their codebases rapidly. Certainly this happens. Don't let my counter cases fool you. It is a rare codebase that doesn't involve some open source software, whether it is merely in the form of a standard library or a widget library, or is full of the stuff. This is by design; if every program had to write every instruction down to the operating system, or the machine itself, there would be no programs. The iterative building process, programs on top of libraries on top of the operating system, is so productive that I can't imagine someone ignoring it. Even for the smallest embedded systems, designers are using the GNU compilers to create great programs for their devices: compile, flash, and go.

Libraries, System Calls, and Widgets

Here we begin to see how open source ideals have changed proprietary development. When proprietary software developers create a program, they may use free software, created or derived libraries, widget libraries, and tools. This includes developers targeting proprietary operating systems such as Windows and OS X. Developers creating software, whether for OS X, Unix/Linux, or Windows, commonly use free tools to do so. They almost always use free libraries in the creation of their programs and often use free user interface elements during the creation of their systems.

Some might think I'm indulging in some mission creep for free software here, assigning a larger role to it than it maybe should enjoy. I'm not. I'll take it even further: if there hadn't been free tools like the GNU compiler collection, the industry would have been forced to create and release them. Otherwise, the computer industry as we know it would not exist and would certainly not be as large as it is right now. This is not to imply that companies somehow owe something to the free software community. However, companies do help out when they can reap a long-term benefit. IBM understands this, as do Novell, Google (my employer), and many others. Even Microsoft uses and releases code under a variety of licenses, including the GPL (its Unix services for Windows) and BSD (Wix), but Microsoft is conflicted both internally and externally, so it's not as easy for it to embrace open source.

Am I saying that without free tools, the compiler would try to charge a per-program fee? No, I think that if free tools hadn't arrived and commoditized the compiler, other competitive concerns would have kept the price of software development tools accessible and cheap. That said, I think free tools played a big part. Free and open source software changed expectations. Microsoft and Intel make no attempts to prevent developers from using their compilers to create free software or software that is counter to their corporate goals. Client licenses, a common fixture in the email/workflow market, are unheard of for mainstream development tools.

If there is one thing about free software that is downright scary to proprietary development shops, it may be this: software that is licensed per client almost always comes under attack from free software. This is forcing in the software industry a shift away from such per-client licenses in all but the most specialized verticals—for instance, the software that runs an MRI machine, or air traffic control software, both of which are so specialized as to not count, because every client is custom. The grand irony here is that in some industries, such a high cost is attached to developing software that some are forming very open source–looking consortiums to solve common software development problems.

Distributed Development

Distributed development is more than just a fad or even a trend. Organizations and companies large and small are using diverse, globally distributed teams to develop their software. The free software development movement showed the world how to develop internationally. Well before SourceForge.net became a site that every programmer had heard of, projects working together over the Internet or far-flung connected corporate networks developed much of the software that we use today.

In fact, the tools they developed to do that are now considered the baseline standard for developerd everywhere. What company in its right mind doesn't mandate that its programmers use some form of version control and bug tracking? I ask this rhetorically, but for a long time in the software business, you couldn't make this assumption. Small development shops would back up their data, for sure, but that's not version control.

Distributed development is about more than just version control. It's also about communications and bug tracking and distribution of the end result of software.

Understanding Version Control

Programming is an inherently incremental process. Code, then build, then test. Repeat. Do not fold, spindle, or mutilate.[5] Each step requires the developer to save the program and run it through a compiler or interpreter. After enough of these cycles, the program can do a new thing or an old thing better, and the developer checks the code into a repository, preferably not on his machine. Then the repository can be backed up or saved on a hierarchical storage system. Then, should a developer's workstation crash, the worst case is that the only work lost is that done since the last check-in.

5　This sentence is famous for being printed on punch cards, an early way of providing computers with data. If they were folded, spindled, or mutilated, they jammed the readers—which makes one speculate what the punch card programmer used for version control. The answer is right there in front of you: as the cards went through revisions, they swapped out cards and retained the old, original cards.

What is actually stored from check-in to check-in is the *difference* from one version to the next. Consider a 100-line program, in which three lines in a program read:

```
for (i=1; i < 1; i++) {
    printf("Hello World\n");
}
```

and one link needs to be changed to:

```
for (i=1; i < 100; i++) {
    printf("Hello to a vast collection of worlds!\n");
}
```

which would then be checked back in. The system would note that only one line had changed and store only the difference between the two files. This way, we avoid wasting storage on what is mostly the same data.

The value of having these iterations can't be overstated. Having a previously known, working (or even broken) copy can help in the event of an editing problem, or when you're trying to track down a bug that simply wasn't there a revision ago. In desperate cases, you can revert to a previous version and start from there. This is like the undo option in your favorite word processor, but one that persists from day to day.

Version control isn't used just in development. I know of IT shops and people who keep entire configuration directories (*/etc*) in version control to protect against editing typos and to help with the rapid setup of new systems. Some people like to keep their home directory in a version control system for the ultimate in document protection. There is even a wiki project that sits on top of the Subversion version control system.

Additionally, good version control systems allow for branching—say, for a development and a release branch. The most popular version control system that many open source projects use is CVS.

CVS

CVS, the concurrent versioning system, allows developers all over the world to work on a local copy of a codebase, going through the familiar "code, build, test" cycles, and check in the differences. CVS is the old standby of version control, much in the same way RCS/SCCS was before it. There are clients for every development environment and it is a rare professional developer who hasn't been exposed to it.

Since it is easy to use and install and it enjoys wide vendor support, CVS continues to be used all over the world and is the dominant version control platform.

Subversion

Only the rise of Subversion has brought real competition to the free version control space. With a much more advanced data store than CVS and with clients available for all platforms, Subversion (SVN) is also very good at dealing with binary data and branching. Both are things that CVS isn't very good at. SVN is also very efficient to use remotely, and CVS is not; CVS was designed for local users and remote use was tacked on later. Additionally, SVN supports a variety of access control methods, supporting any authentication scheme that Apache does (Subversion is an Apache project), which includes LDAP, SMB, or any the developers wish to roll for themselves.

What About SourceSafe?

SourceSafe isn't really version control. Local version control, whether CVS or Source-Safe, is just backup, requiring a level of hardware reliability that simply doesn't exist on a desktop. Since SourceSafe is not designed to be used remotely, you take the life of your codebase in your hands when you use it. There are some SourceSafe remoting programs out there if you must use SourceSafe, but I can't recommend them so long as decent, free SVN and CVS plug-ins exist for Visual Studio.

The Special Case of BitKeeper

BitKeeper, which was written by Larry McVoy, was chosen by Linus Torvalds to use for version control for the Linux kernel. For the Linux kernel, BitKeeper was a very good choice, given the kinds of problems that arise with Linux kernel development. Written for distributed development, BitKeeper is very good at managing multiple repositories and multiple incoming patch streams.

Why is this important? With most version control systems, all your repositories are slaves of one master and resolving differences between different slaves and masters can be very difficult.

The only "problem" with the kernel team's use of BitKeeper was that BitKeeper was not a free software program, although it was available for the use of free software developers at no charge. I say *was* because Larry McVoy recently decided to pull the free version, thus making it impossible for Linux kernel developers to work on the program without paying a large fee.[6]

A great number of developers lamented the use of a proprietary tool for free software development, and the movement off BitKeeper, while disruptive, is a welcome change.

BitKeeper is a tool designed with the open source software model in mind. It has found success among large proprietary development houses specifically because the problems that faced the kernel team in 2001 are the same ones that increasingly face proprietary development shops. All of these teams, not just those working on open source development projects, now face multiple, far-flung teams that are engaged in collaborative development and struggle to do it effectively.

---

6   The kernel team is in the process of moving off of BitKeeper as of this writing.

Collaborative Development

You have a developer in Tokyo, a team in Bangalore, a team in Zurich, and a shop in Seattle, all working on the same codebase. How can you possibly keep the development train from coming off the rails? Communication!

IRC/IM/Email

One might imagine that only now, with the advent of IM and VoIP, can developers keep up with each other. In fact, developers have stayed in touch in something approximating real time since the early days of Unix, when they began to have a great variety of communications tools to use.[7] Early on, two developers on the same machine used the Write or Talk Unix programs, which allowed for a simple exchange of text between users. This grew into Internet Relay Chat (IRC) and then Instant Messenger (IM).

Email itself plays the most important role in development. It is the base packet of persistent knowledge that distributed developer teams have. Wikis are also taking hold as repositories of information.

VoIP

Strangely (to nondevelopers) voice simply hasn't caught on as a terrific tool for ongoing developer communications. While a regular conference call is useful for keeping everyone moving in the same direction, the idea of vocal input while developing would drive many coders away screaming. The phone isn't evil, but maintaining an uninterruptible flow can be very important to developer productivity. Phones also do not create a logfile or other transcript that can be referred to later. Don't take my experiences for gospel here. Read the book *Peopleware*[8] for more information about this. Everywhere I've ever worked, the one constant has been developers wearing headphones, but listening to music, not other developers yammering in their ears.

SourceForge

The online site SourceForge.net is the largest concentration of open source projects and code on the planet. SourceForge boasts some 100,000 projects and 1 million registered developers, and people use its integrated version control, project web hosting, file release mechanism, bug control, and mailing lists to write a vast amount of software. Pulling together these features on a free platform for open source developers proved to be a revolutionary concept. Before, people were left implementing this themselves with Bugzilla (a bug-tracking mechanism) and CVS or some other version control/bug-tracking facilities.

---

7   In fact, the Unix "write" command allowed hackers in the 1970s to communicate in a fashion not so different from IM.

8   Tom Demarco and Timothy Lister, *Peopleware* (New York, NY: Dorset House Publishing Company, 1999).

SourceForge represents, for a lot of people, the next stage in developer environments. VA Software, the company that runs SourceForge through its Open Source Technology Group (OSTG) subsidiary, sells this sort of solution into the enterprise, as does the Brisbane, California–based Collab.net.

Software Distribution

While free software developers know how to code, what about getting the code in front of the user? In the early days of the Free Software Foundation (FSF), the answer was to send out tapes and disks to users who wanted the tools, for a reasonable fee. Now that so many people have connections to the Internet, boxed software is beginning to show its age, but software producers are really just now learning from open source how to distribute software in this way.

Dependencies

When you compile a piece of software, you sometimes end up relying on libraries that you must call from your program to do some task. If you try to run the program without the expected complement of libraries, it cannot run or it may run poorly. Open source developers have created some very smart packaging and installation systems and filesystem methods that can make this a more tractable problem. Once they created these packaging systems and combined them with the Internet, they got online updating. The irony is that, in a lot of ways, Linux and Unix were schooled in this by Windows. A common complaint regarding Linux when comparing it to Windows and OS X is that software can be very difficult to install. One could argue that Windows isn't all that easy to install either, but since Windows is preinstalled on most computers, this is an argument that often falls on deaf ears.

I don't think Linux developers have learned to do installation well yet. There are some standouts, but for the most part, installation ease is still a work in progress. One thing free and proprietary share is the appreciation for and development of online updating systems. This is something Linux distributions get very right. In short, once Linux is installed on your machine, it can be very easy to keep it up to date.

Online Updating/Installation

Online updating is a terrific way of getting software onto your machines. More importantly, it is a terrific way to maintain a secure system over time. Since Linux distributions don't have to worry about software license ownership, it is very easy for the software to determine whether to download a patch or fix, and thus many Linux distributions have systems to facilitate this. Proprietary software development houses such as Microsoft are still trying to figure this out. It is a hard problem when you mix it with licensing concerns. Additionally, when it's done wrong, you can literally crash thousands, or in the case of Microsoft and Apple, millions of machines, so it is really critical to do well. That the Debian and Fedora Core Linux distributions do this at all is quite a feat.

Want a sticky issue? Do you trust your software vendor to allow it to automatically update your software? For some, this question is heard in these ways:

- Do you trust Microsoft to update your operating system?
- Do you trust a bunch of bearded Unix programmers to update your system?

How you react to these questions has a lot to do with the realities of how difficult the problem is, how successful previous auto-updates have been in the past, and how trusting you are—which brings up the subject of the next section.

How Proprietary Software Development Has Changed Open Source

Open source isn't magic, and developers aren't magicians. No developer is immune to security problems and bugs creeping into his code.

Bugs/Security

Free, open, proprietary, closed….Bugs happen. I think open source means fewer bugs, and people have written tens of thousands of words explaining how they agree or disagree with me. One thing I know I'm right about is that both *kinds* of code have bugs. Bugs persist longer in closed codebases, and their closed nature keeps bugs persistent.

If I may paraphrase Socrates, "An unexamined codebase is dead," and by *dead* I mean killed by the hostile environment that is viruses, worms, crackers, and Trojans. Like bugs, security flaws happen in both free and closed software. As a project matures, it must assemble a mantle of testing and quality assurance (QA) techniques that are vital to its ongoing health. I think open source development has learned much from the processes that proprietary software development houses have come up with to support their paying customers.

Testing and QA

As projects mature, so do the testing suites around them. This is a truism for free and for closed software codebases, but the research around this originated in commercial software/hardware and in academia, and open source software has been a ready consumer of this information. The most popular talk I attended lately was in unit testing for Python at the O'Reilly Open Source Conference. The room was packed, with people sitting in the aisles. Testing is huge and is required for any project, free or not.

Project Scaling

Scaling is hard. Whether we're talking about development group size, bandwidth, space, or whatever, scaling any programming project is nontrivial.

Software development has its limits. Product teams can't grow too fast or too large without one of two things happening—either disintermediating technology or project ossification. Fred Brooks's seminal book, *The Mythical Man-Month*, covered this in depth, and the existence of F/OSS development methodologies doesn't change that. In fact, the tools and changes free software has brought to prominence are all around disintermediation and disconnected collaboration.

F/OSS isn't magic. It isn't breaking the speed of light. Most projects, with some notable exceptions, are composed of small teams, with one to three people doing the vast majority of the coding. If you care about project size, you would be well served by reading the findings of the Boston Consulting Group's study of open source software developers on SourceForge. This revealing study analyzes project metrics and motivations. For one thing, you see that projects almost always comprise fewer than five active developers. Many projects have only one developer.

So, what am I talking about when I say *disintermediating technology*? Look at it this way. Imagine that one person decides to create a cake from scratch. He'd have to start with a cake mix and some milk and an egg, right? No! He'd need chocolate, milk, flour, yeast, water, and the other ingredients, right? No! Just for the milk, he'd need a cow, some food and water for the cow, a bench, a milk bottle, a chiller, a pasteurizer, a cap and a rag, some bag balm, and so on, right? Well, you're getting closer. The point is that we accept interfaces all the time, and the successful project finds these interfaces, formalizes them, and spreads the work out along these lines.

We accept power at 120 volts at 60 hertz alternating current. We don't generate the power ourselves. We accept that we don't need to dig for oil, refine it, and pour the refined gas into our cars. We use interfaces with different systems all the time. Programs, too, have interfaces, and the success of a program is in how it manages these interfaces.

Proprietary or not, a successful program is one that interfaces effectively between systems and teams working on these systems. Microsoft doesn't have 5,000 engineers working on Windows. It has them working on the kernel, the printing subsystem, the windowing system, the voice synthesis module, and other components. More importantly, it has groups that work on interfacing between the systems so that they (theoretically) work as a whole. Likewise for the Linux kernel; Linus interacts with a number of captains who control different subsystems, including networking, disk drives, memory, CPU support, and so on. Fractionation, when possible, is key, and when not possible, disastrous—which is why groups working to integrate the whole and making sure the interfaces are appropriate can make all the difference in the success or failure of a project.

This interface management is something that free software has done very well. Many commercial developers would be well served to learn from open source's interface management practices.

Control

Control is something customers and end users have never had over their code. You don't *buy* proprietary software, you rent it, and that rental can be rescinded at any time. If you read the end-user license agreements (EULAs) that accompany proprietary software, you may be left with the feeling that you are not trusted and not liked all that much. For instance, in Microsoft Word's EULA, there is this charming note:

> You may not copy or post any templates available through Internet-based services on any network computer or broadcast it in any media.

So, if you were to take a standard Microsoft Word template (which all templates are derived from) and make one that is suited to your business as, say, a publisher, you would be in violation of your EULA with Microsoft, and thus vulnerable to its lawfirm.

Controlling your software destiny is something I consider extremely important. Take, for instance, my employer, Google. We are able to fix and change the Linux kernel to fit our very specific needs. Do we have to check with Linus or one of his lieutenants before, during, or after we change the network stack? No. If we were running NT on our machines, we would be unable to get such changes made, and were we to enter into a deal where Microsoft would incorporate our requested changes, we would in effect be informing a competitor of our development strategy.

Another example is a recent service pack from Microsoft, which featured a firewall and antivirus package. This package, which is turned on by default after service pack installation, was aimed at stopping the viruses and Trojans endemic to the Windows experience. Funnily enough, it considered iTunes a virus and presented a fairly confusing message asking the user to authorize the program's use of the network.[9] That Microsoft's own media player, which has common network access methods, *wasn't* impeded is telling.

Your computer is not your own; you only borrow that which makes it useful, and when that is taken away, you are left with nothing but a toxic pile of heavy metals and aluminum.

I think this is a subtle but important part of open source's popularity. Many people and companies are interested in controlling their own destiny, and Linux and other open source programs make this possible.

Intellectual Property

Free-software developers believe in intellectual property, probably more so than people who never consider open source software. Developers creating open source have to believe, as the entire structure of the GPL, BSD, MPL, and other licenses depends on the existence of copyright to enforce the clever requirements of those licenses.

---

9   iTunes has a nifty sharing mechanism whereby users stream music to other iTunes users over the network. It's pretty neat.

When you hear people criticizing free-software developers as guiltless communists or pie-in-the-sky dreamers, it is worth remembering that without copyright, there can't be free software.

Discussions concerning intellectual property and free software usually revolve around two issues: patents and trademark. Software can be patented, and things can be trademarked. Exactly how these intersect with free software is complex. Can a piece of software which is patented be released under the GPL and still hold to the letter of the license? Can a program name be trademarked and then released under the BSD and still be a meaningful release? Legal opinion and precedence thus far provide no definite answer.

Open source developers are learning, though, paying attention to the current events around intellectual property and how it affects them.

The reality of intellectual property is something modern developers are almost required to learn. Learning the laws concerning software is the way to protect themselves from those who might send the feds out to arrest them when they come to the United States. I know that sounds like I'm typing this with tinfoil on my head, but I am not kidding.[10]

The problem with this learning process is that it does take time away from coding, which is not good and is a net loss for free software—which may indeed have been the whole point.

Some Final Words

While open source software is about freedom and licenses, it is nonetheless true that open source costs less, under many circumstances, than proprietary software. This is an important aspect of free software. Additionally, it has to be cost competitive against other free products, just as software that costs money must compete against an open source/free offering.

---

10 I wish I were, but I'm not. It happened to Russian developer Dmitry Sklyarov, who intended to discuss his reverse engineering of the Adobe PDF file format at the DefCon developers conference in Las Vegas. Upon landing at McCarran International Airport, he was met by the FBI, which placed him under arrest under the auspices of the Digital Millenium Copyright Act on behalf of Adobe Corporation. As a result, Linux kernel developers no longer have a substantial meeting in the United States, choosing instead to meet in Canada and Australia, two countries that do not have similar laws and rarely extradite for intellectual property–related crimes of this nature. Developers felt this was necessary because the Linux kernel uses code that was reverse engineered. Reverse engineering, by the way, is what made Dell, Phoenix, AMI, AMD, EMC, and a large number of other companies both possible and profitable.

Free Things Are Still Cheaper Than Expensive Things

When I say "competes against other free products," I'm talking about pirated copies of Windows, Office, SQL Server, Oracle, and many others competing against Linux, OpenOffice, MySQL, Postgres, and other best-of-breed free software applications. These applications are doing very well in environments that have little regard, legally or culturally, for software licenses.

Free things have a velocity all their own, and people forget that. I'll leave you with a little anecdote from when I was working for a large law firm in Washington, DC. I was still in college studying computer science, and I ran the law firm's email network during the day. This was 1996 or so, and TCP/IP was clearly the big winner in the network format wars versus NetBIOS and SNA, to a degree that no one could have appreciated. I was in the elevator with one of the intellectual property attorneys at the firm—a fairly technical guy—when he said something like: "You know, if TCP/IP had been properly protected and patented, we could have rigged it so that every packet cost money; they really missed the boat on that one."

Where would the Internet be if this was true? I don't know, but I do know one thing: the Internet would not be running TCP/IP. So, enjoy the freedom of open source software. It is there for you!

A Tale of Two Standards

> It was the best of protocols, it was the worst of protocols, it was the age of
> monopoly, it was the age of Free Software, it was the epoch of openness, it was
> the epoch of proprietary lock-in, it was the season of GNU, it was the season of
> Microsoft, it was the spring of Linux, it was the winter of Windows....

Samba is commonly used as the "glue" between the separate worlds of Unix and
Windows, and because of that, Samba developers have to intimately understand the
design and implementation decisions made in both systems. It is no surprise that
Samba is considered one of the most difficult Free Software projects to understand
and to join, outclassed in complexity only by the voodoo black art of Linux kernel
development. Samba really isn't that hard, however, once you look at the different
standards implemented in the two systems (although some of the decisions in Windows can cause raised eyebrows).

In developing Samba, we're creating a bridge between the most popular standards
currently deployed in the computing world: the Unix/Linux standard of POSIX and
the Microsoft-developed de facto standard of Win32. In this chapter, I will examine
these two standards from an application programmer's perspective. In doing so, I
thought it might be instructive to look at the reasons why each of them exists, what
the intention for creating the particular standard might have been, and how well they
have stood the test of time and the needs of programmers. A historical perspective is
very important, as we look to the future and decide what standards we should

encourage governments and businesses to support, and what effect this will have on the software landscape in the early 21st century.

> Standard: (noun) A flag, banner, or ensign, especially. An emblem or flag of an army, raised on a pole to indicate the rallying point in battle.[1]

The POSIX Standard

POSIX was named (like many things in the Unix software world) by Richard Stallman. It stands for Portable Operating System Interface–X, meaning a portable definition of a Unix-like operating system API. The reason for the existence of the POSIX standard is interesting and lies in the history of the Unix family of operating systems.

As is commonly known, Unix was created in 1969 at AT&T Bell Labs by Ken Thompson and Dennis Richie. Not originally designed for commercialization, the source code was shipped to universities around the world, most notably Berkeley in California. One of the world's first truly portable operating systems, Unix soon splintered into many different versions as people modified the source code to meet their own requirements. Once companies like Sun Microsystems and the original, prelitigious SCO (Santa Cruz Organization) began to commercialize Unix, the original Unix system call API remained the core of the Unix system, but each company added proprietary extensions to differentiate their own version of Unix. Thus began the first of the "Unix wars" (I'm a veteran, but I don't get disability benefits for the scars they caused). For independent software vendors (ISVs), such proprietary variants were a nightmare. You couldn't assume that code that ran correctly on one Unix would even *compile* on another.

During the late 1980s, in an attempt to create a common API for all Unix systems, and fix this problem, the POSIX set of standards was born. Because no one trusted any of the Unix vendors, the Institute of Electrical and Electronics Engineers (IEEE) shepherded the standards process and created the 1003 series of standards, known as POSIX. The POSIX standards cover much more than the operating system APIs, going into detail on system commands, shell scripting, and many other parts of what it means to be a Unix system. I'm only going to discuss the programming API standard part of POSIX here because, as a programmer, that's really the only part of it I care about on a day-to-day basis.

Few people have actually seen an official POSIX standard document, as the IEEE charges money for copies. Back before the Web became really popular, I bought one just to take a look at the real thing. It wasn't cheap (a few hundred dollars, as I recall). Amusingly enough, I don't think Linus Torvalds ever read or referred to it when he was creating Linux; he used other vendors' references to it and manpage descriptions of what POSIX calls were supposed to do.

---

1   *http://www.dictionary.com.*

Reading the POSIX standard document, however, is very interesting. It reads like a legal document; every line of every section is numbered so that it can be referred to in other parts of the text. It's *detailed*. *Really* detailed. The reason for such detail is that it was designed to be a complete specification of how a Unix system has to behave when called from an application program. The secret is that it was meant to allow someone reading the specification to completely reimplement their own version of a Unix operating system starting from scratch, with nothing more than the POSIX spec. The goal is that if someone writes an application that conforms to the POSIX specification, the resulting application can be compiled with *no* changes on any system that is POSIX compliant. There is even a POSIX conformance suite, which allows a system passing the tests to be officially branded a POSIX-compliant system. This was created to reduce costs in government and business procurement procedures. The idea was that you specified "POSIX compliant" in your software purchasing requests, the cheapest system that had the branding could be selected, and it would satisfy the system requirement.

This ended up being less useful than it sounds, given that Microsoft Windows NT has been branded POSIX compliant and generic Linux has not.

Sounds wonderful, right? Unfortunately, reality intruded its ugly head somewhere along the way. Vendors didn't want to give up their proprietary advantages, so each pushed to get its particular implementation of a feature into POSIX. As all vendors don't have implementations of all parts of the standard, this means that many of the features in POSIX are *optional*—usually just the one you need for your application. How can you tell if an implementation of POSIX has the feature you need? If you're lucky, you can test for it at compile time.

The GNU project suffered from these "optional features" more than most proprietary software vendors because the GNU software is intended to be portable across as many systems as possible. To make their software portable across all the weird and wonderful POSIX variants, the wonderful suite of programs known as GNU autoconf was created. The GNU autoconf system allows you to test to see whether a feature exists or works correctly before you even compile the code, thus allowing an application programmer to degrade missing functionality gracefully (i.e., *not* fail at runtime).

Unfortunately, not all features can be tested this way, as sometimes a standard can give too much flexibility, thus causing massive runtime headaches. One of the most instructive examples is in the `pathconf()` call. The function prototype for `pathconf()` looks like this :

```
long pathconf(char *path, int name);
```

Here, `char *path` is a pathname on the system and `int name` is a defined constant giving a configuration option you want to query. The constants causing problems are:

```
_PC_NAME_MAX
_PC_PATH_MAX
```

`_PC_NAME_MAX` queries for the maximum number of characters that can be used in a file-name in a particular directory (specified by `char *path`) on the system. `_PC_PATH_MAX` queries for the maximum number of characters that can be used in a relative path from the particular directory. This seems fine until you consider how Unix filesystems are structured and put together. A typical Unix filesystem looks like Figure 3-1.

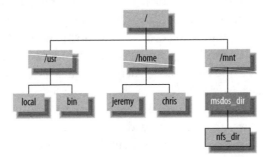

Figure 3-1. *Typical Unix filesystem*

Any of the directory nodes, such as `/usr/bin` or `/mnt`, could be a different filesystem type, not the standard Unix filesystem (maybe even network mounted). In Figure 3-1, the `/mnt/msdos_dir` path has been mounted from a partition containing an old MS-DOS-style FAT filesystem type. The maximum directory entry length on such a system is the old DOS 8.3 maximum of 11 characters. But below the Windows directory could be mounted a different filesystem type with different maximum name restrictions—maybe an NFS mount from a different machine, for example, on the path `/mnt/msdos_dir/nfs_dir`. Now the `pathconf( )` can accommodate these restrictions and tell your application about it—*if you remember to call it on every single possible path and path component your application might use!* Hands up, all application programmers who actually do this....Yes, I thought so. (You at the back, put your hand down. I know how you do things in the U.S. Star Wars missile defense program, but no one programs in ADA anymore, plus your tests never work, OK?) This is an example of something that looks good on paper but in practical terms almost no one would use in an actual application. I know we don't in Samba, not even in the "rewritten from scratch with correctness in mind" Samba4 implementation.

Now let's look at an example of where POSIX gets it spectacularly wrong, and why this happens.

First Implementation Past the Post

Any application program dealing with multiple access to files has to deal with file locking. File locking has several potential strategies, ranging from the "lock this file for my exclusive use" method, to the "lock these 4 bytes at offset 23 as I'm going to be reading from them soon" level of granularity. POSIX implements this kind of functionality via the fcntl( ) call, a sort of jack-of-all-trades for manipulating files (hence "fcntl → file control"). It's not important to know exactly how to program this call. Suffice it to say that a code fragment to set up such a byte range lock looks something like this:

```
int fd = open("/path/to/file", O_RDWR);
```

Now, set up the struct flock structure to describe the kind of byte range lock we need:

```
int ret = fcntl(fd, F_SETLKW, &flock_struct);
```

If ret is zero, we got the lock. Looks simple, right? The byte range lock we got on the region of the file is advisory. This means that other processes can ignore it and are not restricted in terms of reading or writing the byte range covered by the region (that's a difference from the Win32 way of doing things, in which locks are mandatory; if a lock is in place on a region, no other process can write to that region, even if it doesn't test for locks). An existing lock can be detected by another process doing its own fcntl( ) call, asking to lock its own region of interest. Another useful feature is that once the file descriptor open on the file (int fd in the previous example) is closed, the lock is silently removed. This is perfectly acceptable and a rational way of specifying a file locking primitive; just what you'd want.

However, modern Unix processes are not single threaded. They commonly consist of a collection of separate threads of execution, separately scheduled by the kernel. Because the lock primitive has a per-process scope, this means that if separate threads in the same process ask for a lock over the same area, it won't conflict. In addition, because the number of lock requests by a single process over the same region is not recorded (according to the spec), you can lock the region 10 times, but you need to unlock it only once. This is sometimes what you want, but not always: consider a library routine that needs to access a region of a file but doesn't know if the calling processes have the file open. Even if an open file descriptor is passed into the library, the library code can't take any locks. It can never know if it is safe to unlock again without race conditions.

This is an example of a POSIX interface not being future proofed against modern techniques such as threading. A simple amendment to the original primitive allowing a user-defined "locking context" (like a process ID) to be entered in the struct flock structure used to define the lock would have fixed this problem, along with extra flags allowing the number of locks per context to be recorded if needed.

But it gets worse. Consider the following code:

```
int second_fd;
int ret;
struct flock lock;

int fd = open("/path/to/file", O_RDWR);
```

```
/* Set up the "struct flock" structure to describe the
kind of byte range lock we need. */

lock.l_type = F_WRLCK;
lock.l_whence = SEEK_SET;
lock.l_start = 0;
lock.l_len = 4;
lock.l_pid = getpid();

ret = fcntl(fd, F_SETLKW, &lock);

/* Assume we got the lock above (ie. ret == 0). */

/* Get a second file descriptor open on
the original file. Assume this succeeds. */

second_fd = dup(fd);

/* Now immediately close it again. */

ret = close(second_fd);
```

What do you think the effect of this code on the lock created on the first file descriptor should be (so long as the close( ) call returns zero)? If you think it should be silently removed when the second file descriptor is closed, congratulations—you have the same warped mind as the people who implemented the POSIX spec. Yes, that's correct. Any successful close( ) call on any file descriptor referencing a file with locks will drop all the locks on that file, even if they were obtained on another, still-open file descriptor.

Let me be clear: *this behavior is never what you want.* Even experienced programmers are surprised by this behavior, because it makes no sense. After I've described this to Linux kernel hackers their responsse have been that of stunned silence, followed by "but why would it do that"?[2]

The reason is historical and in my opinion, reflects a flaw in the POSIX standards process, one that hopefully won't be repeated in the future. By talking to longtime BSD hacker and POSIX standards committee member, Kirk McKusick (he of the BSD daemon artwork), I finally tracked down why this insane behavior was standardized by the POSIX committee. As he recalls, AT&T took the current behavior to the standards committee as a proposal for byte range locking, as this was how their current code

---

2   To discover if this functionality was actually correctly used by any application program or if anything really depended on it, Andrew Tridgell, the original author of Samba, once hacked the kernel on his Linux laptop to write a kernel debug message if ever this condition occurred. After a week of continuous use, he found one message logged. When he investigated, it turned out to be a bug in the exportfs NFS file exporting command, whereby a library routine was opening and closing the /etc/exports file that had been opened and locked by the main exportfs code. Obviously, the authors didn't expect it to do that either.

implementation worked. The committee asked other ISVs if this was how locking should be done. The ISVs who cared about byte range locking were the large (at the time) database vendors, such as Oracle, Sybase, and Informix. All these companies did byte range locking within their own applications, and none of them depended on, or needed, the underlying operating system to provide locking services for them. So their unanimous answer was "we don't care." In the absence of any strong negative feedback on a proposal, the committee added it "as is" and took as the desired behavior the specifics of the first implementation, the brain-dead one from AT&T.

The "first implementation past the post" style of standardization has saddled POSIX systems with one of the most broken locking implementations in computing history. My hope is that eventually Linux will provide a sane superset of this functionality that can be adopted by other Unixes and eventually find its way back into POSIX.

OK, having dumped on POSIX enough, let's look at one of the things that POSIX really got right and that is an example worth following in the future.

Future Proofing

One of the great successes of POSIX is the ease in which it has adapted to the change from 32-bit to 64-bit computing. Many POSIX applications were able to move to a 64-bit environment with very little or no change, and the reason for that is abstract types.

In contrast to the Win32 API (which even has a bit-size dependency in its very name), all of the POSIX interfaces are defined in terms of abstract datatypes. A file size in POSIX isn't described as a "32-bit integer" or even as a C-language type of unsigned int, but as the type off_t. What is off_t? The answer depends completely on the system implementation. On small or older systems, it is usually defined as a signed 32-bit integer (it's used as a seek position so that it can have a negative value), and on newer systems (Linux, for example) it's defined as a signed 64-bit integer. As long as applications are careful to cast integer types only to the correct off_t type and use these for file-size manipulation, the same application will work on both small and large POSIX systems.

This wasn't done all at once, because most commercial Unix vendors have to provide binary compatibility to older applications running on newer systems, so POSIX had to cope with both 32-bit file-sized applications running alongside newer 64-bit-capable applications on the new 64-bit systems. The way to make this work was determined by the Large File Support working group, which finished its work during the mid-1990s.

The transition to 64 bits was seen as a three-stage process. Stage one was the original old 32-bit applications; stage two was seen as a transitional stage, where new versions of the POSIX interfaces were introduced to allow newer applications to explicitly select 64-bit sizes, and stage three was where all the original POSIX interfaces default to 64-bit clean.

As is usual in POSIX, the selection of what features to support was made available using compile-time macro definitions that could be selected by the application writer. The macros used were:

```
_LARGEFILE_SOURCE
_LARGEFILE64_SOURCE
_FILE_OFFSET_BITS
```

If _LARGEFILE_SOURCE is defined, a few extra functions are made available to applications to fix the problems in some older interfaces, but the default file access is still 32 bit. This corresponds to stage one, described earlier.

If _LARGEFILE64_SOURCE is defined, a whole new set of interfaces is available to POSIX applications that can be explicitly selected for 64-bit file access. These interfaces explicitly allow 64-bit file access and have 64 coded into their names. So, open( ) becomes open64( ), lseek( ) becomes lseek64( ), and a new abstract datatype called off64_t is created and used instead of the off_t file-size datatype in such structures as struct stat64. This corresponds to stage two.

_FILE_OFFSET_BITS represents stage three; this macro can be undefined or set to the values 32 or 64. If undefined or set to 32, it corresponds to stage one (_LARGEFILE_SOURCE). If set to 64, all the original interfaces such as open( ) and lseek( ) are transparently mapped to the 64-bit clean interfaces. This is the end stage of porting to 64 bits, where the underlying system is inherently 64 bit, and nothing special needs to be done to make an application 64-bit aware. On a native 64-bit system that has no older 32-bit binary support, this becomes the default.

As you can see, if a 32-bit POSIX application had no embedded dependencies on file size, simply adding the compile-time flag -D_FILE_OFFSET_BITS=64 would allow a transparent port to a 64-bit system. There are few such applications, though, and Samba was not one of them. We had to go through the stage-two pain of using 64-bit interfaces explicitly (which we did around 1998) before we could track down all the bugs associated with moving to 64 bits. But we didn't have to rewrite completely, and I consider that a success of the underlying standard.

This is an example of how the POSIX standard was farsighted enough to define some interfaces that were so portable and clean that they could survive a transition of underlying native CPU word length. Few other standards can make that claim.

Wither POSIX?

The POSIX standard has not been static; it has managed to evolve (although some would argue too slowly) over time. A major step forward was the establishment of the Single Unix Specification (SUS), which is a superset of POSIX developed in 1998 and adopted by all the major Unix vendors and shepherded by the Unix standards body, "the Open Group." It was a great leap forward when this specification was finally made available for free on the Web from the Open Group web site at *http://www.unix.org*. It certainly saved me from having to hunt down cheap POSIX specifications in secondhand bookshops in Mountain View, California.

The expanded SUS now covers such issues as real-time programming, concurrent programming via the POSIX thread (pthread) interfaces, and internationalization and localization, but unfortunately it does not cover file Access Control Lists (ACLs). Sadly, that specification was never fully agreed on, and so has never made it into the official documents. Interestingly enough, the SUS also doesn't cover the GUI elements, because the history of Unix as primarily a server operating system has meant that GUIs have never been given the priority necessary for Unix to become a desktop system.

Looking at what happened with ACLs is instructive when considering the future of POSIX and the SUS. Because ACLs were sorely needed in real-world environments, individual Unix vendors, such as SGI, Sun, HP, and IBM, added them to their own Unix variants. But without a true standards document, they fell into their old evil ways and added them with different specifications. Then along came Linux....

Linux changed everything. In many ways, the old joke is true: Linux is the Unix defragmentation tool.[3] As Linux became more popular, programs originally written for other Unixes were first ported to it, and then after a while were written for it and then ported to other platforms. This happened to Samba. Sun's SunOS on a SPARC system was, at first, our primary user platform, but after five years or so we rapidly migrated to Linux on Intel x86 systems. We now develop almost exclusively on Linux, and from there port to other Unix systems.

This means the Linux interfaces are starting to take over as the most important standards for Unix-like systems to follow, in some ways supplanting POSIX and the SUS. The ACL implementation for Linux was added into the system, at first via a patch by Andreas Grünbacher, held externally to the main kernel tree. Finally it was adopted by the main Linux vendors, SuSE (now Novell) and Red Hat, and has become part of the official kernel. Other free Unix systems such as FreeBSD quickly followed with their own implementation of the last draft of the POSIX ACL specification, and now there are desktop GUI and other application programs that use the Linux ACL interfaces. As this code is ported to other systems, the pressure is on them to conform to the Linux APIs, not to any standards document. Sun has announced that its Solaris 10 on Intel release will run Linux applications "better than Linux" and will be fully compatible at the system call level with Linux applications. This means Sun must have mapped the Linux ACL interface onto the Solaris one. Is that a good thing?

In a world where Linux is rapidly becoming the dominant version of Unix, does POSIX still have relevance, or should we just assume Linux is the new POSIX?

---

3   This was inspired by novice system administrators coming to Unix from the Windows platform for the first time and asking "where is the system defragmentation tool?", the concept of a filesystem designed well enough not to need one being outside their experience.

The Win32 (Windows) Standard

Win32 was named for an expansion of the older Microsoft Windows interface, renamed the Win16 interface once Microsoft was shipping credible 32-bit systems. I have a confession to make. In my career, I completely ignored the original 16-bit Windows on MS-DOS. At that time, I was already working on sane 32-bit systems (68000 based), and dealing with the original insane 8086 segmented architecture was too painful to contemplate. Win32 was Microsoft's attempt to move the older architecture beyond the limitations of MS-DOS and into something that could compete with Unix systems—and to a large extent Microsoft succeeded spectacularly.

The original 16-bit Windows API added a common GUI on top of MS-DOS, and also abstracted out the lower-level MS-DOS interfaces so that application code had a much cleaner "C" interface to operating system services (not that MS-DOS provided many of those). The Win32 Windows API was actually the "application" level API (not the system call level; I'll discuss that in a moment) for a completely new operating system that would soon be known as Windows NT ("New Technology"). This new system was designed and implemented by Dave Cutler, the architect of Digital Equipment Corporation's VMS system, long a competitor to Unix. It does share some similarities with VMS. The interface choice for applications was very interesting, sitting on top of a system call interface that looks like Figure 3-2.

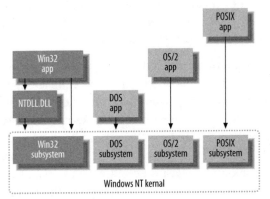

Figure 3-2. *Architecture of the Win32 API*

The idea behind the Windows NT kernel was that it could host several "subsystem" system call interfaces, providing completely different application behavior from the same underlying kernel. It was meant to be a completely customizable operating system, providing different kernel "personalities" any ISV might require. The DOS subsystem and the (not-shown) 16-bit Windows subsystem were essential, as they provided backward compatibility for applications running on MS-DOS and 16-bit Windows; the new operating system would have gathered little acceptance had it not been able to run all the old MS-DOS and Windows applications.

The OS/2 subsystem was designed to allow users of text mode OS/2 applications (which was at one time a Microsoft product) to port them to Windows NT.

The two interesting subsystems are the original POSIX subsystem and the new Win32 subsystem. The POSIX subsystem was added, as the POSIX standard had become very prevalent in procurement contracts. Many of these valuable contracts were available only to systems that passed the POSIX conformance tests. So Microsoft added a minimal POSIX subsystem into the new Windows NT operating system. This original subsystem was, I think it's fair to say, deliberately crippled to make it unuseful for real-world applications: applications using it had no network access and no GUI access, so although a POSIX-compliant system might be required in a procurement contract, there usually was no requirement that the applications running on that system also had to be POSIX compliant. This allowed new applications using the Microsoft-preferred Win32 subsystem to be used instead. All might not have been lost if Microsoft had documented the internal subsystem interface, allowing third-party ISVs to create their own Windows NT kernel subsystems, but Microsoft kept this valuable information to itself (there was an exception to this, which I'll discuss shortly).

So, let's examine the Win32 standard API, the interface designed to run on top of the Win32 kernel subsystem. It would be logical to assume that, like the POSIX system calls, the calls defined in the Win32 API would closely map to kernel-level Win32 subsystem system calls. But that would be incorrect. It turns out that, when released, the Win32 subsystem system call interface was completely undocumented. The calls made from the application-level Win32 API were translated, via various shared libraries (DLLs in Windows parlance)—mainly the *NTDLL.DLL* library—into the real Win32 subsystem system calls.

Why do this, one might ask? Well, the official reasoning is that it allows Microsoft to tune and modify the system call layer at will, improving performance and adding features without being forced to provide backward compatibility application binary interfaces (or ABIs for short). The more nefarious reasoning is that it allows Microsoft applications to cheat, and call directly into the undocumented Win32 subsystem system call interface to provide services that competing applications cannot. Several Microsoft applications were subsequently discovered to be doing just that, of course. One must always remember that Microsoft is not just an operating system vendor, but also the primary vendor of applications that run on its platforms. These days, this is less of a problem, as there are several books that document this system call layer, and there are several applications that allow snooping on any Windows NT kernel calls made by applications, allowing any changes in this layer to be quickly discovered and published. But it left a nasty taste in the mouths of many early Windows NT developers (myself included).

The original Win32 application interface was, on the surface, very well documented and cheaply available in paper form (five books at only $20 each; a bargain compared to a POSIX specification). Like most things in Windows, on the *surface* it looks great. It covers much more than POSIX tries to standardize, and so offers flexible interfaces for manipulating the GUI, graphics, sound, and pen computing, as well as all the standard system services such as file I/O, file locking, threading, and security. Then you start to program with it. If you're used to the POSIX specifications, you almost immediately notice something is different. The *details* are missing. It's *fuzzy* on the details. You notice it the first time you call an API at runtime, and it returns an error that's not listed anywhere in the API documentation. "That's funny...." you think. With POSIX, all possible errors are listed in the return codes section of the API call. In Win32, the errors are a "rough guide."

The lack of detail is one of the reasons that the Wine project finds it difficult to create a working implementation of the Win32 API on Linux. How do you know when it's done? Remember that Linus, with some help, was able to create a decent POSIX implementation within a few years. The poor Wine developers have been laboring at this for 12 years, and it's still not finished. There's always one more wrinkle, one more undocumented behavior that some critical application depends on. Reminds me of Samba somehow, and for very similar reasons.

It's not *entirely* Microsoft's fault. It hasn't documented its API because it hasn't needed to. POSIX was documented in detail due to need: the need of the developers creating implementations of the standard. Microsoft knows that whatever it makes the API do in the next service pack, that's still the Win32 standard. "Wherever you go, there you are," so to speak.

However, the Win32 design does some things very well; security, for instance. Security isn't the number one thing people think of when considering Windows, but in the Win32 API, security is a very great concern. In Win32, every object can be secured, and a property called a Security Descriptor, which contains an ACL, can be attached to it. This means objects—such as processes, files, directories, and even Windows—can have ACLs attached. This is much cleaner than POSIX, in which only objects in the filesystem can have ACLs attached to them.

So, let's look at a Win32 ACL. As in POSIX, all users and groups are identified by a unique identifier. On POSIX, it's a uid_t type for users, and a gid_t type for groups. In Win32, both are of type SID or security identifier. A process or thread in Win32 has a token attached to it that lists the primary SID of the process owner and a list of secondary group SID entries this user belongs to. Like in POSIX, this is attached to a process at creation time and the owner can't modify it to give himself more privileges. A Win32 ACL consists of a list of SID entries with an attached bit mask identifying the operations this SID entry allows or denies. Sounds reasonable, right? But the devil is in the details (see Figure 3-3).

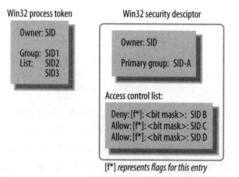

Win32 process token

Owner: SID

Group: SID1
List: SID2
SID3

Win32 security desciptor

Owner: SID

Primary group: SID-A

Access control list:

Deny: [f*]: <bit mask>: SID B
Allow: [f*]: <bit mask>: SID C
Allow: [f*]: <bit mask>: SID D

[f*] represents flags for this entry

Figure 3-3. *Win32 access control*

Each SID entry in an ACL can be an allow entry or a deny entry. Their *order* is important. Reorder a list of entries and swap a deny entry with an allow entry, and the meaning of the ACL can change completely. POSIX ACLs don't have that problem because the evaluation algorithm defines the order in which entries are examined. In addition, the flags defining the entry (marked as [f*] in Figure 3-3) control whether an entry is inherited when the ACL is attached to a "container object" (such as a directory in the filesystem) and may also affect other attributes of this particular entry.

The bit mask enumerates the permissions that this entry allows or denies. But the permissions are (naturally) different, depending on what object the ACL is attached to. Let's look at the kinds of permissions available for a file object:

DELETE
> Delete the object.

READ_CONTROL
> Read the ACL on an object.

WRITE_DAC
> Write the ACL on an object.

FILE_READ_DATA
> Read from the file.

FILE_READ ATTRIBUTES
> Read file metadata.

FILE_READ_EA
> Read extended attributes (if the file has any).

FILE_WRITE_DATA
> Write to the file.

FILE_WRITE_EA
> Write extended attributes (if the file has any).

`FILE_EXECUTE`
   Open for execute (why do we need the `.EXE` tag then?).

`SYNCHRONIZE`
   A permission related to an open file handle, not the file.

And this is one of the simpler kinds of permission-bearing objects in Win32.

If the Win32 API treats security so seriously, why does Windows fail most security tests in the real world? The answer is that most applications ignore this wonderful, flexible security mechanism because it's just *too hard to use*—just like the problem with the POSIX `pathconf( )` call. No one can use the security mechanism correctly; applications would degenerate into a mess. It doesn't help that Microsoft, having realized the APIs controlling security were too difficult to use, keeps adding functions to simplify this mess, sometimes also adding new APIs with a new service pack. In addition, as Microsoft has moved in the "Active Directory" world, it has extended the underlying semantics of the security mechanism,adding new flags and behaviors.

Try taking a look at the "file security dialog" in Windows 2000. It's incomprehensible. No one, especially a system administrator, can keep track of this level of detail across their files. Everyone just sets one default ACL on the root of a directory hierarchy and hopes for the best. Most administrators usually want to do two simple things with an ACL: allow group X but not user Y, and allow group X and also user Z. This is just about comprehensible with POSIX ACLs, although those are near the limit of complexity that people can deal with. The Win32 security system is orders of magnitude more complex than that; it's hopelessly overdesigned. Computer scientists love it, as it's possible to do elegant little proofs of how secure it is, but in the real world, it's simply too much to deal with effectively—great idea, adding ACLs to every system object, but a real shame about the execution.

Just to spread the blame around, the networking "experts" who designed the latest version of Sun's network filesystem, NFS version 4, fell in love with this security mechanism and decided it would be a great idea to add it into the NFSv4 specification. They probably thought it would make interoperability with Windows easier. Of course, they didn't notice that Microsoft had been busily extending the security mechanism as Windows has developed, so they standardized on an old version of the Windows ACL mechanism, as Microsoft documented it (not as it actually works). So now, the Unix world has to deal with this mess—or rather, a new network filesystem with an ACL model that is *almost*, but not quite, compatible with Windows ACLs, and that is completely alien to anything currently found on Unix. I sometimes feel Unix programmers are their own worst enemies.

The Tar Pit: Backward Compatibility

Now, as an example of where Win32 got things spectacularly wrong, I want to look at a horror from the past that unfortunately got added into the Win32 interfaces due to the MS-DOS heritage. My pet hate with Win32 is the idea of "share modes" on open files. In my opinion, this one single legacy design decision has probably done more than any other to hold back the development of cluster-aware network filesystems on Win32 systems.

Under POSIX, an open( ) call is very simple. It takes a pathname to open, the way in which you want to access or create the file (read, write, or both with various create types), and a permission mask that gets applied to files you do create. Under Win32, the equivalent call, CreateFile( ), takes seven parameters, and the interactions among them can be ferociously complex. The parameter that causes all the trouble is the ShareMode parameter, which can take values of any of the following constants OR'ed together:

FILE_SHARE_READ
   Allow others to open for read.

FILE_SHARE_WRITE
   Allow others to open for write.

FILE_SHARE_NONE
   Don't allow any other opens.

FILE_SHARE_DELETE
   Allow open for delete intent.

To make these semantics work, any Windows kernel dealing with an open file has to know about every other application on the system that might have this file open. This was fine back in the single-machine MS-DOS days, when these semantics were first designed, but it is a complete *disaster* when dealing with a clustered filesystem in which a multitude of connected file servers may want to give remote access to the same file, even if they serve out the file read-only to applications. They have to consult some kind of distributed lock management system to keep these MS-DOS-inherited semantics working. While this can be done, it complicates the job enormously and means cluster communication on every CreateFile( ) and CloseHandle( ) Zcall.

This is the bane of backward compatibility. This idea of "share modes" arbitrating what access concurrent applications can have to a file is the cause of many troubles on a Windows system. Ever wonder why Windows has a mechanism built in to allow an application to schedule a file to be moved, but only after a reboot? Share modes in action. Why are some files on a Windows server system impossible to back up due to "another program is currently using this file" errors? Share modes again. There is no security permission that can prevent a user from opening a file with, effectively, "deny all" permissions. If you can open the file for read access, you can get a share mode on it, by design. Consider a network-shared copy of Microsoft Office. Any user must be able to open the file `WINWORD.EXE` (the binary file containing Microsoft Word) to execute it. Given these semantics, any user can open the file with `READ_DATA` access with the `ShareMode` parameter set to `FILE_SHARE_NONE` and thus block use of the file, even over the network. Imagine on a Unix system, being able to open the `/etc/passwd` file with a share mode and deny all other processes access. Watch the system slowly grind to a halt as the other processes get stuck in this tar pit....

World Domination, Fast

I've heaped enough opprobrium on Win32. Let's give it a break and consider something the designers really did get right, and one of the advantages it has over POSIX. I'm talking about the early adoption of the Unicode standard in Win32. When Microsoft was creating Win32, one of the things it realized was that this couldn't just be another English-only, American- and European-centric standard. It had to be able to not only cope with, but also encourage, applications written in all world languages (never accuse Microsoft of thinking small in its domination of the computing world).

Given those criteria, its adoption of Unicode as the native character set for all the system calls in Win32 was a stroke of genius. Even though the Asian countries aren't particularly fond of Unicode, because it merges several character sets they consider separate into one set of code points, Unicode is the best way to cope with the requirements of internationalization and localization in application development.

To allow older MS-DOS and Win16 applications to run, the Win32 API is available in two different forms, selectable by a compiler `#define` of `-DUNICODE` (it also helps if you own the compiler market for Windows, as Microsoft does, as you can standardize tricks like this). The older code-page-based applications call Win32 libraries that internally convert any string arguments to 16-bit Unicode and then call the *real* Win32 library interface, which, like the Windows NT kernel, is Unicode only.

In addition, Win32 comes with a full set of library interfaces to split out the text messages an application may need to display into resource files so that ISVs can easily have them translated for a target market. This eases the internationalization and localization burdens considerably for vendors.

What is more useful, but not as obvious, is that making the Win32 standard natively use Unicode meant developers were immediately confronted with the requirements of multilingual code development. Many applications written in English-speaking (or Western European eight-bit character set–compatible) countries are badly written, making the assumption that a character will always fit within one byte. The early versions of Samba definitely made that mistake and retrofitting multibyte character set handling into old code is a real bear to get right. I know, because I was the person who first had to work on this for Samba (later I got some much-needed help from Andrew), so I may be a little touchy on this subject.

Whenever I did Win32 development, I *immediately* designed with non-English languages in mind, and wrote everything with the abstract type TCHAR (one of the few useful abstract types in Win32), which is selectable at compile time using the Unicode defined to be either `wchar_t` with Unicode turned on, or `unsigned char` with Unicode turned off. Getting yourself in the right multibyte character set mindset from the beginning eliminates a whole class of bugs that you get when having to convert a quick "English-only" hacked-up program into something maintainable for different languages. POSIX has been catching up over the years with the `iconv( )` functionality to cope with character set conversions, and Sun designed `gettext( )` interfaces for localization, but Win32 had it all right from the start.

Wither Win32?

As with POSIX, the Win32 standard has not remained static over time. Microsoft has continued to develop and extend it, and has the advantage that anything it publishes immediately becomes the "standard," as is the case with all single vendor–defined standards.

However, Microsoft is attempting to deemphasize Win32 as it moves into its new .NET environment and the new world of "managed code." Managed code is code running under the control of an underlying virtual machine (called the Common Language Infrastructure, or CLI, in .NET) and can be made to prevent the direct memory access that is the normal mode of operation of an API designed for C coding, such as Win32 or POSIX. Free Software is also making a push into this area, with the Mono project, which implements the Microsoft C# language and .NET-managed code environment on Linux and other POSIX systems.

Even if Microsoft is as successful as it hopes to be in pushing ISV programmers to convert to .NET and managed code using its new C# language, the legacy of applications developed in C using the Win32 API will linger for decades to come. ISV programmers are an ornery lot, especially people who have mastered the Win32 API, due to its less-than-complete documentation.

What seems to happen over the years is that experienced Win32 programmers gain a sort of folk knowledge about the Win32 APIs—i.e., how they really work versus what the documentation says. I often hang out on Usenet Windows discussion groups, and the attitudes of the experienced Windows programmers are very interesting: they usually hate telling novices how stuff works. It's almost as if having learning Windows is a badge of honor, and they don't want to make earning that too easy for the neophytes. They exude an air of "they must suffer as I did."

As Microsoft becomes less interested in Win32 with the release of its new Longhorn Windows client and the move to managed code, is it possible for Microsoft to lose control of it? The POSIX standard is so complete because it was designed to allow programmers reading the standards documents to re-create a POSIX system from scratch. The Win32 standard is nowhere near as well documented as that. However, there is hope in the Wine project, which is attempting to re-create a version of the Win32 API that is binary compatible with Windows on Intel x86 systems. Wine is, in effect, a second implementation of the Win32 system, making it closer to a true vendor-independent standard. Efforts taking place at companies such as CodeWeavers and Transgaming Technologies are very promising; I just finished playing the new Windows-only game *Half-Life 2* on my desktop Linux system, using the Wine technology. This is a significant achievement for the Wine code and bodes well for the future.

Choosing a Standard

BETWEEN TWO EVILS, I ALWAYS LIKE TO TAKE THE ONE I'VE NEVER TRIED BEFORE.

—*Mae West*

So, what should we choose when examining what standards to support and develop applications for? What should we recommend to businesses and governments that are starting to look closely at the open source/free software options available?

It's important that businesses and governments selecting standards-based products pay attention to *open* standards. No more of the Microsoft Word .DOC format standard (which suffers from the same problem as Win32 in terms of it being single-vendor controlled). No de facto vendor standards, no matter how convenient. They need to select standards that are at the same level as POSIX—namely, standards to the level that other implementations can be created from the documentation. It's simple to tell when a standard meets that criterion because other implementations of it exist.

The interesting thing is that both POSIX and Win32 standards are now available on both systems. On Linux, we have the POSIX standard as native, and the Wine project provides a binary-compatible layer for compiled Win32 programs that can run many popular Win32 applications. Perhaps more interestingly for programmers, the Wine project also includes a Linux shared library, *winelib,* which allows Win32 applications to be built from source code form on POSIX systems. What you end up with is an application that looks like a native Windows application, but can be run on non-Intel platforms; something that early versions of Windows NT used to support, but now is restricted to x86-compatible processors. Taking your Win32 application and porting it using *winelib* is an easy way to get your feet wet in the POSIX world, although it won't look like a native Linux application (this may be a positive thing if your users are used to a Windows look and feel).

If you've already gone the .NET and C# route, using the Mono project may enable your code to run on POSIX systems.

On Windows, there is now a full POSIX subsystem, supported by Microsoft and available for free. Earlier I alluded to Microsoft's reluctance to release information on how to create new subsystems for the Windows NT kernel, but it turns out that earlier in its history Microsoft was not so careful. A small San Francisco–based company, Softway Systems, licensed the documentation and produced a product called OpenNT (later renamed Interix), which was a replacement for Microsoft's originally crippled POSIX subsystem. Unfortunately, OpenNT didn't sell very well; someone cruelly referred to it as having "all the application availability of Linux, with the stability of Windows." As the company was failing, Microsoft bought it (probably to bring the real gem of the Windows kernel subsystem interface knowledge back in-house) and used it to create its Services for Unix (SFU) product. SFU contains a full POSIX environment, with a software development kit allowing applications to be written that have access to networking and GUI APIs. The applications written under it run as full peers with the mature Win32 applications, and users can't tell the difference.

Recently Mcrosoft made SFU available as a free download to all Windows users. I like to think the free availability of Samba had something to do with this, but maybe I'm flattering the Samba team too much. As I like to say in my talks, "If you're into piloting Samba on Linux in your organization, you're paying too much for your Microsoft software." But what this means is that if you want to write a completely portable application, the one standard you can count on to be there and fully implemented and supported on Windows, Linux, Solaris, Apple Mac OS X, HP-UX, AIX, IRIX, and all the other Unix systems out there is POSIX.

So, if you'll excuse me, I'm going to look at porting parts of Samba to Windows....

Ben Laurie

CHAPTER 4

Open Source and Security

More than two years ago, in a fit of frustration over the state of open source security, I wrote my first and only blog entry[1] (for O'Reilly's Developer Weblogs):

> June and July were bad months for free software. First Apache chunked encoding vulnerability,[2] and just when we'd finished patching that, we get the OpenSSH hole.[3] Both of these are pretty scary—the first making every single web server potentially exploitable, and the second makes every remotely managed machine vulnerable.

> But we survived that, only to be hit just days later with the BIND resolver problems.[4] Would it ever end? Well, there was a brief respite, but then, at the end of July, we had the OpenSSL buffer overflows.[5]

> All of these were pretty agonising, but it seems we got through it mostly unscathed, by releasing patches widely as soon as possible. Of course, this is painful for users and vendors alike, having to scramble to patch systems before exploits become available. I know that pain only too well: at The Bunker,[6] we had to use every avail-

---

1   *http://www.oreillynet.com/pub/wlg/2004.*
2   *http://cve.mitre.org/cgi-bin/cvename.cgi?name=CVE-2002-0392.*
3   *http://cve.mitre.org/cgi-bin/cvename.cgi?name=CVE-2002-0639.*
4   *http://cve.mitre.org/cgi-bin/cvename.cgi?name=CVE-2002-0651.*
5   *http://cve.mitre.org/cgi-bin/cvename.cgi?name=CAN-2002-0656.*
6   Back in those days, The Bunker belonged to A.L. Digital Ltd., and it wasn't called The Bunker Secure Hosting.

able sysadmin for days on end to fix t'he problems, which seemed to be arriving before we'd had time to catch our breath from the previous one.

But I also know the pain suffered by the discoverer of such problems, so I thought I'd tell you a bit about that. First, I was involved in the Apache chunked encoding problem. That was pretty straightforward, because the vulnerability was released without any consultation with the Apache Software Foundation, a move I consider most ill advised, but it did at least simplify our options: we had to get a patch out as fast as possible. Even so, we thought we could take a little bit of time to produce a fix, since all we were looking at was a denial-of-service attack, and let's face it, Apache doesn't need bugs to suffer denial of service—all this did was make it a little cheaper for the attacker to consume your resources.

That is, until Gobbles[7] came out with the exploit for the problem. Now, this really is the worst possible position to be in. Not only is there an exploitable problem, but the first you know of it is when you see the exploit code. Then we really had to scramble. First we had to figure out how the exploit worked. I figured that out by attacking myself and running Apache under gdb. I have to say that the attack was rather marvelously cunning, and for a while I forgot the urgency of the problem while I unravelled its inner workings. Having worked that out, we were in a position to finally fix the problem, and also, perhaps more importantly, more generically prevent the problem from occurring again through a different route. Once we had done that, it was just a matter of writing the advisory, releasing the patches, and posting the advisory to the usual places.

The OpenSSL problems were a rather different story. I found these whilst working on a security review of OpenSSL commissioned by DARPA[8] and the USAF.[9] OpenSSL is a rather large and messy piece of code that I had, until DARPA funded it, hesitated to do a security review of, partly because it was a big job, but also partly because I was sure I was going to find stuff. And sure enough, I found problems (yes, I know this flies in the face of conventional wisdom— many eyes may be a good thing, but most of those eyes are not trained observers, and the ones that are do not necessarily have the time or energy to check the code in the detail that is required). Not as many as I expected, but then, I haven't finished yet (and perhaps I never will, it does seem to be a never-ending process). Having found some problems, which were definitely exploitable, I was then faced with an agonising decision: release them and run the risk that I would find more, and force the world to go through the process of upgrading again, or sit on them until I'd finished, and run the risk that someone else would discovered them and exploit them.

---

7   A hacker (or group of hackers, it is not known which).
8   The United States Defense Advanced Research Projects Agency, responsible for spending a great deal of money on national security—in this case, for a thing known as CHATS, or Composable High Assurance Trusted Systems.
9   Yes, I do mean the United States Air Force.

In fact, I dithered on this question for at least a month—then one of the problems I'd found was fixed in the development version without even being noted as a security fix, and another was reported as a bug. I decided life was getting too dangerous and decided to release the advisory, complete or not. Now, you might think that not being under huge time pressure is a good thing, but in some ways it is not. The first problem came because various other members of the team thought I should involve various other security alerting mechanisms—for example, CERT[10] or a mailing list operated by most of the free OS vendors.[11] But there's a problem with this: CERT's process is slow and cumbersome and I was already nervous about delay. Vendor security lists are also dangerous because you can't really be sure who is reading them and what their real interests are. And, more deeply, I have to wonder why vendors should have the benefit of early notification, when it is my view that they should arrange things so that their users could use my patches as easily as I can. I build almost everything from original source, so patches tend to be very easy to use. RPMs[12] and ports[13] make this harder, and vendors who release no source at all clearly completely screw up their customers. Why should I help people who are getting in the way of the people who matter (i.e., the users of the software)?

Then, to make matters worse, one of the more serious problems was reported independently to the OpenSSL team by CERT, who had been alerted by Defcon.[14] I was going, and there was no way I was delaying release of the patches until after DeFcon. So, the day before I got on a plane, I finally released the advisory. And the rest is history.

So, what's the point of all this? Well, the point is this: it was a complete waste of time. I needn't have agonised over CERT or delay or any of the rest of it. Because half the world didn't do a damn thing about the fact they were vulnerable, and because of that, as of yesterday, a worm is spreading through the Net like wildfire.

Why do I bother?

Two years later, I am still bothering, so I suppose that I do think there's some point. But there are interesting questions to ask about open source security—is it really true that "many eyes" doesn't work? How do we evaluate claims about the respective virtues of open and closed source security? Has anything changed in those two years? What is the future of open source security?

---

10 CERT is an organization funded to characterize security issues and alert the appropriate parties— a job they do not do very well, in my opinion.

11 Apparently, I'm not one, so I'm not on this list.

12 One of those recursive definitions programmers love: RPM Package Manager, a widely used system for distributing packaged open source software, particularly for various flavors of Linux.

13 FreeBSD's package management system. Also used by other BSDs.

14 DefCon is a popular hacker's convention, held annually in Las Vegas.

Many Eyes

Although it's still often used as an argument, it seems quite clear to me that the "many eyes" argument,[15] when applied to security, is not true. It is worth remembering what was originally said: "Many eyes make all bugs shallow" (Eric S. Raymond). I believe this is actually true, if read in the right context. Once you have found a bug, many eyes will, and indeed, do, make fixing it quick and easy.

Security vulnerabilities are no different in this respect—once they are found, they are generally easy to track down and fix (the Apache chunked encoding vulnerability was the hardest I've ever had to track, and even that took only one long day's work). But vulnerabilities aren't like bugs in that sense—until they are discovered. Once you find them, you have a recipe for making the software behave unexpectedly. Until that time, what do you have? A piece of software that does what you expect.

The idea that bugs and security vulnerabilities are really the same thing is quite wrong—and it's an idea that I suspect has been perpetrated by the reliability community,[16] sensing a new source of funding. Software is reliable if it does what is expected when operated as expected. It is secure if it does what is expected *under all circumstances*. This is a very critical difference, indeed. Nonsecurity bugs have a significant qualitative difference from security bugs—people don't go out of their way to find bizarre things to do to make the software go wrong just for the fun of it. And if they do, and it's not a security hole...well, yes, that's interesting, and we'll fix it one day but, in the meantime, you didn't *need* that functionality, so just stop poking yourself in the eye and it will stop hurting.

What has happened is that advocates of open source have taken the "many eyes" argument to mean that because the source is available, many people will examine it for weaknesses. This simply isn't true: most people never look at the source at all (until it doesn't work), and even if they do, most do not have the experience to find the problems. The argument simply does not hold water, and it's time we, as a community, abandon it.

However, there is an important sense in which the "many eyes" theory holds a grain of truth: those who *want* to look at the source to check for vulnerabilities, can. The interesting question is whether those who want to look the the code are generally the good guys or the bad guys. But this is a question I will come to later, when I compare open and closed source.

---

15   The argument is that if enough people look at the code, bugs (and hence security issues) will be found before they bite you.

16   Academics who study the reliability, as opposed to the security, of computer systems.

Open Versus Closed Source

Since I wrote my rant, Microsoft has decided that security is important (at least for sales), and as a result, there's been a sudden increased interest in the truth of the claim that open source is "more secure" than closed source—and, of course, the counterclaim of the opposite.

But this claim is not easy to examine, for all sorts of reasons. First, what do we mean by "more secure"? We could mean that there are fewer security bugs, but surely we have to take severity of the bugs into account, and then we're being subjective. We could mean that when bugs are found, they get fixed faster, or they damage fewer people. Or we might not be talking about bugs at all. We might mean that the security properties of the system are better in some way, or that we can more easily evaluate our exposure to security problems.

I expect that, at some point, almost everyone with a serious interest in this question will choose one of these definitions, and at some other point a completely different one.

Who Is the Audience?

It is also important to recognize that there are at least two completely different reasons to ask the question "is A more secure than B?" One is that you are trying to sell A to an audience that just wants to tick the "secure" box on their checklist, and the other is because you actually care about whether your product/web site/company/whatever is secure, and are in a position to have an informed opinion.

It is, perhaps, unkind to split the audience in this way but, sadly, it appears to be a very real split. Most people, if asked whether they think the software they use should be secure will say, "Oh yeah, security, that's definitely a good thing, we want that." But this does not stop them from clicking Yes to the dialog box that says "Would you like me to install this Trojan now?" or running products with a widely known and truly dismal security record.

However, it is a useful distinction to make. If you are trying to sell to an audience that wants to tick the security box, you will use quite different tactics than if the audience truly cares about security. This gives rise to the kind of analysis I see more and more. For example, *http://dotnetjunkies.com/WebLog/stefandemetz/archive/2004/10/11/28280.aspx* has an article titled "Myth debunking: SQL Server vs. MySQL security 2003–2004 (SQL Server has less bugs!!)." The first sentence of the article gives the game away: "Seems that yet again a MS product has less bugs that (sic) the corresponding LAMP[17] product." What is this telling us? Someone found an example of a closed source product that is "better" at security than the corresponding open source

---

17   LAMP stands for Linux, Apache, MySQL, Perl (or PHP) and is common shorthand for the cluster of open source commonly used to develop web sites.

one. Therefore, all closed source products are "better" at security than open source products. If we keep on saying it, it must be true, right?

Even if I ignore the obviously selective nature of this style of analysis, I still have to question the value of simply counting vulnerabilities. I know that if you do that, Apache appears to have a worse record than IIS recently (though not over longer periods).

But I also know that the last few supposed vulnerabilities in Apache have been either simple denial-of-service (DoS) attacks[18] or vulnerabilities in obscure modules that very few people use. Certainly I didn't even bother to upgrade my servers for any of the last half-dozen or so; they simply weren't affected.

So, for this kind of analysis to be meaningful, you have to get into classifying vulnerabilities for severity. Unfortunately, there's not really any correct way to do this. Severity is in the eye of the beholder. For example, my standard threat model (i.e., the one I use for my own servers, and generally advise my clients to use, at least as a basis) is that all local users[19] have root,[20] whether you gave it to them or not. So, local vulnerabilities[21] are not vulnerabilities at all in my threat model. But, of course, not everyone sees it that way. Some think they can control local users, so to them, these holes matter.

Incidentally, you might wonder why I dismiss DoS attacks; that is because it is essentially impossible to prevent DoS attacks, even on perfectly functioning servers, since their function is to provide a service available to all, and simply using that service enough will cause a DoS. They are unavoidable, as people subject to sustained DoS attacks know to their pain.

Time to Fix

Another measure that I consider quite revealing is "time to fix"—that is, the time between a vulnerability becoming known and a fix for it coming available. There are really two distinct measures here, because we must differentiate between private and public disclosure. If a problem is disclosed only to the "vendor,"[22] the vendor has the leisure to take time fixing it, bearing in mind that if one person found it, so will others—

---

18 In a DoS attack, the attacker prevents access by legitimate users of a service by loading the service so heavily that it cannot handle the demand. This is often achieved by a distributed denial of service (DDoS) attack, in which the attacker uses a network of "owned" (i.e., under the control of the attacker and not the legitimate owner) machines to simultaneously attack the victim's server.
19 That is, people with user accounts on the machine, rather than visitors to web pages or people with mail accounts, for example.
20 Root is the all-powerful administrative account on a Unix machine.
21 A local vulnerability is one that only a local user can exploit.
22 A term I am not at all fond of, since, although I am described as a "vendor" of Apache, OpenSSL, and so forth, I've never sold any of them.

meaning "leisure" is not the same as "forever," as some vendors tend to think. The time to fix then becomes a matter of negotiation between vendor and discloser (an example of a reasonably widely accepted set of guidelines for disclosure can be found at *http://www.wiretrip.net/rfp/policy.html*, though the guidelines are not, by any means, universally accepted) and really isn't of huge significance in any case, because the fix and the bug will be revealed simultaneously.

What is interesting to measure is the time between public disclosures (also known as zero-days) and the corresponding fixes. What we find here is quite interesting. Some groups care about security a lot more than others! Apache, for example, has never, to my knowledge, taken more than a day to fix such a problem, but Gaim[23] recently left a widely known security hole open for more than a month. Perhaps the most interesting thing is that whenever time to fix is studied, we see commercial vendors—Sun and Microsoft, for example—pitted against open source packagers—Red Hat and Debian and the like—but this very much distorts the picture. Packagers will almost always be slower than the authors of the software, for the obvious reason that they can't make their packages until the authors have released the fix.

This leads to another area of debate. A key difference between open and closed source is the number of "vendors" a package has. Generally, closed source has but a single vendor, but because of the current trend towards packagers of open source, any particular piece of software appears, to the public anyway, to have many different vendors. This leads to an unfortunate situation: open source packagers would like to be able to release their packages at the same time as the authors of the packages. I've never been happy with this idea, for a variety of reasons. First, there are so many packagers that it is very difficult to convince myself that they will keep the details of the problem secret, which is critical if the users are not to be exposed to the Bad Guys. Second, how do you define what a packager is? It appears that the critical test I am supposed to apply is whether they make money from packaging or not![24] This is not only blatantly unfair, but it also flies in the face of what open source is all about. Why should the person who participates fully in the open source process by building from source be penalized in favor of mere middlemen who encourage people *not* to participate?[25]

Of course, the argument, then, is that I should care more about packagers because if they are vulnerable, it affects more people. I should choose whom I involve in the release process on the basis of how many actual users will be affected, either posi-

---

23 A popular open source instant messaging client.
24 Of course, not all packagers make money, but I've only experienced this kind of pressure from those that do.
25 This is because vendors tend to encourage users to treat them as traditional closed source businesses—with their own support, their own versions of software, and so forth—instead of engaging the users with the actual authors of the software they are using.

tively or negatively, depending on whether I include the packager or not, by my choice. I should also take into account the importance of these users. A recent argument has been that I should involve organizations such as the National Infrastructure Security Co-ordination Centre (NISCC), a UK body that does pretty much what it says on the tin, and runs the UK CERT (see *http://www.niscc.gov.uk* for more information) because they represent users of more critical importance than mere mortals. This is an argument I actually have some sympathy with. After all, I also depend on our infrastructure. But in practice, we soon become mired in vested interests and commercial considerations because, guess what? Our infrastructure uses software from packagers of various kinds, so obviously I must protect the bottom line by making sure they don't look to be lagging behind these strange people who give away security fixes to just anyone.

If these people really cared about users, they would be working to find ways that enable the users to get the fixes directly from the authors, without needing the packager to get its act together before the user can have a fix. But they don't, of course. They care about their bank balance, which is the saddest thing about security today: it is seen as a source of revenue, not an obligation.

Incidentally, a recent Forrester Research report claims that packagers are actually quite slow—as slow as or slower than closed source companies—at getting out fixes. This doesn't surprise me, because a packager generally has to wait for the (fast!) response of the authors before doing its own thing.

Visibility of Bugs and Changes

There is argument that lack of source is actually a virtue for security. Potential attackers can't examine it for bugs, and when vulnerabilities are found, they can't see what, exactly, was changed.

The idea that vulnerabilities are found by looking at the source is an attractive one, but is not really borne out by what we see in the real world. For a start, reading the source to anything substantial is *really* hard work. I know—I did it for OpenSSL, as I said earlier. In fact, vulnerabilities are usually found when software misbehaves, given unusual input or environment. The attacker follows up, investigating why that misbehavior occurred and using the bug thus revealed for their own evil ends. The "chunked encoding" bug I mentioned earlier is a great example of this. This was found by the common practice of feeding programs large numbers of the same character repeatedly. When Apache was fed *A* lots of times, it ended up treating it as a count of characters in hex, and it came out negative, which turns out to be a Bad Thing. In this case, all that was needed was eight characters, but the problem was found by feeding Apache several thousand.[26]

---

26 This particular method is popular because it is so easy: `perl -e "print 'A'x10000" | target`.

So, not having the source might slow down an attacker slightly, but given the availability of excellent tools like IDA (a very capable disassembler) and Ollydbg (a powerful [and free] debugger), not by very much.

What about updates? The argument is that when source is available, the attacker can compare the old and new versions of the source to see what has changed, and then use that to craft software that can exploit unfixed versions of the package. In fact, because most open source uses version control software, and often has an ethos of checking in changes that are as small as possible, usually the attacker can find just the exact changes that fixed the problem without any clutter arising from unrelated changes.

But does this argument hold water? Not really, as, for example, Halvar Flake has demonstrated very clearly with his Binary Difference Analysis tool. What this does is take two versions of a program, before and after a fix, disassembles them, and then uses graph isomorphisms to work out what has changed. I've seen this tool in action, and it is very impressive. Halvar claims (and I believe him) that he can have an exploit out for a patched binary in one to eight hours from seeing the new version.

Review

Another important aspect to security is the ability to assess the risks. With closed source, this can be done only on the basis of history and reputation, but with open source, it is possible to go and look for yourself. Although you are not likely to find bugs this way, as I stated earlier, you can get a good idea about the quality of the code, the way it has been written, and how careful the author is about security. And, of course, you still have history and reputation to aid you.

Who's the Boss?

Finally, probably the most important thing about open source is the issue of who is in control. When a security problem is found, what happens if the author doesn't fix it? If the product is a closed source one, that generally is that. The user is doomed. He must either stop using it, find a way around the problem, or remain vulnerable. In contrast, with open source, users are never at the mercy of the maintainer. They can always fix the problem themselves.

It is often argued that this isn't a real choice for end users; usually end users are not programmers, so they cannot fix these problems themselves. This is true, but it completely misses the point. Just as the average driver isn't a car mechanic but still has a reasonably free choice of who fixes his car,[27] he can also choose a software maintainer to fix his software for him. In practice, this is rarely needed because (at least for any widely used software) there's almost always someone willing to take on the task.

---

27 This is a metaphor that is rapidly going out-of-date, as car manufacturers make cars more and more computerized and harder and harder for anyone not sanctioned by the manufacturer to work on. Who knows—perhaps this will lead to an open source culture in the car world.

Digression: Threat Models

I mentioned threat models earlier. Because not all my readers will be security experts, it is worth spending a moment to explain what I mean. When you evaluate a threat to your systems, you have to have a context in which to do it. Simply saying "I have a security hole" tells you almost nothing useful about it. What you want to know is how bad it is, how fast you have to fix it, what it will cost if you don't fix it, and what it will cost if you do fix it.

To make that assessment, there are various things you need to know. The obvious ones are what systems you are running; what the value of each component is; what impact the vulnerability will have on each component; how likely you are to be attacked; and so forth. But less obvious is the question of whether you actually care about the attack at all—and this is where threat models come in. They characterize what you have already assumed yourself to be vulnerable to and how you are vulnerable it.

So, as I mentioned, my threat model is that local users have root. Because root can do, essentially, anything she wants, this means that any vulnerability that can only be exploited by a local user, *no matter what it is, and no matter how bad*, is irrelevant to me. They could do that already.

Threat models can get quite complicated, and you may well find that when a new vulnerability comes along, you have to consider what your model actually is, because you don't already know. For example, suppose there's an attack on the domain name service that allows it to be faked. Do you care? Was that something you assumed had to be correct when you built your system, or is incorrectness merely a nuisance?

Anyway, I don't want to turn this chapter into a textbook on security, so suffice it to say that threat models are important, everyone's is different, and you can't evaluate the impact of vulnerabilities without one—which means, really, that the whole question of which is better is one only you can answer.

The Future

> PREDICTION IS DIFFICULT, ESPECIALLY ABOUT THE FUTURE.
>
> —*Niels Bohr/Mark Twain*[28]

There are two futures: the one we should have, and the one we're going to get. I'll talk about the one we should have first, because it's more fun, more interesting, and definitely more secure.

---

28 Apparently it's difficult about the past too—we don't know which of these people said this!

Today's operating systems and software are based on decades of experience with developing software that was run by nice guys on machines over which they controlled access relatively easily (whether as users or nonusers interacting with the machine or software in some way). This was a world where your biggest security threat was a student playing a prank. We learned a great deal about how to write software that did clever things, was easy to use, and had pretty interfaces.

Unfortunately, we learned almost nothing about how to write secure software. And in the meantime, we built up a huge amount of insecure software. Worse, we used insecure languages to write the insecure software in. And worse even than that, we used languages thath there's no real prospect of securing. And we continue to use them, and the same insecure operating systems we wrote, with ever-increasing teetering towers of software piled on top of them.

So, in my Brave New World, we get smart enough to scrap all this and use an idea invented in the 1960s: capabilities. Unfortunately, academics decided very early on that capabilities had all sorts of problems, and this has prevented their widespread adoption. Mark Miller and Jon Shapiro, in "Paradigm Regained: Abstraction Mechanisms for Access Control" (*http://www.erights.org/talks/asian03/paradigm-revised.pdf*), have very effectively debunked these criticisms, though I have to admit to being bemused by how anyone could believe them in the first place, since they are so easily solved.

In any case, there are still some of us around who believe in capabilities, and I entertain the fond hope that we may start using them on a larger scale. The foremost project using capabilities at the moment is the E language (*http://www.erights.org*), which, as well as being a capability language from the ground up, has some very nice features for distributed computing, and is well worth a look. Unfortunately, I do not believe a language with such esoteric (and ever-changing) syntax will ever be widely used. It seems that privilege belongs to a very few. Perhaps more promising from the point of view of likelihood of adoption is my own nascent CAPerl (think "Kapow!") project, which adds capabilities to Perl. Although this is far less elegant and satisfying, it has the virtue of looking almost exactly like Perl to the experienced programmer, and so I do have some hope that it might actually get used. I don't have a web site for it yet, so I invite you to Google for it.

No discussion of capabilities in the 21st century would be complete without mentioning EROS (*http://www.eros-os.org*). Funnily enough, EROS is short for Extremely Reliable Operating System, since its author, Jon Shapiro, thought that was what was important about it when he started writing it. Now, though, we are far more interested in its security properties than in its reliability. EROS, like E, implements capabilities from the ground up. More importantly, it runs on PCs. Unfortunately, it seems it is a project that won't be finished. Work is, however, starting soon on the second attempt.

Of course, if I really think this will happen, I'm on crack. Not enough people care enough about security to contemplate throwing everything away and starting again (make no mistake, that's what it takes). But I can (and do) hope that people will start writing new things using capabilities. And I hope that drawing them to your attention will assist that.

Now I'll move on to what I think will really happen. Certainly people have become more aware of security as an issue, and the increasing use of open source in corporate environments also increases the pressure on security. It seems likely that this will drive open source toward better ways to deliver updates faster. I don't think it is actually possible to drastically improve open source's record on fixing security issues. I believe that by any measure, that open source is ahead of closed source. But the flow from author to end user is not yet a smooth one.

Interestingly, the fix for that is strongly related to the fix for another widely acknowledged problem with open source: package management systems. We do not yet have the ultra-smooth systems to handle installation and update of systems in a way that makes it a no-brainer for end users. Open source and closed source present interestingly different problems. Open source packages of any complexity tend to depend on other open source packages, usually with a completely different set of authors and release cycles. Managing installation in this environment is much harder than in the closed source situation, where one vendor—even one that buys components from others—is in control of the whole package. I think the open source world is moving toward better package management, and this will automatically improve the end user's management of security.

However, for corporate environments, this probably makes little difference. In such situations, there are almost always elaborate procedures for rolling out new versions which are almost unchanged when using open source. Even so, clearer visibility of dependencies and, therefore, what needs to be upgraded when a fix comes out, would be useful.

I also hope that better package management would reduce the dependency of users (at least, if they choose to have their dependency reduced) on packagers. Although packagers, in theory, add value, they also add latency. Perhaps worse, they damage the open source model by introducing dozens of slightly different versions of each package, through the widespread practice of applying patches to the packages instead of contributing them back to the original authors, which reduces the effectiveness of community development by splitting the community into many smaller subcommunities.

As always, there is a price to be paid for better package management. Automated updates are a fantastic vector to mount automated attacks. We know well how to prevent such attacks using public key cryptography, but once more, the complexity of multiple authors introduces problems of key management to which there aren't really good answers, at least, so far.[29]

One thing that does seem certain is that the increasing trend of concern about security by end users will continue. The seemingly never-ending rise of spam, adware, and Trojans, if nothing else, has put it on everyone's agenda, and that doesn't seem likely to change.

Interesting Projects

I've already mentioned some projects in passing, but no chapter on open source security would be complete without mentioning some of the more interesting projects out there. I'll start with the obvious ones and move on to the more esoteric. This list probably reflects my current obsession with privacy and anonymity:

OpenSSL
> Well known, but still essential. This library implements most known cryptographic algorithms, as well as the SSL and TLS protocols. It is very widely used in both free and non-free software, and at the time of this writing was in the final stages of obtaining FIPS-140 certification. *http://www.openssl.org.*

Apache 2
> Of course, we've all known and loved Apache for years. Finally, Apache 2 has HTTPS support out of the box. *http://www.apache.org.*

Mozilla
> A suite of web browser, mail, and news reading software, and related utilities. You probably don't think of this as security software, but it is probably second only to Apache in the number of financial transactions it protects. And it does it with a minimum of fuss. What's more, it isn't plagued with its closed-source rivals' fondness for installing evil software you never intended to install! *http://www.mozilla.org.*

GnuPG
> Implementing the OpenPGP standard under the GPL. Primarily used for email, but also the mainstay for validation of open source packages (using, of course, public key cryptography). *http://www.gnupg.org.*

Enigmail
> Small, but (almost) perfectly formed. This is a plug-in for the increasingly popular (and, of course, open source) email client, Thunderbird, providing a nicely streamlined interface for GnuPG. *http://enigmail.mozdev.org.*

---

29 I should perhapsat this point plug KeyMan, a package I designed to solve this problem, but since it has singularly failed to take off, that might be inappropriate.

CVE

Common Vulnerabilities and Exposures. This is a database of security problems, both commercial and open source. The idea is to provide a uniform reference for each problem, so it's easy to tell if two different people are talking about the same bug. *http://cve.mitre.org*.

TOR

The onion router. Onion routing has been a theoretical possibility for a long time, providing a way to make arbitrary connections anonymously. Zero Knowledge Systems spectacularly failed to exploit it commercially, but now it has come from a most unlikely source: the U.S. Navy. The Navy's funding recently ran out, but the Electronic Frontier Foundation stepped up to take over. Well worth a look. *http://tor.eff.org*.

Conclusion

In the end, it seems to me there's little to be sensibly said that, from the viewpoint of security, truly differentiates between open and closed source. The points I believe are critical are my ability to review the code for myself and my ability to fix it myself when it is broken. By "myself" I do, of course, include "or anyone of my choice." What I don't believe in—at all—is the often-quoted but never-proven "many eyes" theory.

In the digression on threat models, I mentioned that the only person who can really answer the question of whether open source is better for security is you. Leave the camp of people who think security is a good thing that we should all have more of, and join the camp of people who have thought about what it means to them, what they value, and so, what they choose.

CHAPTER 5

Michael Olson

Dual Licensing

Over the past decade, there have been many attempts to commercialize open source software. One common strategy has been to create services businesses, which offer consulting and support to users of open source. Another strategy has been to build hybrid businesses, which distribute open source platforms with proprietary add-ons, and which make money by licensing the add-ons.

A third strategy, and the focus of this chapter, is called *dual licensing*. Companies that use dual licensing provide a single software product under two different licenses. One license, which imposes open source terms, is available to a certain class of users. A second license, with proprietary terms, is available to others.

Business and Politics

This chapter is about business. Software is deep in the modern economy: it provides the mechanism for the flow of capital around the world, and it is itself a good that can be produced, bought, and sold. Whenever something interesting happens in the world of software, business leaders pay attention.

Open source is interesting. It enforces new rules for use and distribution of software products. It changes the economics of software production. It impacts the way that companies can capitalize on the software they control.

At the same time, though, open source has no business agenda. Open source is about freedom in the political sense. It is about peer review and scientific collaboration. Open source licenses take no position whatsoever on the profitability of business models. Open source is not antibusiness; it has no opinion.

Businesspeople, of course, have opinions on open source. Some years ago, when open source was still new to the business community, it was unfamiliar, and that unfamiliarity bred fear, confusion, and opposition. More recently, as quality open source software products have proven their value to all kinds of businesses, that opinion has shifted. Most businesspeople who have thought about open source at all are guardedly interested in it, and want to find ways to use it in their companies to be more efficient, to spend less, or to earn more.

An informed opinion on open source is important. Just as software is a powerful economic force, open source is a powerful force in the development and deployment of modern software systems. The Internet, including the World Wide Web, and a wide range of the services that run on it (for example, e-commerce sites such as Amazon.com; information retrieval services such as Google; and portals such as Yahoo!) exist *because* of open source software. Open source is a genie that is too big for its bottle. Now that it is out, it will not fit back in. If you do not put it to work for you, it is likely to wreak bad magic on your business.

My own opinion on open source is simple. It is one tool among many that can, when used sensibly, create business value. I run a business based on open source, but my agenda is commercial, not political. I understand the politics behind open source, and I appreciate and respect many smart people in the open source community. When I sit down at my desk, though, I am more interested in the difference between income and expenses on my profit and loss statement. To the extent that open source helps move that difference in the right direction, I care deeply about it.

The open source business strategy about which I know the most—and the focus of this chapter—is called *dual licensing*. Dual licensing is a way to make a single software product available under two different licenses. One is an open source license, and encourages sharing and collaboration. The other is a more conventional proprietary license, and permits secrecy and competition—which promote the creation of proprietary value. Dual licensing is a way to give a single product to open source users on open source terms, and to paying proprietary customers on conventional proprietary terms.

Open Source: Distribution Versus Development

Open source can include a distribution strategy or a development strategy. While people often think of the development strategy (many programmers working on a common project) when they think of high-profile projects like Linux, we are instead going to focus on the distribution strategy.

Open source is just a way to put product in many users' hands inexpensively. Dual-licensing businesses do not use collaborative development to build their products. In fact, as we will see, that production strategy is poisonous to dual-licensing businesses.

A dual-licensing business can take advantage of the cheap and ubiquitous Internet to distribute its products at low costs. Open source licensing promotes use at much lower cost, with much less friction, than an expensive marketing campaign could do. Dual-licensing businesses can distribute software to more people, more cheaply, than their proprietary competitors.

At the same time, dual licensing permits these businesses to generate revenue by licensing the software to certain users for a fee. Software licensing revenue is good revenue—because you can make and license a second copy of a piece of software for essentially no additional cost, businesses and the financial markets like licensing revenue. Selling support or services, by contrast, imposes new costs with every deal, because a business must have the capacity to answer the telephone for every new customer it captures. As a result, licensing-based businesses, including dual-licensing businesses, generally get higher valuations and can raise capital more cheaply than businesses based exclusively on services.

A Primer on Intellectual Property

Dual licensing, when you first encounter it, can seem like a parlor trick. How can I possibly charge money for software that you can get for free?

There is, in fact, no sleight of hand. Nobody gets tricked. Dual licensing requires some care, and the same diligence that most businesspeople bring to their jobs, but it turns out that governments around the world want businesses to be able to do this. They have constructed legal infrastructures that permit all sorts of businesses to make money with confidence. All it takes is to behave like a business and comply with the law.

Understanding the mechanism here is easier with some background in intellectual property law. Most software vendors make money by charging fees for an intangible product—algorithms and their implementations as expressed in computer code. All these businesses, including dual-licensing businesses, exist because of crucial legal concepts regarding intellectual property—concepts such as ownership and license grants.

Ownership

Ownership of real property is easy to understand—my ball, your house, her island. Ownership of something intangible like the expression of ideas, is harder to understand. Over several centuries, governments have created rules about who may and may not make copies of artifacts like books and music. This "copyright" law balances the interests of the creator of a work—an author or a composer—with the interests of consumers who purchase those copies. Copyright law clarifies issues and resolves disputes among authors and readers. Copyright law has been adapted to apply to software, balancing the rights of software authors with those people who license and use their computer programs.[1]

Copyright law worldwide generally says that the *creator* of an *expression* of an idea owns that expression. From the point of view of a computer programmer, this means that whoever writes a program owns the program. If the programmer copied parts of it from somewhere else, of course, he was not the creator of those parts, and he does not own them.

Unless the owner expressly assigns his rights to someone else, he owns his creation. Most employment agreements include just such an assignment provision, so the code a programmer writes for his job instantly becomes the property of his employer. The Free Software Foundation, an important organization in the free software movement, requires such assignment for any contribution to projects that it manages.

Without an explicit assignment, there can be many different owners of a large software product. This is exactly the case with many open source projects, including Linux. If 10 developers collaborate on an open source package, each of the 10 owns the pieces she produced. None of them owns the entire work.

Licensing

Except in very rare circumstances, no one ever buys computer software. The person who created the work is its owner; a purchaser only buys certain rights to use the copy of the software in his possession. This distinction is critical to businesses that build software products. These businesses do not sell *ownership*. Instead, they sell *licenses* to the software, where "licenses" generally means a collection of rights to copy, run, and use the software product.

The owner is allowed, under the law, broad latitude to set conditions on use of the software. If someone wants a license to use the software, the owner can require that person to do any number of different things in exchange for the license.

Most proprietary software vendors require payment of a fee. This has been a remarkably successful business strategy for many decades, producing a number of billion-dollar software companies.

Open source software licensing does not require payment of a fee. The owners of open source software do impose other conditions on their open source licensees.

---

1   The bodies of law dealing with trade secrets and patents are also both important to businesses, but are beyond the scope of this discussion. Patents in particular complicate open source software development and distribution, and are something of a lightning rod in the debate between proponents and opponents of open source.

Those conditions vary, depending on the open source license in use, but two broad categories are most common:

- *Reciprocal* licenses require that any recipient promise to contribute back any changes or additions to the software. Reciprocal licenses are coercive; they essentially enforce sharing. The most common reciprocal license in use is the GNU General Public License, or GPL.

- *Academic* licenses usually require very little—often, just acknowledgment of the original owner's work on the software. Academic licenses encourage reuse of other programmers' work by making it available on liberal terms. The most popular academic license in use is the Berkeley Software Distribution, or BSD, license.

In both the proprietary and open source cases, the owner of the software sets the conditions that others must meet to use the software.

The owner is allowed to set different conditions for different people. This same situation applies to real property: you might allow your brother to sleep in your spare bedroom for free, but probably would not let me do that on the same terms. At best, I could hope to pay you rent for use of the room. Likewise, the owner of a piece of software can grant different rights, on different terms, to different people.

This is common in proprietary software businesses. For example, if you buy a computer with an operating system and a word processor installed, you are most likely permitted to use those packages, but not to make copies of them for others to use. By contrast, the computer manufacturer is allowed to copy, install, and distribute both pieces of software to you and to others. The company that owns the operating system and word processor chooses to grant the computer manufacturer broader rights than you have, because the owner's goal is to make money in business, and creating a distribution channel that generates license fees is a good way to do that.

The fundamental point to remember is that open source licensing rests on the same foundation that proprietary software businesses use for their own licensing: copyright law and the notions of ownership and licensing that inform it. In both the proprietary and open source cases, the *owner* grants *licenses* to use the software under certain conditions.

One main benefit of being an owner is having the right to set those conditions.

Dual Licensing

Dual licensing exploits these attributes of copyright law to construct profitable businesses using open source software. The software's owner distributes a single product under both an open source license and a proprietary license. Open source licensees pay no fees, but make certain promises. Proprietary licensees have different rights, and pay a fee for that consideration.

By making the software available on open source terms, the owner creates a large and inexpensive distribution channel. Generally, users can download the software on the Internet for free. In many cases, the software is bundled with third-party offerings, and is distributed by others not formally affiliated with the owner. Open source distribution makes the software ubiquitous, putting it in the hands of many users at very low cost.

Reciprocity

Open source, as it has emerged from the academic world, emphasizes the importance of reciprocity. Fundamentally, everyone in the community is allowed to share the benefits of the code that others have written. Everyone is expected to share their work, however. If you give me your work for free, I reciprocate and give you my work on the same terms. This emphasis on reciprocity encourages collaboration and allows the community to build better software than any small group could on its own.

In a dual-licensing business, the open source license demands reciprocity. If a customer uses the software under the open source license, his own additions—which may well include the code for his own product—must be open source, as well.

Under the proprietary license, by contrast, reciprocity is not required. The customer's own code, and any changes or additions he makes to the original product, can remain proprietary.

This distinction between the two licenses is crucial. Sharing source code is a virtue in the open source community, but is a serious competitive disadvantage in many commercial settings. Proprietary software vendors would much rather pay money than give away the competitive advantage inherent in their source code.

Warranty

A warranty is a promise. These promises are the bedrock of most commercial relationships. Software vendors that use proprietary licenses to do business with their customers and suppliers use warranties to spell out clearly what each party expects from the other.

As a software vendor, I might warrant that my software product will work as documented, and that I will fix any errors if it does not. I will generally warrant that I wrote the software, and that I am the person who owns it. I may ask my customer to warrant that he will not give my software away for free to anyone else. If either of us breaks one of these promises, our licensing agreement will usually spell out the steps that the other party can take to fix the problem. If we cannot fix the problem, either of us can sue the other in court for breakingthe promise.

People do not generally make promises casually. There are real costs in keeping your word. For example, if I have promised that my software works, and it does not, I need to spend time, and possibly money, to fix it.

Virtually all open source licenses—and certainly all those used by dual-licensing businesses—are "as is" licenses, with no warranties, no promises, and no recourse in the event of problems. The user gets the software as is, and assumes all risk in using it. This allows the business to minimize business risk from users of its software who did not pay a license fee. Put simply, if the user does not pay for the license, the owner will not use his money to protect the user in the case of a problem.

Proprietary licenses, by contrast, generally include clear warranties. A company can afford to make promises to its paying customers, because it can use some of the money paid to defray the costs of keeping the promise.

Depending on how a customer plans to use a software package, he may or may not be able to tolerate the "as is" nature of an open source license. If the customer is willing to risk problems in the software, and to fix those problems himself, the open source license is fine. On the other hand, if the customer plans to ship the software to customers of his own, he may want the assurance of a committed partner behind him. Likewise, if the customer runs mission-critical infrastructure on open source software, he may need the confidence that commercial warranties provide.

Competitive Issues

Many vendors that consider dual-licensing strategies are concerned with competition.

Naturally, open source distribution presupposes that there are no valuable proprietary techniques in the software that need to be kept secret. If an owner distributes a software package in source form, competitors can read the source code, just like any other person who receives a copy.

This seems at first like it should be a major concern, but in point of fact, the amount of innovation in the software market is much smaller than is generally supposed. Especially in relatively mature sectors of technology, such as operating systems and databases, the techniques and algorithms are well understood by all the suppliers The combination of particular techniques that any single vendor uses may be somewhat interesting to its competitors. All of the vendors in a market hire from the same pool of technical talent, and even poach employees from one another. As a result, most vendors know pretty well how their competitors' products work.

Proprietary vendors will naturally claim that their products *are* novel and innovative, and must be kept secret. The problem is that there is no way to test this assertion, since doing so requires examination of the source code, and that is precisely what the vendors cannot permit, yet still protect their market positions.

In some cases, these claims are no doubt true. The strategy of concealment, however, does not distinguish between ugliness and beauty, and may be used to hide either.

A different competitive concern arises from confusion between ownership and licensing. Some vendors considering dual licensing worry that their competitors will download the open source version of the product and license it under proprietary terms to others. In effect, this would allow a competitor to use the vendor's own product to compete with him.

Copyright law, of course, prohibits this. Only the owner has the right to grant proprietary licenses to the software. In fact, no competitor can copy pieces of the work into its own products, since those products would then be forced to comply with the terms of the open source license.

There is another, subtler, force that protects dual-licensing vendors from unscrupulous competition. Open source software is preemptive: the existence of a quality open source package makes it very difficult or impossible to introduce a for-fee competitor. The cost of building such a competitive offering has to be recovered somehow. If a no-cost offering is available, selling an alternative for a fee will likely fail. The result is that very few businesses are willing to challenge open source software in the market once it is established.

The converse, of course, is not true. Open source frequently, and aggressively, challenges established proprietary software.

Even in cases where proprietary alternatives exist, dual licensing creates competitive advantage. The open source product is ubiquitous as a result of an inexpensive distribution channel. In addition, the fact that many users are able to use the software at no charge actually takes money away from purely proprietary vendors in the market. If a user can choose between a good open source product for free, and a competitive proprietary product for a fee, the for-pay competitive vendor is likely to lose the sale. Dual-licensing vendors are not paid by their open source users. Neither, however, are their competitors.

Ownership

Only the owner of a work has the standing to offer it to different users under different licenses. As a consequence, dual licensing works only if there is a single, well-defined owner of a work. Otherwise, customers have no single entity with whom to negotiate their license terms. In practice, projects that use the open source development model (a large community of independent programmers collaborating on a work) are poor candidates for dual licensing. There are just too many contributors to contact for permission every time a new customer wants to license the software under non-open source terms.

As a result, dual-licensing businesses do not use the open source development model. Instead, they invest in the development of the open source software and do not accept contributions from the community at large. Dual-licensing businesses rely on open source as a distribution strategy, not as a production strategy.

Ownership is an issue in the larger open source community, of course. As a practical matter, assessing the provenance of contributions to an open source project is hard. Individual contributors are often judgment-proof, and ensuring that they have not misappropriated the intellectual property they contribute to the project requires real diligence. Projects manage this problem the same way that proprietary vendors do—they put experienced managers over coders to review contributions, and they rely on mature and experienced contributors to build the product.

In fact, this ownership issue cuts both ways. Proprietary vendors are equally at risk from employees who take pieces of open source software and incorporate it into the proprietary products that their employers build.

The answer for both proprietary and open source developers, of course, is to set up formal policies and standards on intellectual property use. Diligence is simply part of the job. Both proprietary and open source developers can and do write large, sophisticated products without stealing from others.

Practical Considerations

Dual licensing works. Using it, businesses can generate revenue based on license fees paid for copies of software. Because the public markets view this revenue as inherently lower margin than services revenue, dual-licensing companies have attractive valuations and can raise capital at parity with proprietary software vendors.

Attractive Margins

Every sustainable business must consistently earn more money than it spends. This is often hard.

One of the key considerations in designing a new business is the margin that the revenue stream can produce. Margin is, at base, the percentage of revenue that is profit. If you pay $100 to build a widget, and you pay a salesperson $100 to sell it, the widget costs you $200. If you charge $400 for it, you have an attractive margin—50%. If you charge $205 for it, you have a very thin margin—2.5%.

Most technology businesses sell licenses, or services, or a mixture of the two. For example, a company might license its product to customers, and might also sell consulting services to help customers integrate the product into existing IT infrastructures.

Licensing businesses generally have attractive margins. This is because, while it may cost $1 million in payroll to produce a new software program, the second copy of that program is essentially free. Having sunk the cost into building the product, the company can sell as many copies as it likes without paying the developers to start all over again. As a result, selling a new copy of an existing product really just requires paying the sales team that sells it.

Services businesses, by contrast, have much slimmer margins. This is because, to deliver the service, the business must have the people on the payroll who can do the work that is inherent in the services contract. Selling more consulting engagements requires that you hire more consultants. Thus, the costs that a services business incurs on every contract include not just the cost of selling it, but also the cost of fulfilling it.

Of course, very few companies are purely licensing businesses. Most provide at least customer support. However, building attractive margins is easier in a business with a significant licensing component.

Dual-licensing businesses offer two different revenue streams. Proprietary customers pay licensing fees, giving the business a high-margin licensing revenue stream. Both proprietary and open source users may choose to buy consulting or technical support, giving the business a lower-margin services revenue stream. This diversity in the revenue streams can allow the company to weather temporary slowdowns in either its licensing or its services businesses.

Capital

Over the past several years, the success of open source business models, and of dual-licensing models in particular, has caught the attention of the investment community. As investors have learned more about open source and dual licensing, they have begun to make investments—sometimes substantial investments—in new businesses. In addition, established companies have begun to look at open source as a strategic business advantage, and not just as a low-cost way to bring technology in-house. Major technology vendors have demonstrated their willingness to acquire open source businesses at prices that reward the principals and original investors in those businesses.

Venture capitalists and other sources of early-stage funding for companies look for a few key attributes in open source businesses.

First, of course, the company must have a credible, defensible business model. It is, generally, easier to raise money for a company that uses dual licensing, than for one planning to make money exclusively from services. The big risk for a new services business is that it will demonstrate the viability of a new services opportunity, and thus attract the attention of a large established services player. Popular open source packages have enormous installed bases. Businesses exist today that earn tens of millions of dollars supporting and training those users. That is enough money to interest even a large, established company.

The real problem with providing only services for an open source package is that the strategy is hard to defend against competitors. While it is true that the original authors of an open source package have an advantage in supporting it—after all, they know the most about how it works—freely available source code permits anyone else to learn the product internals and offer the same service.

Investors and acquirers are generally willing to pay less for a pure services business than for a licensing business, or for a business that combines the two sources of revenues. After all, costs grow in proportion to revenue growth. Every services sale requires staffing to meet the service commitment. Selling licenses for software, on the other hand, is much cheaper; making another copy of an existing product is free.

Second, investors look for open source packages with a solid market presence. The investor typically wants to see a track record of successful deployment to prove that a market exists. Thus, it is very hard to raise money to build a brand-new open source package from the ground up. Indeed, virtually all of the open source businesses funded in the past decade have tried to commercialize on the value of existing open source packages.

This creates a chicken-and-egg problem. The open source package must exist before it can be funded. Somehow, though, the package must get written in the first place. Generally, the open source packages that serve as the foundation for successful businesses were labors of love in the beginning, created by developers working in their spare time for free. Only when they succeeded in building a credible product could they raise money.

Third, and significantly, investors always look at the team in which they are asked to invest. As a rule, no investor will back an unbalanced organization—all engineering, or all marketing, for example. A fundable open source business must combine the technical expertise of a solid group of engineers with experienced management. In particular, the company must be properly staffed to market and sell the licenses or services that the business will offer. This point may seem obvious, but in practice, it is very hard to put together a solid team with the ability to make and execute a business plan that demands both solid product development and execution against a financial plan.

Once an open source business is successful in the market, it can choose to go in a number of directions.

If the business is generating profits, the owners may choose to continue to run it as a privately held company, earning a return on their effort and investment in the form of dividends. As a rule, venture-backed businesses do not have this option. The venture capital community wants to liquidate its investment at a profit within a few years of making the initial investment, and will press management to sell the company at some point. Absent that pressure, however, a strong, profitable, and growing business is a wonderful thing to own.

More commonly, small businesses are acquired by larger businesses, and folded into the bigger company. This rewards the original investors, as well as the principals who started the company and helped to make it successful. The acquisition will most often pay off in cash, stock in the bigger company, or a mixture of the two.

Established companies are willing to buy open source businesses for a variety of reasons. For example, open source technology is often ubiquitous in the market, and controlling that technology can give the big company important advantages over its competitors. In addition, an open source platform can create opportunities to build new proprietary product offerings that run on top of the open source, which creates new service and licensing revenue. The open source business's revenue may, on its own, be enough to capture the attention of the larger company.

The reasons that any particular acquirer chooses to buy any other business depend deeply on the peculiar circumstances of each, so there is no cookbook for building a company to sell. In general, though, a solid customer base, a track record of consistent growth, and a profitable revenue stream are good things.

The last way that small, privately held businesses transform themselves is to offer their stock for sale on the public markets. This is a much less common practice in today than it was six years earlier; a business must, in general, have very high revenues to make this transition. In addition, the demands on the management team in a publicly traded company are very different from, and in many ways more onerous than, the demands on a private company's team. The requirements for reporting financial information to the public markets and the need to manage the public investment community's expectations dramatically change the way that a company president works. As a result, many companies are forced to change their executive teams before offering themselves for sale on the public market.

Choosing Licenses

One of the most important decisions in a dual-licensing business is what the terms will be for both the proprietary and the open source license. This issue is much bigger for the open source license, because that license is never negotiated. People simply download the software and accept the terms. As a result, a mistake in the open source terms is a mistake every single time the software is distributed.

Academic licenses are poorly suited to dual-licensing businesses, but reciprocal licenses generally work well. An academic license simply requires acknowledgment of the owner. Most rational businesspeople would much rather make that acknowledgment than hand over precious capital for use of software. A reciprocal license, by contrast, requires that the customer's own intellectual property be given away under an open source license. There is a very large population of potential customers who are more interested in protecting their intellectual property than in saving money.

As a result, the only kind of open source license that makes sense for a dual-licensing business is a very strong reciprocal license. The best example, of course, is the GNU GPL. Choosing the GPL gives you the benefit of the work already done by the Free Software Foundation, or FSF, in drafting the license and defining the key terms. In addition, the FSF has a strong vested interest in demonstrating the enforceability of the GPL, so, in the event of a dispute over unauthorized use of your dual-licensed product, you have access to a seasoned legal team that knows the subject well

It is almost certainly a mistake to try to draft your own open source license for a dual-licensing business. Doing so requires that you understand and apply the fundamental concepts of open source development and distribution in a new legal license. This is work you do not need to do if you choose an existing license. More importantly, writing a new license from scratch creates the opportunity to make bad mistakes. Finally, drafting a new license will require you to educate the open source and proprietary software business communities about the terms of your own license. This will create friction and inhibit adoption. You want developers concentrating on your software, not on your license.

Need and Pain

The interplay between the software product and its open source license is probably the single most important business issue for dual-licensing companies. The software product must create *need* in the market. Ideally, it will be so attractive to customers that they simply have to use it. At the same time, the open source license must cause enough pain that some users would rather pay money than endure the pain.

My company makes a product called Berkeley DB. It is an established product with a good reputation, but the only way for our customers to use it is to combine it with software they write themselves. That act of combination gives us leverage, because the resulting work is derived from our software, and we thus can dictate the terms under which the derivation must be licensed.

Our open source license requires that the entire work be released in open source form. Open source users, of course, can do this, but the requirement is poisonous to most proprietary vendors. Our dual-licensing strategy lets the proprietary customers pay for a proprietary license and keep their own intellectual property secret.

Linux is, once again, a good example of a product for which dual licensing would not work. Ignoring the ownership issues raised earlier, no customer *needs* to create and redistribute derivative versions of Linux. Customers simply want to install and use the operating system on their computers. None of those actions is forbidden by the GPL, so there is no pain.

Of course there are businesses making money distributing Linux, but they are doing so in different ways. Dual licensing works with only certain sorts of software, with very specific ownership characteristics.

Any vendor considering dual licensing must consider both technology and licensing when designing a business model. The software technology must be constructed so that users need to do something specific—for example, combine it with their own intellectual property and distribute the combined work to others—to use it. The open source license must make this activity painful to at least some customers with money. These customers must be willing to pay enough money to avoid that pain to make the business profitable.

On the other hand, the open source license must not be painful to the open source community, or it will undermine the benefits of the cheap and ubiquitous distribution channel that open source licensing provides. Open source users must experience only pleasure in their use of the software, or the product will fail to penetrate the market.

Measuring the Market

Businesspeople generally measure markets in terms of dollars—the amount of money that customers in the market will spend for a product or service. Dual-licensing businesses need a different metric, because they distribute their software to a combination of paying and nonpaying customers. The result is that a dual-licensing business can have a much larger installed base than a measure of dollars spent would suggest.

Dual-licensing businesses need to look at this issue pragmatically. Put bluntly, a dual- licensing business is never going to get all the money on the table. Some users would rather meet the open source terms than pay money for the software. A dual-licensing business must balance the size of its installed base, and the concomitant opportunity to sell more proprietary licenses, with the money that could be extracted by charging for every use.

The long-term goal of the business is to maximize profits and to grow. By foregoing some revenue in the short term, a dual-licensing business may make its product ubiquitous, which creates additional long-term opportunity. Though you might never get all the money on the table, you can find yourself sitting at a much bigger table this way.

There are two other benefits to a large installed base, even if not everyone in it pays a license fee.One, noted earlier, is that the earth is scorched for competitors: the only way to compete on price with a free product is to pay people to use a competitive one. This is an expensive way to capture customers.

The other, of course, is the opportunity to sell services independently of license fees. A large installed base may be willing to pay for consulting and support, even if it is not willing to pay for the software. Generally, dual-licensing businesses offer these services and book revenue profitably as a result. As a rule, however, dual licensing is more valuable for the license fees that proprietary users pay than for the service fees that open source users are willing to pay.

Piracy

Because open source software is widely available, it is easy for software pirates to make and distribute copies in ways that violate the open source license for the code. Significantly, it is easy for a user to download the software under an open source license, and then to use it in ways that only a paid proprietary license would permit.

Piracy is not, of course, a problem unique to open source or dual-licensing companies. Every software vendor—indeed, every software developer who distributes a product—runs the risk that unscrupulous users will pirate copies of the product. Individual vendors and umbrella organizations like the Business Software Alliance battle the problem continually.

Piracy is a very real business problem for dual-licensing vendors. The two lines of defense are diligence in protecting intellectual property rights, and consistent and clear explanations of the terms under which the software is distributed.

Diligence in protecting intellectual property rights is relatively simple. Anytime a dual-licensing vendor learns of a misappropriation of its software, it must pursue and resolve the issue. This is no different from the rules that apply to a proprietary software vendor, of course.

Consistent and clear messaging on the license terms is equally important. Open source licensing is now generally well understood by the software industry. By emphasizing the conditions that apply to open source use, and making clear the difference between its paid and proprietary licenses, a dual-licensing vendor can educate its users and capture business before problems crop up.

In general, no responsible business or consumer wants to misappropriate the intellectual property of another. The penalties are large, the risks are unacceptable, and it simply is not fair. Most consumers of software want to obey the law.

Dual-licensing vendors are in exactly the same position as purely proprietary vendors. Digital distribution means that copying is easy. There are pirated copies of both proprietary and dual-licensed software in the world. Vendors of both have the same law behind them and the same recourse against pirates.

The Social Contract

Open source is much more than an ingredient in a business model, of course. The movement predates the dual-licensing model by decades, and its long history has produced a complex and nuanced present. Any business that proposes to use open source technology, including dual-licensing businesses, must understand and participate in the open source movement as it exists today.

The open source community generally cares a great deal about reputation; companies, as well as individuals, gain status by being smart and by contributing to the good of the community. Contribution certainly includes developing open source technology for use by others, and any dual-licensing business will do that. Contribution can also mean, for example, promoting and explaining open source technology in the press and at industry meetings—a task for which businesses are often better suited than individuals—and supporting worthy open source projects with staff time and money.

It is unusual in business for an outside group that pays a company no money and that has no legislative authority to have as much influence over a business and its policies as the general open source community does over open source businesses. If a company alienates the open source community, the main advantages of open source distribution—ubiquity and a large installed base—can disappear in a flood of bad press and ill will on developer discussion lists. Executives at dual-licensing businesses must speak clearly to their open source constituencies. They must show the community the respect it deserves in order to earn the community's respect in return.

Besides merely making friends with the open source community, dual-licensing businesses must pay attention to issues that are unique to the business model.

An unwritten social contract among open source developers says that the community generally will produce software that benefits the community most. This is in some sense driven by Darwin—the people who write open source code work on the problems they consider most important, so if there is widespread pain around a particular issue, there will be widespread attention to addressing it.

The social contract, of course, is not perfect. For example, many open source projects are more poorly documented than their proprietary competitors, because very few programmers enjoy writing documentation, and few will do it without being paid. While proprietary enterprise software is not always more polished than open source, it is generally the case that the drudgery in proprietary development gets more attention than it does in open source projects. A programmer working as a volunteer often lacks the patience to do the exhaustive testing and debugging, interface cleanup, and so on, that commercial licensees of software have come to expect.

The introduction of companies focused on profits into this mix alters the social contract of open source. The change is, in some respects, for the better, but it is a change nonetheless.

Businesses driven by profits will invest to maximize those profits. A large customer willing to pay a significant price for a new feature or for changed behavior in a product will get more attention from a business than will a large number of nonpaying users who all want some different feature or behavior. In business, customers vote with their money. In pure open source, contributors vote with their programming time.

A dual-licensing business will, and should, pay attention to the requirements of its paying customers. The business needs to balance the needs of its paying customers with the needs of the nonpaying open source user community. After all, the company does not want to alienate its open source users and thereby lose the benefits of its open source distribution.

As a rule, this issue requires attention, but seldom causes real problems. Paying customers are usually interested in speed, reliability, or enhancements that make a product more powerful. Those improvements are all useful and interesting to open source users as well, so it seldom happens that the open source community loses and the paying customer wins.

From the point of view of the paying customer, the ability of money to influence the product roadmap is actually a benefit. Many companies considering adopting open source technology are concerned that they have no way to influence the development community to solve real business problems for them. Because dual-licensing companies care about income and profits, customers know that they can get the vendor's attention the old-fashioned way: by pulling out a checkbook.

Trends and the Future

Dual licensing is an innovative strategy that combines open source distribution with proprietary licensing. The combination confers competitive advantages on businesses that use it. These advantages include a low-cost distribution channel, powerful product marketing, and a high-margin, scalable revenue stream.

Of course, dual licensing is just one tool that businesspeople can use to build sustainable enterprises in a world where technology and economic forces are in constant change. The rest of this chapter examines dual licensing in that larger context.

Global Development

The flow of capital and information across borders has transformed the world from a collection of economic islands into a more integrated global economy. That trend will surely continue, and will likely accelerate, over the next decade. At the same time, security concerns, and particularly worries about terrorism, create a strong political incentive to encourage the development of a prosperous middle class with an economic stake in the future in emerging economies.

Worldwide economic development is necessarily influenced by information technology. Leaders of emerging nations are clearly interested in building clean, sustainable IT industries. Doing so requires a significant investment in education, but also an investment in enabling technology.

Dual licensing provides a way for nascent knowledge economies to grow. Because dual-licensed software is available for use at no charge under open source terms, emerging businesses can preserve precious capital by choosing to comply with the terms of the open source license.

China is an instructive example. In 2004, the Chinese government announced its intention to invest in the development of a version of Linux tailored for use in the country, with language and character set support and other new features. The economic advantage conferred by a low- or no-cost operating system in a country where an enormous number of computers will be installed soon is obvious. Between 2005 and 2010, China is likely to become the world's largest consumer and producer of open source software. Consumption will increase earlier, and faster, than production. Demand for high-quality open source and dual-licensed products will be very high.

Although there is little or no short-term revenue for dual-licensing businesses here, they do establish a long-term competitive advantage. As these small businesses in emerging economies grow, and as they build value in expertise or new hardware and software products, they can choose to pay some of their new capital for the more permissive terms of the proprietary license. The cost of switching encourages customers to stay with the product that they know, and the one built into their own products.

This same strategy applies to developed economies. Software vendors are generally interested in the education market, in part because they want the next generation of software consumers trained to use their products. Many vendors encourage university researchers and students to use their products so that the students will prefer those products when they graduate and eventually recommend software purchases to their new employers.

Dual-license vendors have an advantage in both cases, relative to proprietary vendors. Because the open source license terms permit use at no charge under reasonable conditions, individual business owners, as well as researchers and students, can choose the open source software easily. They do not need to negotiate a special low- or no-cost introductory license to use the product. There is much less friction in the distribution of open source software than in most proprietary software distribution, and the lower friction translates into higher adoption.

Open Models

Dual licensing is, at base, the combination of a venerable business strategy—licensing software for money—with a relatively new open source distribution strategy. This combination is interesting and valuable on its own, but it is by no means the only such combination that is possible.

The global Internet makes the distribution of content much cheaper and easier than it ever was before. At the same time, it eliminates barriers to sharing among widely dispersed individuals and companies. The effect is to create new opportunities for the collaborative creation of intellectual property.

Some examples of this collaborative development were all but unknown just a few years ago. Weblogs, or blogs, are common today. They have emerged as an alternative source for news and information, replacing older media such as newspapers and radio, especially for fast-breaking stories. Similarly, the publication and production of scientific journals is changing. Single-vendor control over high-profile journals, and the vendor's ability to dictate pricing to the market, is eroding because researchers can collaborate and publish their research online more easily than ever before. Finally, some web sites encourage user contributions to make themselves more valuable to visitors. An excellent example is the book and other product reviews on Amazon.com. Amazon's visitors share their reviews with one another because they benefit from that sharing, but Amazon itself gains an enormous advantage because of the depth and breadth of those reviews.

In all three of these cases, old-fashioned ideas, like news distribution, journal publication, and product reviews, have been transformed by the open, collaborative processes that the Internet encourages. Licensing is as much an issue here as in software distribution—ownership of the content, and the right to distribute it online, must be considered carefully.

Collaboration and open distribution continue to transform the way that businesses operate. That transformation necessarily damages some established businesses, especially those where a middleman controlled the flow of goods or information, and was able to extract a fee from the flow. The Internet allows individuals to bypass that middleman—the well-known "disintermediation" strategy—and to share and publish on their own.

Any new business built on a disintermediation strategy must consider the law. Unlawful distribution of copyrighted music files is theft, not sharing, but that fact alone does not mean that the existing music retailing industry makes sense in a world with cheap ubiquitous bandwidth. The next generation of businesses must find ways to use the legal system to protect intellectual property, even while they take advantage of the powerful collaborative properties of the global Internet.

The Future of Software

Open source has transformed the way that software is produced. Now, with dual licensing, open source is changing the way that proprietary software is distributed.

Dual licensing will never replace either pure proprietary or pure open source strategies. There are compelling reasons for both to exist.

Purely proprietary distribution allows companies to invest significantly in new development and to be rewarded for that investment. Particularly at the edge of information technology, where the newest products and services are built, businesses will use proprietary licensing to protect their competitive position and to extract maximum value from their efforts.

Purely open source development and distribution, on the other hand, reduces costs of core infrastructure for consumers, and powerful market forces encourage investment in that cost reduction. Open source has been most successful in those sectors of the market where technology is mature and stable—operating systems, databases, web servers, and middleware. This is not to say that open source development is not innovative, but its impact has been largest in the cases where it has commoditized products in mature markets.

The net effect of open source distribution, from the point of view of the consumer, is to reduce the total cost of software licensing. Dual licensing does nothing to reverse this trend; dual-licensed software will exert downward price pressure on established markets just as purely open source software does, though the net reduction in costs will be lower since some consumers continue to pay for software licenses.

Dual licensing will continue to grow in popularity as new and established businesses apply the strategy to new opportunities. Competitive pressure ensures that businesses will look for ways to gain advantages over one another, and building a hybrid business offers advantages in many cases. Other novel hybrid strategies will, no doubt, appear in the future.

CHAPTER 6

Ian Murdock

Open Source and the Commoditization of Software

It is said that the only things certain in life are death and taxes. For those of us in the IT industry, we can add one more to the list: commoditization. The question is, how do we deal with it, particularly if we are IT vendors and not simply IT consumers, for whom commoditization is an unquestionably positive event?

Commoditization is something that happens to every successful industry eventually—success attracts attention, and there are always competitors willing to offer lower prices to compensate for lesser-known brands or "good enough" quality, as well as customers to whom price means more than brand, quality, or anything else the high-end providers have to offer.

Often, to remain competitive at lower price points, the low-end provider employs a strategy of imitation—for example, investing less in research and development than its high-end peers, and instead relying on the high-end providers to "fight it out" and establish standards and best practices it can then imitate in its own products.

This strategy works because success also breeds interoperability. Unless a company monopolizes a market (a temporary condition, given today's antitrust laws), an industry eventually coalesces around a series of de facto standards that govern how competing products work with each other, or how consumers interact with like products from different vendors. In other words, given time and a large enough market, every industry naturally develops its own lingua franca.

This kind of natural standardization is good for consumers and for the world as a whole. Few people, for example, would know how to type if every typewriter used a different layout for its keys, and the telephone wouldn't be in widespread use today if each carrier's network couldn't talk to any of its competitors' networks. And where would we be today without the descendants of typewriters and telephones—namely, computer keyboards and telecommunications?

Of course, from any incumbent's point of view, an ideal world would allow, say, the market leader in typewriters to own the layout of its product's keys, so anyone who learned to type using its product would face huge barriers to switching to a competitor's product. Fortunately, the layout of a typewriter's keys and similar interoperability features are very difficult proprietary positions to enforce, so once a standard way of inter-operating emerges, all vendors are free to imitate that standard in their own products.

The moral of the story is that standardization, and thus commoditization, are both natural market forces as well as key events in human history. When an industry matures and competing products become more or less interchangeable commodities, this allows new industries to build atop them to create new and innovative products that would not have otherwise been possible if the industries they built upon had not standardized. In the case of typewriters and telephones, it is clear that the industries they enabled—the computer industry, e-commerce, etc.—greatly exceed the size of the industries that enabled them, both economically and in their contribution to human progress.

So, how do incumbent firms fight commoditization? Another moral of the story is that they *shouldn't*. The forces of commoditization, being natural market forces, cannot be beaten. Yet time and time again, incumbent firms fight them. First, the challengers are ignored or dismissed as cheap knockoffs, unsuitable for any but the least-demanding customer. Then they are ridiculed for lacking imagination and innovation. Then, invariably, they are imitated—but by this point, it is too late, as the market has fundamentally changed, and the incumbent finds itself unable to compete because the challengers were built for a commodity market and the incumbent was not. In very simple terms, this is Clayton Christensen's *Innovator's Dilemma* at work.

This chapter argues that the open source movement is just another commoditization event and that, like other commoditization events, it represents a disruptive shift in the software industry as well as an opportunity for entrant firms to unseat the established firms against seemingly overwhelming odds. That being said, commoditization does not equate to certain death to the established firms if they have the vision to see beyond the disruptive events that may befall them in the short term and can adapt themselves to the new commodity environment. Above all, this chapter aims to convey that commoditization is a natural and unstoppable force that is good for everyone involved—if that force is allowed to develop on its natural course.

Commoditization and the IT Industry

The computer industry managed to escape the forces of commoditization for the first 20 years or so of its life—a natural occurrence given the industry was young enough and small enough that standards had not yet had the opportunity to emerge. In the first two decades of the industry, computer manufacturers delivered an end-to-end solution to the customer, from the hardware on up through the operating system software that ran the hardware to the applications that ran on top of the operating system. Every layer of the stack—and, most importantly, the interfaces between them—was proprietary to the computer vendor. As a result, every computer spoke a different "language," and it was difficult to get different types of computers to "talk to each other" and interoperate.

Because of these incompatibilities, the initial choice of hardware implicitly tied the buyer to an operating system; in turn, the operating system dictated what applications the buyer would be able to use. Over time, the high cost of computing technology made it financially impractical for the buyer to move away from the incumbent vendor because previous investments in that vendor's technology would have to be discarded. The combination caused users to become "locked in" to a single vendor.

However, as the industry matured, the dynamic changed. Entrant firms such as Apple, Apollo, and Sun saw the opportunity to create products that targeted an entirely new class of computing consumer—the individual user—that could not afford the mainframes and minicomputers sold by established firms such as IBM, DEC, and Data General.

By focusing on "good-enough" quality and lower prices, and by tapping into years of consumer frustration caused by batch processing, timesharing, and incompatibility between proprietary stacks, the new "personal computing" products were received enthusiastically and began to appear in offices and dens everywhere.

The strategies employed by one entrant firm in particular and one established firm in particular would forever change the computer industry. The latter, ironically, would lead directly to the commoditization of the hardware industry, and the former would lead directly to the ongoing commoditization of the software industry.

On the hardware side, IBM sought to stem the rising tide of Apple by introducing its own personal computing product, the IBM PC. Because of internal cost structures designed around multimillion-dollar mainframe products as well as an aggressive product launch timeline, IBM decided to use off-the-shelf parts for the IBM PC instead of following its traditional approach of developing proprietary components in house.

On the software side, Sun sought to attain a competitive advantage against the proprietary stacks of the mainframe and minicomputer vendors by basing its workstation products on the Unix operating system. Unix was already hugely popular in academia and corporate research labs, so this approach gave Sun instant access to a large portfolio of compatible applications as well as an enormous user base already familiar with the operating system that shipped on its products.

In other words, Unix was an *open system*—that is, a system based on open standards. Unix variants from different groups (for example, AT&T Unix and BSD Unix, the two variants in widespread use in the early 1980s) were largely based on the same APIs. Because of this, applications could be easily ported from one version of Unix to another, and users familiar with one version of Unix could easily learn to operate a different version.

Decommoditization: The Failure of Open Systems

The impact of Sun's decision was the first to be felt. Open systems quickly became popular because of the compatibility they offered—a completely foreign notion at the time. Users adopted systems based on open standards because doing so allowed them to move freely among products from different vendors, avoiding the lock-in common in the proprietary world. Soon, numerous companies—including some of the mainframe and minicomputer vendors—launched Unix-based workstations to compete with Sun's, and Unix became big business.

As the Unix market grew, the competition for customers became fierce. "Compatibility among products," which helped the Unix vendors win converts from the proprietary world, changed from an asset to a liability. In an attempt to imitate the lock-in strategies that had served the mainframe vendors so well for so many years, the Unix vendors themselves began adding incompatible features to their respective products. This ultimately fragmented the market and alienated customers. By the late 1980s, Unix was no longer a lingua franca for the workstation market, but a veritable tower of Babel.

Meanwhile, IBM's decision to use off-the-shelf parts in the IBM PC inadvertently created the industry's first open hardware platform. It was not long before a new wave of entrants, such as Compaq, Dell, and Gateway, realized they could build products that were 100% compatible with the IBM PC, thus gaining access to a large base of applications and users, much as Sun had done by adopting Unix. On the component side, two companies experienced the biggest windfall from IBM's decision: Intel and Microsoft. As the clone market emerged, both companies found an entire market to sell to, not just a single company—a much larger opportunity, even if that single company was IBM.

At this point, the events set in motion by IBM and Sun intersected. As the Unix vendors were competing vigorously with each other through the introduction of proprietary extensions to Unix, thereby "decommoditizing" the lowest level of the software stack, the fully commoditized PC waited in the wings. As PCs became more powerful, they began to replace workstations, and as PCs continued their march upmarket, the market power of the PC vendors (and, thus, the vendors of their constituent components) increased dramatically. In particular, the new ubiquity of the PC helped Microsoft's Windows operating system replace Unix as the lingua franca of not just the new PC-based workstation market, but also of the entire computer industry.

Why did Unix fail while the PC has succeeded beyond anyone's wildest expectations, particularly those of its progenitor, IBM? Both began life as open systems—as ecosystems of sorts—and both grew enormously popular because of their open nature. On the Unix side, though, each vendor tried to own the ecosystem by itself, and, in the end, all they collectively managed to do was destroy it. Meanwhile, on the PC side, the ecosystem won out in the end, for the betterment of all who embraced that ecosystem; and, most importantly, the existence of that ecosystem enabled the creation of other ecosystems above it. For example, without a truly open platform in every office and den, the Internet would not have been able to take root, and it too evolved into an ecosystem that has spawned countless products, services, industries, and ecosystems that were previously unimaginable.

Linux: A Response from the Trenches

It was into this environment that Linux emerged in the early 1990s. At first the mere hobby project of a young college student, Linux captured the imagination of those who could best be described as the "collateral damage" of the Unix wars. Two features of Linux made it appeal to this large group of users and developers: its compatibility with Unix, with which they were intimately familiar; and that it was licensed under the GNU General Public License (GPL), which not only allowed the scores of Unix refugees to contribute to its development, but also guaranteed that Unix-style fragmentation could never happen to the result of the community's work, at least at the source-code level.

Linux grew by leaps and bounds during the 1990s. As with previous challengers, it was first ignored, then ridiculed, by the incumbents, primarily Microsoft, which had masterfully used its position as the de facto standard operating system to expand into numerous additional markets and gain additional—even unprecedented—market power. Unlike so many companies that had come before it, Microsoft wielded the forces of commoditization expertly. By offering its products at lower prices than its competitors could afford to offer them, Microsoft preemptively commoditized many of the markets in which it competed, depending on high volume to make its products profitable and making it impossible for challengers to undercut it.

As Microsoft's power grew, so did the desire of Microsoft's competitors to counter it. By the late 1990s, it was clear that Linux was a powerful force, and many of the industry's largest companies began to see it as a competitive weapon. These companies also recognized that the power behind Linux wasn't so much its technology as its licensing and development model, by now referred to as "open source"—and in particular, the open source model's ability to "out-commoditize" Microsoft.

The fundamental question is this: why is Linux (and the open source movement it helped launch) able to out-commoditize Microsoft? Because it, like the PC, the Internet, and the other open systems and open standards we take for granted today, is more of an ecosystem than a technology. Indeed, Linux builds above those previous ecosystems—without open, commoditized hardware, and without the Internet to enable the open source development model to work, Linux would not exist today.

Microsoft may wield the forces of commoditization more expertly than any company that has come before it, but its platform is not an ecosystem. By definition, an ecosystem is an environment to be shared, not owned. Linux is positioned to become the lingua franca of the lowest level of the software stack, if we never forget it is an ecosystem and not a product to be owned. Looking at the lessons of the past, if it remains an ecosystem, we all win. If not, we destroy it.

"So, How Do You Make Money from Free Software?"

If the open source movement represents the commoditization of software, how can the challengers of today's software industry utilize its commoditizing power to unseat the incumbents, Microsoft in particular? Perhaps more importantly, if this strategy succeeds, is there money to be made in a software industry that has been commoditized? Finally, are there lessons that can be applied from past commoditization events, particularly the events that reshaped the hardware industry in the 1980s?

For a textbook example of how to turn the commoditization of an industry into business advantage, one need look no further than Dell Computer. Dell, of course, was one of the companies that started life in the mid-1980s to build IBM clones. Dell's initial claim to fame was "build to order," taking advantage of the fact that a PC was not really a product in itself, but rather, an assemblage of numerous products that any individual with a moderate amount of skill could assemble himself—a direct lineage from IBM's decision to base the original IBM PC on off-the-shelf parts.

Unlike some of its competitors, Dell saw itself for what it truly was: an assembler of off-the-shelf components and a distributor of these components in a form its customers found useful—namely, a complete PC. Dell gave its customers choice—not an overwhelming amount of choice, but enough choice to give those with the skill to build their own PCs reason to buy from Dell instead of building themselves. Its competitors, on the other hand, attempted to mold the PC into a monolithic, unchangeable product, a collection of specific components from specific vendors with the occasional bit of proprietary technology added to the mix—a thinly veiled attempt to decommoditize the PC standard and own it all to themselves.

To accommodate its new approach to selling hardware, Dell had to develop a new kind of business model. Over the years, the Dell model became more about the assembly of product than the final product of that assembly process. Dell became remarkably good at assembling components from a multitude of suppliers into cohesive wholes, and in negotiating with those suppliers to get the lowest possible price. It stuck to commodity components, allowing the market to pick winning technologies and resisting the temptation to invest heavily in the R&D required to play the proprietary lock-in game its competitors were playing. It employed unusual tactics on the sales side, most notably selling directly to the consumer instead of going through wholesalers and resellers, each of whom took a substantial slice of the profit margin.

As a result of its streamlined processes and lower cost structure, Dell was able to sell PCs at a much lower price than its competitors could. As the PC market grew, and as the market commoditized further with each failed proprietary extension to the PC standard, Dell's position grew stronger. As the PC began to move upmarket, it simply became less expensive to "outsource" the assembly of the PCs to a supplier that specialized in assembling them, and Dell was extremely well positioned to play this new role. Today, as the commoditization of PCs extends to other parts of the hardware market—servers, storage, printers, handheld devices—Dell continues to be extremely well-positioned, and its entry into a new market is often taken as impending doom for that market's established firms.

The First Business Models for Linux

So, what lessons can we learn from Dell as open source commoditizes the software world? Namely, that operating in a commodity market calls for entirely different business models than the business models that have preceded them. Beyond general conclusions such as this, what specific lessons are there to be learned from Dell's success? As a start, we will look at the lowest layer of the software stack, the operating system, and attempt to draw parallels between Dell's successful strategy and the strategies of today's open source operating system vendors—namely, the Linux distribution companies.

To millions of users around the world, "Linux" is an operating system. They're right, of course, but the reality is far more complex than that. First of all, Linux proper is just the kernel, or core, of the operating system—the rest of the software that comprises the "Linux operating system" is developed independently from the kernel, by different groups that often have different release schedules, motivations, and goals.[1]

Traditional operating systems are built by cohesive teams, carefully coordinated groups of product managers, project managers, and programmers at companies and universities. In contrast, Linux is built by thousands of individuals—hackers and hobbyists and professional programmers—some paid to work on specific projects but the majority simply working on what interests them. And the reality is even more involved: Linux is not just a single system, but hundreds of subsystems, programs,

---

1   To avoid confusion, I will use the term *Linux* to refer to the operating system, following standard usage. When referring to just the Linux kernel, I will say "the Linux kernel."

and applications, themselves developed by their own communities of individuals around the world.

So, who glues all this mishmash together into something that actually looks like an operating system? Since almost the inception of the Linux community, this has been the job of the "Linux distribution," a curious term in itself for those coming from broader computing circles accustomed to operating systems being built by cohesive teams, or at least teams of cohesive teams.

A Linux distribution is a collection of software (typically free or open source software) combined with the Linux kernel to form a complete operating system. The first distributions (HJ Lu's boot/root diskettes, MCC Interim) were very small affairs, designed simply to help bootstrap the core of a Linux system, on which the user (typically a Linux hacker himself, eager to get into writing some code) could compile the rest of the system by hand and as needed.

A second generation emerged (SLS, Slackware, Debian) that aimed to expand the breadth and depth of software shipped by the first-generation distributions, including software typical end users of Unix systems might find useful, such as the X Window System and document formatting systems. In addition, the second-generation distributions attempted to be easier to install than the first, as they were targeted not at Linux hackers eager to get into writing code, but rather, at the ever-expanding collection of end users Linux was just beginning to attract at the time.

As Linux's user base grew, many in the Linux community began to sense a business opportunity, and the first Linux companies were formed: Red Hat, Caldera, SuSE, and many others whose names have long been forgotten. These companies formed around the concept of selling commercial distributions to the expanding Linux user base. A third generation of Linux distributions was born.

The commercial opportunity was ripe, as the primary means of acquiring Linux to that point had been the Internet, and up to that point, the primary users of Linux had been students at universities, where Internet access was plentiful. However, in the broader population where Linux was beginning to get noticed, potential Linux users were lucky to have dial-up access to online systems such as CompuServe. Combined with the rising popularity of CD-ROM drives and the growing size of distributions to incorporate more and more software to appeal to a wider and wider audience, the first business models for Linux were born.

These business models served the first Linux companies well through most of the 1990s and, indeed, this is where the term "Linux distribution" originated—the companies themselves were little more than assemblers and distributors of Linux software, including the Linux kernel, the GNU compiler toolchain, and the other software that came with a typical Linux system. As the typical Linux user became less and less of a technologist and more and more of a traditional end user, the focus of the distributions shifted from simple assembly and distribution to making the distributions easier to install and use.

Linux Commercialization at a Crossroads

Of course, as distributors of a commodity (for, after all, any company could easily become a Linux distributor—all the software being distributed was free), these new Linux distribution companies lacked the "proprietary advantage" every business needs to survive, not to mention thrive. So, following time-honored tradition, many of the Linux companies kept their "value add" proprietary in an attempt to better compete with each other.

For a time, one company took a different approach: Red Hat. After a brief flirtation with proprietary extensions, Red Hat announced its products would include only open source software. Why? It listened to what the market was telling it. The scores of Unix refugees, now occupying important positions in the companies that were adopting Linux in droves, had already been down that path; furthermore, the giants of the industry now supporting Linux, which by now included virtually all of the companies that had participated in Unix's destruction and had seen the consequences, saw Linux as a commodity platform that could recapture the position they had collectively handed to Microsoft in the early 1990s. As a result, Red Hat emerged as the market-leading supplier of Linux software.

However, as the Linux market continued to grow, and as it began to take a place at the core of the computer industry, Red Hat bumped up against its own ceiling, caused by lack of proprietary advantage—other companies were beginning to take in billions of dollars per year in revenue from Linux-based sales, while Red Hat seemed to have hit its peak at $100 million or so.

To counter this, Red Hat came up with a strategy that was still in keeping with its "100% open source" market position. Instead of focusing on selling Linux as a boxed product, it would sell software updates to those boxed products in the form of annual subscriptions. This strategy by itself proved inadequate, as the software updates it distributed were available for free. So, it combined the new strategy with another maneuver, a redefinition of the "Linux platform" to one it could define and control itself.

Moving away from its traditional, freely redistributable Red Hat Linux product line, it launched Red Hat Enterprise Linux. The key part of the strategy behind Enterprise Linux was that independent software vendors (ISVs) and independent hardware vendors (IHVs) were directed to certify to this new "high-end" Linux platform, while the old Red Hat Linux was relegated to software developers and infrastructure roles. The other key part of the strategy was that Enterprise Linux was no longer freely redistributable—the acquisition of the product was tied to the subscription, and any redistribution of the product caused the subscription to be null and void.

In other words, if Linux users wanted access to the applications and hardware certified to Red Hat's platform, they had to run Enterprise Linux. To run Enterprise Linux, they had to acquire it from Red Hat via the new subscription model, which entailed signing a subscription agreement that forbade them from redistributing it. More precisely, customers were still free to redistribute Enterprise Linux, but in doing so, they lost all support from Red Hat and, most importantly, from the legions of ISVs and IHVs that certified to the Red Hat platform. Red Hat's transformation was complete when it dropped its Red Hat Linux product line altogether in 2003. Red Hat's new model was still in keeping with the letter of the open source movement but no longer with its spirit.

Proprietary Linux?

By any measure of the term, this is proprietary lock-in, albeit proprietary lock-in that does not involve the traditional attainment through source code intellectual property—i.e., proprietary software. In a way, Red Hat has learned from the lessons of Dell's success: it has come up with a clever new business model to match the commodity market in which it competes—operating systems used to be sold as products in boxes or bundled with other products, and Red Hat realized this approach would not be profitable in the new operating system market Linux was helping to create; so it found a new way to sell its operating system products that was profitable.

However, in a very real way, Red Hat's model is also dangerously close to the model employed by the Unix vendors, which had catastrophic consequences. It is attempting to decommoditize the Linux platform, not through proprietary extensions in the form of software, but through a redefinition of the Linux platform to its own ends and the restriction of how that platform can be used and redistributed. Sure, the source code to its platform is still freely redistributable, but with the shift of proprietary position away from source code intellectual property and to third-party relationships and subscription agreements, the rules of the game have changed dramatically here as well.

What's at Stake?

Red Hat's new business model may be helping its revenues in the short term, but is it in Red Hat's best long-term interest—not to mention the best interest of the Linux ecosystem as a whole—if Linux is owned by a single company, or if Linux fragments like Unix did as Red Hat's competitors follow down the proprietary Linux path?

If Red Hat's business model is wrong, what is the right business model for Linux distribution vendors? In my view, the Dell model can be taken a step further than any of the Linux distributors have thought to take it. After all, what are open source technologies but commodity software components, and what are Linux distributions but assemblers of those components into products the end customer finds useful?

Indeed, such an "assembler of commodity software components" business model might fully realize many of the benefits of Linux that the traditional, product-oriented business models of Linux distribution companies have failed to capture: flexibility and choice, without the substantial expertise and financial investment required to adapt a Linux distribution for its own purposes. What if a Linux distribution was a collection of parts that could be mixed and matched to suit the needs of the company buying it instead of a one-size-fits-all, monolithic product like the Linux distributions of today?

As with new business models that have come before it, such an approach would open Linux to new markets, markets that are already using Linux, but for whom today's product-oriented business models are ill-suited: server appliance vendors, set-top box makers, and others to whom Linux is an invisible vehicle for driving their own products. In their world, Linux is a piece of infrastructure, not a product to be owned by Red Hat or otherwise.

Indeed, this is the model being employed by my company, Progeny. Our approach is to embrace the commoditizing effect Linux and open-source software have on the software industry instead of fighting it. Since every company needs a proprietary advantage of some kind, we've chosen to focus on building advantage through our processes, not technology, much as Dell did—in other words, to leverage our expertise in distribution building to help other companies assemble commodity software components from disparate places into cohesive wholes, and to do so in a scalable and flexible way.

Beyond building a better business model around Linux, what's at stake? I contend far more is at stake, for one simple reason: Linux needs to remain a commodity, as it is now a core piece of infrastructural technology at the heart of the computer industry. Indeed, Linux was enabled by the commodity nature of the last infrastructural technology to redefine the IT industry: the Internet.

In "IT Doesn't Matter," which appeared in the May 2003 edition of *Harvard Business Review*, Nicholas Carr points out that infrastructural technologies "[offer] far more value when shared than when used in isolation." What happens if Linux is decommoditized and ends up being the proprietary product of a single company to serve its own purposes? What if the PC or the Internet had been decommoditized? Where would we be today?

Carr's essay provides hope that there is money to be made in infrastructural technologies that have been fully commoditized, and that there's no need to try to own those infrastructural technologies:

> ...the picture may not be as bleak as it seems for vendors, at least those with the foresight and skill to adapt to the new environment. The importance of infrastructural technologies to the day-to-day operations of business means that they continue to absorb large amounts of corporate cash long after they have become

commodities—indefinitely, in many cases. Virtually all companies today continue to spend heavily on electricity and phone service, for example, and many manufacturers continue to spend a lot on rail transport. Moreover, the standardized nature of infrastructural technologies often leads to the establishment of lucrative monopolies and oligopolies.

Carr's essay also provides historical perspective on the commoditization process:

> …infrastructural technologies often lead to broader market changes.[…]A company that sees what's coming can gain a step on myopic rivals. In the mid-1800s, when America started to lay down rail lines in earnest, it was already possible to transport goods over long distances—hundreds of steamships plied the country's rivers. Businessmen probably assumed that rail transport would essentially follow the steamship model, with some incremental enhancements. In fact, the greater speed, capacity, and reach of the railroads fundamentally changed the structure of American industry.

In a commodity world, technologists need to think about innovating in their business models as much as (if not more than) innovating in their technology. Of course, it's a natural trap for the technologist to think about technology alone, but technology is but a small part of the technology *business*. Look for your competition's Achilles' heel, which more often than not is an outdated business model in a changing world, not technology. To attack your competition with technology alone is to charge the giants head on, and this approach is doomed to failure the vast majority of the time.

Businesses operating in a commodity world also need to build business models with the larger ecosystem in mind. It is tempting, once the incumbents have been overthrown through the powers of commoditization, to lapse into the same old proprietary lock-in strategies that served the former incumbents so well. In effect, though, this is decommoditizing the industry, "poisoning the well." It is possible to build a successful business in a commodity market, as Dell and many others before it have shown, and in the long run, it is far better to ride the forces of commoditization than to fight them.

CHAPTER 7

Matthew N. Asay

Open Source and the Commodity Urge: Disruptive Models for a Disruptive Development Process

Open source hastens software's natural trend toward standardization/commodification. While technologically innovative companies will always find ample customer interest, the most important innovations for the next decade of software will come from business model innovation, mostly spawned by open source license requirements. Open source builds a new intellectual property regime centered on the source of code, not source code. Protection, in other words, shifts to "owning" the code creator, rather than the product she creates. Those business models that acknowledge this and leverage it will yield better profits than those that attempt a halfway embrace (or rejection) of open source.

Introduction

We are missing the point. Yes, open source imposes dramatic changes on the software industry, and yes, it is roiling the fortunes of many an established vendor. It will continue to do so, and at an increased pace. Yet despite the sometimes anguished, sometimes giddy reception that open source has provoked in the IT world, open source is not novel. It is not odd.

Open source is simply the software world's mechanism for becoming just like everything else.

All the world's a commodity—or a service to support and distribute commodities: this book that you are reading, the chair that supports you, the restaurant you will eat at tonight—everything—including, increasingly, software, thanks to open source. Open source accelerates the natural progress of software toward commodification, or standardization.

It is critical that IT vendors understand this so that they can deploy (or fight, if they so choose) open source effectively, and more intelligently choose how and where to innovate. Open source does not destroy all value in software innovation; instead, it shifts the control point from the code itself to the creator of the code. In so doing, open source software will not pillage all closed source software. As in other industries, there will continue to be plenty of room for upmarket vendors (e.g., Whole Foods in grocery retailing; Starbucks in coffee; and Nordstom in retail clothing).

That said, there is no room for middling and muddling. Open source will commodify from the bottom up while "upmarket" vendors will dominate "up the stack." Everything else will be a wasteland. Just as Safeway finds itself pummeled by Wal-Mart and Whole Foods so, too, will middle-ground IT vendors find themselves grasping at a dwindling market opportunity.

Open source offers hope, but perhaps not for the reasons normally associated with it. Much has been made about the open source revolution, and with good reason. But perhaps the best reason has little to do with development of source code, and instead has much to do with distribution, marketing, and sales. In other words, what we thought was a software development methodology may have far more importance as a business strategy that undercuts competitors while driving down costs and shifting control to buyers. In such a world, those who understand and leverage open source commodification (or escape it) will thrive—everyone else will be marginalized into economic oblivion. Commodification, the highest stage of capitalism; open source, the highest stage of software.

A Brief History of Software

Once upon a time, software did not matter—hardware did. Software was something that hardware vendors wrote to help them sell hardware. Little more. Software was important because it made hardware operate, but customers understood that they were paying for hardware, and not the software that ran on top of it. (This is still somewhat true of certain areas of the embedded software market.)

As hardware commodified, software grew in importance as a differentiator with these same customers. Not all hardware commodified at the same pace: Solaris servers, for example, handled a workload in a way that commodified hardware could not, and Sun consequently charged a premium. But the real premium increasingly gravitated up the IT stack to the applications that people ran on their hardware. Hardware was important, but only because the applications had to run on something. With the rise of Dell and other commodifiers, however, IT buyers came to care less and less about the "guts" of their computing experiences—they bought systems for the applications they could run, for the productivity they could achieve.

No single company did more to send the applications trend into hyperdrive than Microsoft. Microsoft made software easily consumable by the masses. By simplifying computing, and by doing so at a dramatically lower cost, Microsoft grew the market by competing against nonconsumption and underconsumption, inviting multitudes of average users into the hitherto closed world of computing. Microsoft's Visual Basic lowered the bar of expertise to be a proficient developer, and its Office suite created a market of home and business users who suddenly could create brochures, quality letterhead, etc. Whatever open source developers' feelings about Microsoft, they should acknowledge Microsoft as the natural parent to their own brand of commodification.

For this is what open source is doing: commodifying software. We are now at the point where mainstream software is becoming commodified by the open source community, perhaps pushing all value to the services that support hardware and software. Microsoft, in a sense, is being out-Microsofted.

In this world, customers benefit as vendors focus on solving their business problems, instead of innovating new methods to achieve customer lock-in. Much of today's IT world is composed of expensive, monolithic software "solutions" that end up creating complexity and integration problems, instead of resolving customer problems. That is, today's IT industry is a morass of conflicting standards, complex installations, tepid product interoperability, and expense—all products of the industry's Wild West "level of thinking" in its adolescent years. Increasingly, however, customers are tired of subsidizing the disarray and are turning to open source as a way to get more for less. As open source proliferates, the cost of infrastructure software will plummet, freeing up resources that the CIO can spend on resolving application requirements up the stack.

Vendors, for their part, also benefit from increased use of open source, because it removes the "IP safety blanket." Because code is open, vendors must find innovative ways to satisfy and "lock in" customers. Copyright and patent are fine, but they pit the vendor in an adversarial relationship with the customer, whereas open source control mechanisms tend to force vendors to win by intimately understanding and fixing their customers' business problems. In addition, this commodification of IT will push vendors to move up the stack (and off the stack, into services) to deliver increased customer loyalty/value. Finally, as prices for software drop to match the drop in hardware costs, more buyers will enter the market, increasing the size of the market. Everyone wins.

A New Brand of Intellectual Property Protection

To fully appreciate this trend, it is critical that we better understand the intellectual property regime powering the open source revolution. Intellectual property (IP) law has always been about control. That control benefits creators by holding off would-be competitors long enough to allow the creator to attempt to profit from her innovation. I write a piece of software; I copyright it; I sell it (assuming it is a useful piece of software and I have adequately marketed it so that people know about it). Simple. This has been the software industry's dominant model for decades, and has created a few mammoth software companies that have successfully exploited their IP to generate billions of dollars in revenues. In this model, exclusion (i.e., the ability to keep competitors or customers from copying one's code and distributing it to others) yields profits. In this model, the code itself—locked up and protected—matters most.

In the open source world, at least as defined by the GNU General Public License (GPL), IP continues to play a critical role, but it is a different kind of IP. Dubbed "copyleft," open source IP focuses on keeping code access open rather than closed. And, unlike in the world of proprietary software, the code matters less than the coder—anyone can see the code, but not everyone can replicate the coder's influence on the community to which she contributes her code. By virtue of her contribution, she builds influence in her chosen code community, and this influence translates into a new kind of IP: reputation property instead of intellectual property.

In this new world of open source, reputation property means as much as or more than traditional intellectual property. If I employ the developers on a given project, I have a measure of control over the direction that the open source project will take. But even more importantly, the more developers I employ who work on, say, the PostgreSQL database project, the more likely it is that would-be customers will trust me to be able to support it. Once a company is thought of as the default support vendor for a given project, the harder it becomes to dislodge that vendor. This jibes perfectly with other commodity businesses where brand, price, and service provide the only lock-in, a benevolent lock-in *that customers choose* instead of one that vendors impose.

In this way, the open source code creator exercises a form of control over her creation, and that control translates into *her* (and only her) ability to charge a premium for the software. As an open source creator, then, my options for deriving profit from my creation are not more limited, but they *are* different. Instead of a limited monopoly guarded by law, I have a monopoly guarded by common sense: buyers want to buy from the most qualified source of support. They pay to have access to the source: not the source code, but the source of the code.

This distinction is important. The importance of source code gets trumpeted so often that one would think that every IT buyer on the planet is clamoring for access to source code. They are not. Indeed, Microsoft recently conducted a survey of its customers and found that roughly 60% felt that access to source code was "critical." But when pressed on the matter, 95% said that they would never look at the source, and a whopping 99% said that they would never modify it. (If they did, chances are that such modification would violate their support agreement anyway, whether their vendor was Microsoft, Novell-SuSE, Oracle, Red Hat, etc.) In sum, customers perceive source code access to be important, but are not exactly sure why. As I will detail shortly, the "why" relates to a desire for additional choice and control, and choice and control drive down costs.

Of course, nothing in this chapter should lead one to believe that source code is irrelevant. Source matters. It matters because it lowers switching costs (i.e., the cost of swapping out one vendor's software for a competing vendor's software); because it provides buyers with more control over their IT, as it allows them to shape code to fit their particular needs; and because it provides a mechanism for keeping vendors honest, by forcing them to take responsibility for the quality of their code. In short, source code matters because it shifts control back to the buyer, which forces vendors to offer better software at lower prices. While none of these source code benefits requires the intervention of a vendor, we should not get sucked into the belief that vendors matter but little in the open source world. Instead, open source actually makes vendors more relevant to customers than ever before at dramatically lower prices.

Besides benefiting customers, this GPL licensing scheme offers vendors a way to exercise an incredible amount of control over competitors. By open sourcing my code under the GPL, I push my competitors to follow suit or to increase their R&D efforts to escape commodification. Unfortunately for them, this counterstrategy of a unilateral R&D arms race tends toward paltry results: customers will often opt for the "good enough" product when the price is dramatically lower. Yes, my closed source competitors could simply take my freely accessible source code, "fork" it, and build it into their own products, but they almost never will. Doing so compels them to open source their own software, which they will be disinclined to do. Even if they did so, however, and even if my competitor were not a stodgy old closed source vendor, but rather, an agile open source predator, it would matter little, because open source buyers invariably favor the source of the source, as it were: they trust the creator of the code to support it best.

Open Distribution, Not Source

This has huge implications for the software industry. Disruptive vendors can opt to completely open source their code, relying on reputation property to net them revenues, and further relying on their freely available alternative to competitive products to force competitors to meet them on their home turf. This is not to say that all vendors must adopt an open source strategy, but rather, that they must compete with open source's lower cost structures and superior distribution mechanisms. All must increasingly compete on open source's terms. More detail is needed on why this is so.

The Open Source Weapon

Open source enables a vendor to maximize its market penetration at minimal cost, which is the goal of every IT vendor, but particularly of emerging-growth vendors seeking to displace incumbent vendors. One of the biggest roadblocks to any company's growth is the Bureaucracy Bottleneck—the larger the buyer (and, hence, the larger the opportunity), the more layers of bureaucracy an IT buyer must fight through to try-before-they-buy. Not so with open source, which surreptitiously makes its way into enterprises via free download.

Such distribution fattens a vendor's bottom line without fattening the customer's price tag. MySQL had 10 million downloads in 2003, and by mid-2004 had more than 5 million installations. Of these would-be customers, 5,000 have returned to buy a support contract/license from MySQL, bumping the company's revenues by 100% to $10 million in 2003. The revenue growth is important, but even more so is that it achieved this growth by spending less than 10% of total revenues on sales and marketing activities. By contrast, most public software companies spend 45–50% of total revenues on sales and marketing, and companies of MySQL's size generally spend 21.8% on these activities, according to a study done by Softletter.com. In short, open source creates a small universe of prequalified buyers who seek out the vendor, instead of the other way around, with the vendor's primary marketing costs relating to setting up an FTP server and mostly word-of-mouth-type evangelism to developers.

The savings do not stop there. Whether the open source vendor "borrows" much of its code (e.g., Novell, Specifix, Gluecode) or creates it almost entirely in-house and then open sources it (MySQL, SugarCRM, JBoss), open source delivers development-related cost savings. For the "borrowers," the cost savings are obvious: they leverage a well-developed body of code, most of it written by individuals not on their payroll. For the JBosses and MySQLs of the world, which do 85–100% of their own development work, there is still a significant QA savings from the global pool of testers who submit bug fixes and code contributions (which may or may not be used by the vendor)—cheaper to build, cheaper to sell, cheaper to buy.

Proliferating Open Source Beyond the Enterprise

Today, open source largely confines itself to infrastructure software, in part because this is where the widest computing community resides. Community-based open source projects require a sufficient body of developers with aptitude and interest in a given development problem. But as the IT industry begins to recognize the promise of emerging open source business models, community becomes less critical to the success of a project. As this happens, no area of the software stack will be exempt from open source's influence and intrusion.

Significantly, open source business models will pave the way for open source to conquer the Great Middle Class of IT: the small to medium-size enterprise (SME) market. Open source, as a disruptive methodology in the Clayton Christensen sense, will do more than simply allow startup vendors to compete against established vendors in established markets: it will help to create new markets by competing against under-consumption and nonconsumption. The Internet, and open source's unparalleled use of it, is at this trend's core.

In "Does IT Matter?" Nick Carr offers one example of how this worked in another industry: chocolate. Milton Hershey, founder of the Hershey Company, noticed a gap in the chocolate landscape, a gap the railroad could resolve. Until his observation, transportation deficiencies had forced production to stick close to consumption—there was no quick and refrigerated way to ship chocolate over long distances, creating scores of micromarkets for chocolate. With the advent of the railroad, however, Hershey conceived the idea of a national chocolate market, built it, and owned it.

Today, the Internet parallels the railroad infrastructure of Hershey's era. It offers independent software vendors (ISVs) a similar opportunity to that which Hershey had, yet the majority of ISVs are not capitalizing on it. Yes, some traditional ISVs can and do offer their products for download, but this is a makeshift attempt to leverage the Internet. Open source is an Internet phenomenon—it depends upon the Internet and extends the Internet's utility. Open source should disproportionately benefit from the Internet's distribution mechanism, provided companies understand this fact and act accordingly. As I will show shortly, tomorrow's most successful software vendors will triumph to the extent that they develop models that leverage the Internet as a distribution mechanism, and use open source licensing as the rules-based system to govern that distribution.

So, Why Not Freeware?

Let us assume that all of this is true. Open source is great because it enables upstart competitors to undercut established vendors on price while providing their customers Porsche technology at Pinto pricing. But if open source is so important because it allows me to freely distribute my product over the Internet, why is freeware not equally disruptive?[1] Stated another way, if the source code does not matter, and only distribution matters, why not just give the software away as freeware and charge users who require support? Why offer something (access to source code) that simply does not matter?

The easy way out of this apparent quandary is to allow that while open distribution matters most, open source code access is also important. But this does not get us very far. I will therefore detail the reasons that freeware cannot match open source as a distribution strategy. Most importantly, I will explain how the two matter most when they intersect, making distribution without access as hollow as access without distribution.

---

[1] Larry Augustin originally needled me with this question, for which I thank him.

Don't view. Don't modify. What do you do?

As mentioned earlier, it is an indisputable fact that the vast majority of IT buyers will never view or modify source code, even if offered the ability. There are numerous reasons for this, but the most compelling one is that customers expect to pay for a solution to their problems, and not merely a tool to help them solve their problems. (More on this shortly.) No company can afford the time and human resources necessary to resolve all IT problems; therefore, they take "shortcuts" by buying software that purports to fix certain problems for them. This applies equally well to closed source software and open source software. Most IT buyers just want their software to work, they don't want to have to fiddle with it.

By opting not to view or modify source code, does an IT buyer thereby opt out of any and all of the benefits of access to the source code? Absolutely not. Just because customers do not choose to exercise their rights to view and modify source code does not mean they do not benefit from the right, even when not exercised. On one level, the option to view the source code serves as a surrogate for the actual exercise of this ability. As an example, because I can review the database code that Sleepycat delivers to me, it forces Sleepycat to provide a higher-quality product than closed source vendors would have to offer.

R0ml Lefkowitz of AT&T Wireless gives a tangible example of how this works. In June 2003, R0ml related that he had asked his wife to solicit multiple contractors' bids for a home improvement project. Instead of gathering several bids, however, R0ml's wife procured only one bid. When he asked her why, she responded that she figured the contractor would assume she had collected a number of bids, and so would give her his best bid from the start. The option to exercise choice, then, served her as a useful surrogate for actual choice.

In this way, access to source code motivates the code's vendor to provide a superior product, knowing that it will be open for all to see. It also functions as a security blanket for customers. Hopefully, they will never have to look at the source code. But if Vendor X fails to deliver on its promises, or if it goes out of business, that customer will have the option (unpleasant though it might be) to have some other services firm support the stranded code. Source code access lets buyers rely on their vendors...but not too much. Importantly, the more independent the buyer is from the vendor, the lower the vendor's prices must be. More on this shortly.

As IT buyers have grown comfortable with open source projects, another benefit has emerged. Initially, it is true that buyers will tend to want to avoid tampering with the software they buy. Over time, however, as they grow familiar with a product (closed or open), the buyer's developers will want to make tweaks here or there. They begin to support themselves, in other words, because calling out for support takes unnecessary time (and patience). In a closed source world, however, their ability to tweak the "solutions" they buy is limited. In an open source world, the Amazons of the world have free reign over their IT. Source code access gives customers the ability to experiment with and tailor software to their liking, if they choose, and on their own time table.

Other reasons have been suggested. Developers like to be creative. Like anyone else, they prefer not to have self-expression manacled, and they choose to express themselves in the code they write and modify. On a mailing list in June 2003, Frank Hecker argued that access to source code is critical for developers at the heart of a company's IT infrastructure. Such people want to be able to modify code, and so must have access to source, because they will need to be able to fix any problems they may download into the company. Outside the corporate firewall, Ben Tilly (on the same email list) stipulated that while the majority of developers are not involved in open source, access to source code is critical to that core of developers who do participate. They are the lifeblood of open source communities and are the ones who will openly extend assistance to newbies who just want the code to work without getting their hands dirty.

All of which is a verbose way of repeating the earlier point: source matters, even where it may not directly matter to the end user. Access to source extends benefits to users beyond those chosen few who actually exercise the right to touch source code. Source matters because choice matters. Choice matters for a number of reasons, not the least of which is that choice drives down prices. And choice is amplified by open source, not by freeware, which has its source code closed.

Open source. Open choice. Open wallet.

Many successful software vendors would have us believe otherwise. That is, they want to sell suites of services that take care of all needs, that reduce complexity, and that reduce choice. The primary perpetrator of this strategy is Microsoft, which, as Dana Gardner of Yankee Group notes, wants to "make the end user any offer they can't refuse—to go Windows everywhere" (*http://enterprise-windows-it.newsfactor.com/story.xhtml?story_id=22143*). One major problem with buying into these monolithic visions is that once in, the switching costs to go with another vendor are prohibitive.

By buying into Microsoft or any other vendor that holds out greatly reduced choice as a way to accomplish moderately reduced complexity, a buyer surrenders his IT destiny to that vendor. He upgrades when the vendor wants him to. It gets new technology when the vendor chooses to innovate. (I would argue, and have on several occasions, that Microsoft's market dominance has caused it to stagnate in terms of innovation. When was the last time Microsoft's Office product significantly improved over the last version? And yet the buyer keeps buying, because he finds himself on the Microsoft treadmill.) And he pays whatever the vendor demands, because the he has no other options. He is a prisoner of the vendor's universe, however expansive the vendor pretends that universe to be.

Over time, buyers who condemn themselves to such vendor-controlled realities will pay more for their IT, both in hard costs and in opportunity costs. Open source offers the opposite vision: maximum freedom to shift among vendors (even while staying with the same or similar code base). Open source therefore costs less in the short term and, especially, in the long term.

If we step outside the IT world for a moment, this point will become even clearer. My wife and I recently redid our landscaping, including our cement work. Or, rather, we wisely chose to hire out the work. True, with a Dummy's Guide to Cement (my "source code," as it were), even I might have been able to figure it out and could have completed the project satisfactorily. Had I opted to do the work myself, the cost would have been X. Because I could have done the work myself for X, my cement contractor was only able to charge me 1.5X. Had he bid higher, I would have had strong incentive to perform the work myself. I had access to the source, so his pricing power was curtailed. (In the same way, access to a closed binary, as with freeware, does not accomplish this same effect of driving costs down.)

The cement contractor ended up performing shoddy work and walked off with a portion of our money, for which he had not completed the associated work. Measuring it out, it will cost us 5–10X to hire a lawyer to compel our cement contractor to satisfactorily complete his 1.5X worth of work. I happen to be a lawyer, but not one that has ever actually practiced law, so I am stuck paying the lawyer's fees: $500/hour to recover $1,500+ in payments owed to me by the contractor.

Perverse world, you say? Yes, I suppose so, but the point is that the delta between the cost of me doing my own cement work and the cost of me doing my own legal work is directly proportionate to the skill set involved and the artificial licenses set up by the legal profession to keep would-be attorneys in "would-be land." I am effectively barred from accessing the "source code" of the legal (and medical, among others) profession, which drives up the price that I must pay.

Again, access to source code, whether in software or cement, offers choice, and choice ensures lower prices. It does not matter that most people will never choose to do their own cement work, just as it does not matter that most IT buyers will never choose to view and modify source code. The important thing is that they could if they were so inclined. That "could" is instrumental in dropping prices through the floor.

Such lower prices, then, allow the CIO to spend more money on developers who can further customize software to meet that specific organization's requirements. Imagine that: IT that works for a customer, rather than against it. *That* is innovative.

Such innovation is what open source is all about, and is why it continues to make inroads in the enterprise, in embedded devices, and everywhere else. Open source brings choice, and choice saves money. Freeware does not engender such choice. Meaningful choice is not created by cost-free technology; rather, choice is created by the freedom to manipulate code to one's personal (or corporate) advantage, or to have it widely distributed in such a way that one can benefit from others' exercise of that freedom. Access matters little without distribution, and distribution matters little without access. Together, they spell the commencement of a new age of software innovation, innovation that benefits vendors' and buyers' bottom lines.

Open Source Business Models

All of this may sound plausible enough, but we now need to trace through some real-world business models that vendors use or could use to leverage the benefits of open source to drive revenues and boost profits. Before doing so, it is important to note that an "open source business model" means more than simply supporting Linux as an operating system. In other words, the fact that my CRM system runs on Linux, or that my hardware appliance has Linux as its core, does not make me an open source company. An open source business model means that a vendor somehow engages with the open source community.

With that said, I also need to stress that whatever the importance of open source, not all companies must adopt open source to find success. Open source is the great commodifier, but there will always be those who successfully evade that commodification. Other industries prove instructive on this point.

Take retail, for example. Even as low-cost commodifiers devour middle ground in this market, profits persist up the stack. Wal-Mart is the 8,000-pound gorilla of commodification, cannibalizing groceries, clothing, and just about everything else on which it puts its hands. Still, for all of Wal-Mart's success in low-end fashion, for example, Nordstrom continues to win at the higher-end game. This is more than a case of customer snobbery—it has to do with an experience that Nordstrom delivers (superior customer service) that Wal-Mart is structurally incapable of offering. (This same phenomenon exists in coffee—why are consumers so willing to shell out $4 for a cup of coffee? Because Starbucks has defined a customer experience that transcends Maxwell House at home.)

Another example is the groceries market, as Charles Fishman highlights in the July 2004 issue of *Fast Company* (*http://www.fastcompany.com/magazine/84/wholefoods.html*). In just a few short years, Wal-Mart has become the United States' largest grocery chain, yet the title of "Most Profitable Grocery Chain" and "Fastest Growing Grocery Chain" eludes Wal-Mart (and its European competitor, Aldi). No, those titles go to Whole Foods, with remarkable year-over-year growth in the face of a nationwide 2.5% annual compound growth rate: 17% (2003), 21% (2002), 21% (2001), 23% (2000), and 14% (1999). Whole Foods delivers an upscale grocery experience, offering organic foods and superb quality, and Wal-Mart stocks its shelves according to its modus operandi: decent selection at rock-bottom prices. Both chains have found their respective strategies to be highly profitable—everyone else has gobbled their dust. Whole Foods has registered

$188 million in profits over the last several years, and Food Lion cleared only $150 million with seven times as many stores and five times Whole Foods' revenues. Safeway, for its part, lost $1 billion in the same period on even greater revenues. The takeaway? There is no room in the middle for undifferentiated players. One either commodifies or evades commodification through innovation. Everyone else languishes.

Both Source (a.k.a. Mixed Source) Model

So, the first business model is for the technical innovator that refuses to join open source commodification at all. But what about those companies that opt for a "both source" model, whereby they offer both open source and proprietary software? This model has promise and peril, requiring the vendor to walk a fine line between the model's divergent business requirements (low-end commodification/standardization coupled with high-end specialization). To the extent that a company marries the two, it must do so with a clear understanding of open source complements and substitutes to its proprietary product portfolio.

Both source offers a way to fill in the gaps left by open source, and to charge a premium for this "service," while still delivering open source software. Such a model seems to be ideal for established players that cannot abandon existing customers of closed source products, and blanch at the thought of losing existing profit margins. Of course, whether a vendor can avert the open source "threat" depends upon whether open source has created a viable *substitute* to its product. If so, head-on competition with that open source project is likely futile unless it can move significantly upward in the feature set. Even if it can, competing against free and "good enough" is exceptionally difficult.

A both source strategy makes more sense where the vendor can define and contribute to open source complements. In economic terms, a complement is something that completes a whole; in software terms, it is a component of a software solution. So, just as French fries may be considered a complement to a hamburger, so, too, is Apache a complement to IBM's WebSphere product. Importantly, the more complements that exist for a given product, the more desirable that product becomes for customers, so vendors want as many low-cost, high-quality complements to their products as possible. Oracle likes Linux and x86 hardware because it drives down the total cost of a customer's database solution…without lowering the cost of Oracle's software. Customers, thus, can buy more Oracle software, which gives Larry Ellison more time on his boat.

So far, this sounds like a reasonable defensive strategy for vendors that want to toe-dip into the open source community without getting very wet. But both source also allows vendors to take a scorched-earth agenda against their competitors, by skillfully choosing to build open source complements to their proprietary software, complements which cut directly at those areas that competitors have chosen to retain as proprietary. Of course, such a strategy must bear in mind the distinct possibility that the opposing vendor will then choose to open source pieces of its portfolio that injure the first vendor, creating a lot more open source software, but not necessarily any profits for either company.

Still, it is an open question whether this is a solid strategy against upstart competitors with lower costs who can undercut a proprietary product's margins. In addition, a both source strategy works best for vendors with products that have respectable market share. Slapping open source on or around an also-ran product will not sell it. Good technology, good service, and good sales/marketing sell products. Open source, by itself, is as much a losing strategy as closed source, by itself.

Open source complements to a market leader's products make that proprietary product more valuable by lowering the total cost of the product. Hence, while both source may be the easiest step for a closed source company to make, it will help the vendor only if its products were already competing well against other proprietary products. Again, both source offers no panacea for market losers. The lesson? Companies should adopt the both source strategy when they are on top of their games, instead of when they are losing the final sets of their matches. For market losers, a better bet is to make the difficult transition into a pure-play open source vendor, as defined shortly.

Professional Open Source (a.k.a. Services) Model

The dirty little secret of open source is that the term *open source community* is something of a misnomer. In general, the actual number of contributors to any given project, including the Linux kernel, is tiny. Thus, to "own" an open source project requires little outlay of human resources in terms of numbers, though it may require a significant amount of time to build reputation capital within a given open source community. (Newbies to the Linux kernel, for example, should expect to put in two years or more before they can hope to attain "committer" status in the kernel hierarchy.) Despite the low number of developers required to corner the market on an open source project, the importance of doing so is massive: employing a majority of the developers on a given project roughly equates to intellectual property ownership, as explained earlier.

For this reason, companies that spawn open source projects—e.g., JBoss—are able to completely open source their code without abandoning pricing power. JBoss, for its part, employs roughly 85% of the developers who contribute to the JBoss open source project. JBoss offers its code under the Lesser General Public License (LGPL), which allows users a wide range of action vis-à-vis their code, including the right to fork the JBoss project and start JBoss II.

But no one does that, for reasons already detailed.

Because of the heavy JBoss "ownership" of the committers to the project, the company does not save a great deal of money on development costs. It functions much like any closed source company, except that its development is open for public view and consumption. Any appreciable development savings derive from the bug finds/fixes that JBoss receives from its development community.

Still, the professional open source business model is not really about development savings. Rather, it is about maximizing distribution of one's product; getting it beyond the purchasing firewall/bureaucracy bottleneck to plant the product in the hands of its developer end users so that they can try and then revisit the professional open source vendor for support/service contracts. To get approval to use BEA's Weblogic or IBM's WebSphere, a developer would need to go through a cumbersome process. To use JBoss, she simply needs to click "Click here to download." And while the developer might choose to support herself through newsgroups or other online fora, in production situations she will generally turn to the source of the code (in this case, JBoss). This is the classic open source model, though it is only now starting to be exploited effectively.

Dual-License Model

The dual-license model has been popularized by MySQL, but has been around for some time, most notably deployed by Sleepycat and Trolltech. In the case of a dual-license vendor, that vendor employs not most, but *all,* of the developers who contribute to the code. Because it employs all of the developers, it also owns all of the copyrights to its work. Then, as the owner of the copyrights, it is entitled to license its software under one or more different licenses.

However, the fact that it owns the copyrights and employs the developers begs the question as to what benefit, if any, it derives from its open source status. The answer, as with the professional open source model, lies in distribution strategy. For a dual-license vendor, open source is less a matter of development and more a matter of distribution for open source vendors. Yes, the dual-license vendor derives benefit from outside developers who contribute code (though MySQL, for example, tends to repurpose/rewrite incoming code to help it better fit its existing code base) and bug fixes, but its primary benefit is in the ability to broadcast its product to the world with customers benefiting from lower prices and less lock-in.

Also interesting, though not a benefit touted by the primary adopters of the dual-license model (MySQL, Sleepycat, Trolltech, and now SugarCRM), is the fact that the dual-license strategy provides the customers with a mechanism to buy their way out of the GPL, if consider this desirable. This is of particular benefit in the embedded world where, for example, Linksys might receive GPL'd code from Broadcom and might want a closed source license to that code, so that it will not have to open source the software running its routers and access points (a purely hypothetical example, of course...).

db4objects is promoting its embedded database with precisely this message, one that customers appreciate because however much a vendor may prefer the GPL or another open source license, the fact remains that it may not always be the best fit for a given customer. As such, the dual-license model offers customers a way to pay for the right to choose the license under which to receive software.

ASP Model

Such are the prominent open source models that are easily recognized as such: open source business models deployed by open source companies. But, as Tim O'Reilly has been telling the industry for years, "open source" is a much bigger tent than we may recognize. Tim includes such "infoware" vendors as Google and Amazon in his open source tent. The common denominator between the two? Internet infrastructure powered by open source (Linux and a great deal else), plus an architecture that promotes participation that makes the infoware increasingly valuable.

The enterprise IT industry has also been moving toward a related model for standard enterprise applications, calling it utility computing, on-demand computing, and a range of other names. In this model, IT vendors deliver computing power in a utility fashion: Enterprise Consumer X gets the computing cycles when it needs them, instead of buying all of the hardware/software upfront.

Importantly, customers in this model buy IT (including software) as a service, rather than as a standalone product. As such, customers do not really buy software at all— they buy solutions to their business problems. Whether the "guts" of that solution are open or closed source does not matter anymore. Customers simply pay for value, delivered as a service: SP (service property) rather than IP (intellectual property).

This sounds much like software's ASP model, in which software is delivered to the customer over the Internet, hosted on a central server by the vendor, with customers paying for the value they access over the network. A prominent example is Salesforce.com. Whether the software underpinning the service is closed or open source becomes irrelevant. The requirement to release modifications to a given open source project is triggered by distribution: so long as the code itself is not actually distributed (and only the resultant service dictated by the code is), open sourcing of modifications is not required, but voluntary.

Other Models

The aforementioned business models are the primary models in use today by most open source companies. However, some of the most interesting new companies employ equally interesting (and innovative) business models, generally altering the way open source software is supported. For such companies, the real customer benefit of open source is the availability of source code. But, as noted, major vendors such as Novell and Red Hat, which tie their support contracts to specific product builds, obviate this benefit. Gluecode and Specifix resolve this irony and may point to tomorrow's most successful open source business models.

Managed source model

Gluecode incorporates three levels of source code support into its model. First, Gluecode conglomerates various Apache packages, tests them, and generally makes them play nicely together so that customers need not worry about visiting Apache.com for themselves. Second, Gluecode develops its own proprietary software to extend Apache's reach and thereby provide a compelling solution for portals/business process management (BPM). Third, and unique to Gluecode, the company offers source-level support to its customers, allowing them to check in their code to Gluecode's CVS repository. Gluecode runs the customer's code against test suites to ensure the customer's modifications or additions work properly with Gluecode's open+closed codebase, enabling the customer to become something more: a development partner.

Code-level service model

Specifix does something similar. Focusing on embedded and server deployments of Linux, Specifix allows its customers to modify Specifix's Linux distribution to meet their particular requirements. Instead of invalidating their support agreement with Specifix by doing so, Specifix tracks exactly where the modifications were made and allows the customer to support its own modifications, while continuing to support the original Specifix distribution. The customer may, in other words, opt for the "road less traveled," but Specifix is happy to maintain the more-trafficked road for them, keeping it parallel with the customer's chosen divergence.

Conclusion

Open source propels software toward Commodity Land, a happy place where customers pay for real value and vendors compete on that value, not intellectual property lock-in. Each of the open source business models detailed in this chapter will help to further this trend, making open source mainstream and possibly displacing the traditional, IP-based model as the default.

We are thus on the cusp of a Kuhnian paradigm shift, one that will fundamentally alter the way IT vendors create, sell, and distribute software. Once apparently stymied by the restrictions that open source licensing places on traditional business practices by IT vendors, open source vendors are now finding that open source licensing creates as many opportunities as it closes, changing the nature of software competition for decades to come. This means that incumbent and emerging IT vendors must understand the new rules of engagement to compete effectively. Whether they like to admit it or not, open source will force every software vendor to come to grips with omnipresent, ravenous commodification.

Some may opt for technical innovation over business model innovation, with varying degrees of success. However, such innovators should recognize that while copyright and patent provide potent protections, they also put the vendor in an adversarial relationship with the customer. As such, these traditional intellectual property tools hurt customers as much as (or more than) they do competitors, and will put them at a disadvantage against open source competitors who offer customers choice and value at lower prices. Open source, then, allows vendors to lay waste to their competitors' profit margins by lowering their own costs of distribution, sales, marketing, and development, while simultaneously blessing their customers with increased IT flexibility and a more finely tailored approach to solving their business problems.

Open source, then, offers a new way to innovate, a new way to compete, and a new way to win.

CHAPTER 8

Stephen R. Walli

Under the Hood: Open Source and Open Standards Business Models in Context

People debate regularly whether open source software is "good for business," and how one makes money on something given away "for free." People raise concerns over the commoditization effects of open source,[1] and portray a gloomy road ahead where open source software will "eat its way" up a stack of functionality to the logical conclusion where software has become valueless.

Standards as a commoditization driver have been well understood for quite some time across many industries. A standard exists to enable multiple implementations. The economic argument is that they serve to broaden the market for all producers while fostering price competition (which also fosters production efficiency) for the benefit of consumers. Industry associations of vendors support such work where it expands their market opportunities in complementing areas. Governments support such work because of the "good" economic effects. Seldom does one hear complaints about this commoditization effect, and vendors continue to participate in the development of standards and compete on implementations regardless of that effect.

In this chapter, we will take a look at traditional working definitions of open standards and open source software focusing on the veneer of differences, then step back

---

1 "Will Open Source Middleware Commoditize J2EE?", Nov. 5, 2004, *http://linuxworld.com/story/46957.htm*; visited Nov. 9, 2004. Also: Paula Rooney, "Open Source Will Commoditize Storage, Databases and Security," Jan. 20, 2004, *http://www.storagepipeline.com/hardware/17500067*; visited Nov. 9, 2004.

and look under the hood at a broader business context for the dynamics at work to provide a business model where standards and open source software can be seen in context.

Open Standards

A standard can be a specification, a practice, or a reference model. It is used to define an interface between two (or more) entities such that they can interact in some predictable fashion and to ensure certain minimum requirements are met. Standards exist to encourage and enable multiple implementations.

It is important to put some simple perspective on the standards discussions that follow, as books can be written about this seemingly dry subject. We will look at the context for standards defined by their development and use, a process for developing and maintaining standards, and a set of implementation issues such as intellectual property concerns, conformance and certification concerns. Finally, we'll discuss the history of the concept of "open standards."

Standardization efforts are typically divided into various categories, but the classification systems are often orthogonal. For example:

- Standards can be categorized by the type of development organization—e.g., national or international body, industry and trade associations, and consortia.

- Standards can be viewed as industry voluntary efforts or government-regulated efforts.

- Standards can be thought of as formal de jure—developed specifications, or market-dominant de facto product technologies.

All standards live within a context of development and use. Many formal standards are developed by national bodies or international organizations such as ISO. These standards often define procurement policy for government organizations and large enterprises alike. Industry and trade associations develop standards relevant to their expert and specialized constituencies. In the information technology space, for example, the IEEE has a standards arm, and historically CBEMA (now NCITS) and Ecma International acted as standards development organizations in the U.S. and Europe, respectively, for IT standards. Each of these three organizations was accredited within its national and regional geographies to produce standards that could be later adopted by the relevant nationally or internationally sponsored standards organization to prevent overlapping efforts, and to build on the relevant expertise within different industry groups.

Narrowing the focus even further, consortia of vendors often arise within a specific area of technology within an industry to develop standards and specifications. The consortia often try to build specifications more quickly to expand a particular market, feeling that the more traditional organizations are too slow to deliver standards.

We can categorize standards differently if we bucket them between regulatory versus voluntary standards. Government regulation defines a separate set of concerns over the voluntary work of many organizations within industries. Such government involvement is often driven by economic concerns for the public good (e.g., communications-related standards) or safety issues (e.g., pharmaceutical testing and registration requirements or vehicle safety). Regulatory-based standards will not be discussed further in this chapter because the focus is on the role of standards and open source in market-dynamic areas rather than government-regulated areas.

Another categorization attempts to discuss the difference between de jure standards developed in a consensus-based process and de facto standards. A more accurate statement might be that de facto technology describes a market-dominant product, rather than a specification for interoperability open to all implementers.[2]

Common examples of voluntary information technology standards across this organizational spectrum include SQL, HTML, TCP/IP, and programming language standards like C/C++ and C#.

Standards act as a yardstick against which multiple competing implementations can be judged in the marketplace to make sure that certain basic requirements are met. Vendors compete on implementation beyond the standard to establish competitive differentiation in the market. Ultimately, customers choose the product that does more than simply meet their base requirements. It is this relationship among specification, implementation, and competitive differentiation that provides basic interoperability among vendors, drives competition, and spurs innovation.

All standards organizations have rules about participation, construction, adoption, and amendment. They establish processes for how meetings are carried out to promote fairness of discourse and prevent anticompetitive practices. Standards development organizations also put in place intellectual property rules to ensure participants are aware of the intellectual property landscape with respect to the standard under development.

Most standards bodies require participating holders of essential patents to announce the existence of such patents, and to make them available on "reasonable and nondiscriminatory" (RAND)[3] terms if an implementation of the standard would require a license to the patented technology. Under RAND terms, patent holders cannot dis-

---

2   As we shall see, Clayton Christensen has proposed a situation that says there is market pressure that can change de facto technologies into de jure standards.
3   While each organization's rules are stated somewhat differently and with different levels of formality, a quick look at the governing rules of the IEEE, ISO, IETF, and Ecma International shows a remarkable similarity.

criminate against a particular company or a particular platform. Standards organizations supporting such patent policies ensure that developers interested in delivering standards-based products can do so, while ensuring developers that have invested in a particular invention still have their investment respected.

It is important to remember, however, that no standards development organization can speak for the intellectual property of developers that are not participants in that organization. Standards development organizations structure their patent policies this way because they cannot be the policing organizations nor bear the liability for patent infringement cases from nonparticipants. They are neither funded nor set up to do so. Indeed, if they took on this role, they would likely collapse under the fiscal burden and serve no one.

The interesting thing to observe is that while standards exist to encourage multiple implementations, patents are government-enacted legal tools to protect a single implementation. Patents exist to allow the developing company government-enforced, time-limited legal protection of an invention by preventing others from building the invention. It allows the inventing company to recover the costs of bringing an invention to market in return for publishing the idea for future use by the broader market. A patent is in some regards the antithesis of a standard. Standards are to trade agreements as patents are to tariffs. By definition, they serve different purposes in the economic landscape.

Just as standards organizations are not organized or funded to handle intellectual property liability claims, neither are they typically the conformance certifying agencies for implementations for the standards they produce. Conformance requirements in the standards and specifications are typically simple "claim" style—i.e., you provide the functionality required by the standard and claim conformance to the standard. Organizations that care about conformance then take on the fiscal and legal responsibility of verification around the conformance claims. For example, in the government space in the U.S., the National Institute of Standards and Technology (NIST) developed a procurement process (FIPS[4]) and certification testing process for the standards that it cared to use in those procurements. The government was acting appropriately to protect and serve the public good in federal procurement policy—essentially putting public tax dollars where its mouth was to improve the return on investment. In a commercial setting, The Open Group (née X/Open) as a market consortium handled conformance claims and liability for its specifications. Beyond the testing requirement, warranties of conformance are required and a brand license is signed which is tied to the trademark usage associated with the standards it produced. Companies that wanted to use the trademark on their products in the market had to pay royalties. The X/Open standards were developed through the organization and a company paid for its seat at

---

4   FIPS stands for Federal Information Processing Standards.

the specification setting table through its consortium membership dues. Conformance certification, on the other hand, was funded through the cost of trademark use.

If standards act to define a base functionality to encourage multiple implementations, essentially the greatest common denominator for a specific technology, they help create a commodity. This results in a constant, healthy tension among the standards bodies' participants as they work with each other on the standard, while simultaneously vying for market share with their different products.

The term "open" with respect to standards became a mantra in the late 1980s and early 1990s, and was tied to the concept of "open systems." As Cargill observed, "open systems" was marketing-speak for the idea that if all the vendors would just build their computing products to "open" standards, the consumer would be able to build data processing systems by mixing and matching information processing hardware and software modules in much the same way that one could mix and match stereo components to build the desired system.[5] "Open systems" was a description of the architecture the consumer thought should exist. Unfortunately, the complexity of interconnected data processing systems doesn't lend itself so readily to the metaphor of a single-purpose device (i.e., the stereo system) and the ability for plug compatibility between stereo components to solve all the attendant complexity.

"Openness" became a quality attributed to the standards that would enable open systems. The openness was an attribute of the creation process (the standard was built in some form of public, consensus-based process open to all participants) rather than an attribute of implementations of the standard.

The development model for a standard is unrelated to the development model used for the implementation of that standard. It is equally possible for a standard (open or otherwise) to be implemented in a closed proprietary software product or in an open source software project.

Open Source Software

Open source software (OSS) is a term applied to a collection of software development, licensing, and distribution practices. A lot has been written about OSS over the past decade, as various open source projects gain market importance and the license models demonstrate economic significance. Eric Raymond's original treatise[6] on the development practices remains relevant. The Open Source Initiative (*http://www.opensource.org*) publishes the definition of the requirements a license must meet to be considered an

---

5   Carl Cargill, *Open Systems Standardization: A Business Approach* (Upper Saddle River, NJ: Prentice Hall 1997), 70-71.
6   Eric Raymond, *The Cathedral & the Bazaar* (Sebastopol, CA: O'Reilly Media, 2001). The original essay was published in 1997.

"open source software" license. I will focus on a group of attributes of OSS projects that sets up the economic discussion to come.

OSS projects are interesting "buckets" of technology. Successful OSS projects share a number of attributes.

For instance, distributed communities with good software development practices develop technology packages that satisfy well-defined needs.

- Software quality is a measure of community activity (i.e., the developer customers).
- Contributions reflect the individual economic considerations of the contributor and are based on selfish asymmetric value propositions.

The projects reflect their Unix history of loosely coupled component architectures with well-defined interfaces that make it easy to assemble larger solutions (e.g., the LAMP stack is assembled from Linux, Apache, MySQL, and Perl/Python/PHP).

OSS projects develop software packages in a distributed community where the core developers that inspired the project act as a hub for the evolution of the software as a "benevolent dictatorship." Just like all successful software projects, successful OSS projects support a strong software engineering discipline and ethic at the project's core. Essentially, good software is developed by good software developers.

What makes the software "open source" is the licensing model. While a wide variety of licenses are considered "open source licenses," the basic common denominator (without relisting all the requirements from the Open Source Initiative) is that the software's source code is always freely available and users can modify it without restriction; however, requirements associated with distributing the software may exist. In similar fashion to standards efforts supporting a lack of discrimination (either in participation within the context of their community or in their intellectual property engagement goals), OSS licensing discriminates against no one. Anyone can participate in the community development of the software. Anyone is free to use the software. Anyone can see the source code. Anyone can distribute the software. In each case, requirements may be imposed by the license or reputation that must be earned in the community, which would lead some to not want to participate, but nothing inherent in the process prevents participation, use, or distribution.

An interesting dividing line in the licensing schemes is whether the license is considered "viral." A reciprocal license such as the GNU General Public License (GPL) attaches itself to new software by requiring that if the software is modified and distributed, the license is attached to the new software. This forces the "open" aspect upon new software, keeping the source code publicly available. A company may be wary of publishing the source code to its software, as it may contain trade secrets or other third-party licensed software for which it doesn't have the ability to publish the source. The classic permissive licenses arose in academic settings (e.g., the Berkeley license and the MIT Project Athena license) and had no requirement to associate new work with the license. This class of licenses was very liberal in what was allowed, and a company could easily take software, modify it, and not publish the new source code.

One of the most interesting aspects of OSS development is the economics of the community participation. Surveys have been run and much has been written about the rationale for participation.[7] The "simple" economics is that participants in a community get more than they give. It is a normal selfish asymmetric value proposition. To understand that statement, think about context for a moment. Many people in many walks of life use and value their skill sets differently in different contexts. A writer might be a technical writer or communications writer for a corporation as her paying job, but still use that same collection of writing skills teaching an English as a Second Language class in the evenings, working on a writing project with her child's class at school, and writing a sonnet to a loved one. In each case, she values her skill set differently, and the reward accordingly. Software developers are no different. The interesting aspect of community is that corporations are equally economically rational in their participation. Developers and corporations participate in OSS projects because of the same simple asymmetric value proposition. Many companies participate in OSS projects and draw upon the software to deliver the products and services upon which they base their revenue streams. We will look at this a little more closely in a moment.

Coupling the license and distribution model that ensures the source code is freely available, with a core project team that is disciplined allows for the community effect of OSS development to shine. The community of interest in a particular project can directly contribute changes and bug fixes. While there may be orders of magnitude of difference in the number of bug reports submitted, down to the number of bug reports submitted with proposed fixes, down to the number of "good" fixes that meet the bar defined by the core project team, there is definitely a net gain for the project, both from a testing and a bug fixing point of view, as well as the opportunity to find new talent for the project that wants to participate.

The Real Business Model

Customers view solutions as a network of related "bits" that have to come together in some definable fashion to solve their IT problems. This network can be defined with nodes representing various technology objects and the paths between nodes representing the relationships. This is a very informal network, but very real. For example, a

---

7   Most notable were the surveys by the Boston Consulting Group (*http://www.bcg.com/publications/publication_view.jsp?pubID=935&language=English*, Dec. 15, 2004) and the broader FLOSS survey done at the University of Maastricht (*http://www.infonomics.nl/FLOSS/report/index.htm*, Dec. 15, 2004).

solution for a new retail inventory management system will include nodes representing the existing application systems to which the new retail system must interface, computer resources on which it will run, the programming language environment in which it will be developed and maintained, the staff and their experience and skill sets that will develop and then maintain the new system, databases with which it will need to interact, . The other application systems to which the new retail inventory system will need to interface have their own historical networks. The platform resources may represent a different network view if multiple application systems share the fundamental computing platform. Companies define architectures for their IT functions to attempt to simplify the decisions that need to be made, and often publish these as internal procurement and development standards. History also counts in the network—for example, some shops always buy "Unix" hardware or always program in C or Java, because that is how their resource history has developed.

Turning the discussion around to the vendor-centric product perspective, Geoff Moore defined a model[8] in 1991 for technology adoption that suggests that once a market starts to develop, a company best leads by providing a customer the best "whole product solution." By this he means that the vendor offers its core value product proposition to the customer and then needs to wrap as much around that product as it can to present a "complete" product solution to the customer to meet the customer's broader needs, essentially mapping as much of the customer solution network as possible. Another way to think about this is that the vendor wants to provide as many complements as it can to its core product offering, covering as much of the customer's solution network as is feasible to present the best (most valuable) solution in the customer's eyes.

The business of a vendor would then be to ensure that the complements were as inexpensive as possible, indeed commoditized if possible, so that the whole solution, from the customer's perspective, is as inexpensive as possible—but the lion's share of the revenue would come to the vendor through its core offering. Several business tactics and tools are available to the vendor to try to drive these complement spaces:

- Traditional buy-versus-build strategies can be used to ensure that as much as of the customer's solution is provided through the vendor's own brand, regardless of whether the complement products are offered as add-ons or are bundled directly with the core revenue stream.

- Develop a rich ecosystem of add-ons by encouraging developer and partner networks to provide a richer whole solution to the customer. Publishing proprietary specifications for the complement space enables more partners to develop businesses in the complement spaces.

- Develop tool spaces that help add complements to the complement ecosystem.

---

8   Geoffrey Moore, *Crossing the Chasm* (New York: Harper Collins, 1999).

- Provide certification programs around the core technology to ensure that there are lots of service professionals to help the customers complete and support their solution. Indeed, a company might have its own consulting services arm for parts of a solution, and provide certifications for other parts of a solution.

Taking this view, a company's assets and offerings also form a network of related products and services it matches against the customer's solutions network through the sales and marketing functions. Each node in the network has cost, risk, and revenue models associated with it, and as long as the overall revenue model is greater than the sum of the costs, the company will be profitable.

It is important to remember, however, that no company exists alone in the market to solve the customer's problems. Each vendor in a particular space must have different product networks to allow a differentiation in its sales pitch to the customer. Different vendor companies will also behave differently in their hiring and acquiring strategies to shore up their "whole product offerings."

In addition, it is important to note that one can now look at intellectual property (IP) tools (and by that I mean trademarks, patents, copyrights, and trade secrets) in context. Each of these four legal property types or tools (regardless of legal and geographical jurisdiction) provides a different set of legal protections at different costs. One is far more likely to spend heavily and strategically with IP protection tools in the spaces defining one's core product value proposition or in spaces in which one has the greatest investment, than farther out in the complement spaces of one's product offering network. Indeed, in the complement spaces, a vendor may aggressively publish (or sparingly strategically patent) to ensure that no other vendor can patent in the complement space and raise the prices on that complement.

If we now start to consider open source and open standards in this core-complement context, we see that they are simply additional tools in the tool chest to drive complement spaces. Let's look at each separately for a moment.

Open Source Complements

It becomes very easy for a vendor (OEM, ISV, or systems integrator) to bootstrap a complement product or project space for its core value proposition to its customers using open source software directly. The projects are polished to product readiness either within the company or within the community itself. To "buy" versus "build" as complement strategies for a vendor, we can now add "borrow" and "share." If a vendor joins an existing community, it can polish the OSS project to product readiness to complement its core value proposition to its customer. If it starts its own project, it can be used as the hook to find and engage with new customers around the rest of its core offering.

The engagement in the community is actually a very leveraged conversation directly with people interested in the community's project and then possibly the company's offerings. As people cross the line from community participant and software user to potential customer, they are self-selecting the vendor's services. This is a very efficient way to find new customers. This does not mean one should consider the community as a mass-marketing broadcast channel (it's not), but rather, as a public conversation with one's customers and potential customers. This is not for the faint hearted. Unlike a traditional "Go to Market" plan, the technical people have real-time unmanaged discussions with the customers.[9]

The vendor's challenge becomes ensuring that products remain products and communities are communities. Starting a community project is not that risky if the vendor plays by the rules, staffing it with good software developers that will lead the community well, and understanding that the real return is the conversation they have with customers, and the product complement effect. The "community" at large does not exist to work for free improving a company's products. This mistake is still being made despite the public experiences of the past.

The community leadership is a benevolent dictatorship. Sponsoring the community (or earning your place in an existing community) does give the vendor the opportunity to manage things on its own terms. Software stability is maintained through the community project by the leadership. Project direction is developed by the community leadership and people that have joined the community and earned their position of trust. There may not be a road map with a view three to five years out, as is almost necessary in a product, but the complement space doesn't need the rigor of the core product. Viewpoint becomes important. A customer's view of the need for a road map around a solution may not map to a vendor's view of the need for a product road map.

While a number of relatively small companies are using OSS in their businesses, large vendor participation is very interesting. *[Caveat lector: the following examples are observations from the author and do not represent any direct knowledge of these vendors' business plans or models.]*

IBM has made three big plays: Apache, Linux, and Eclipse. IBM joined the Apache community six years ago, borrowing a web server while selling WebSphere. It joined the Linux community four years ago while managing the commodity curve on the AIX product line and using it as a competitive shot into the Sun server market. Most recently, it has begun a "share" project creating the Eclipse project out of technology it acquired (and then it acquired Rationale).

In joining the Apache community, IBM doesn't need to maintain its own web server team and can focus its efforts on WebSphere instead. In the Linux community, it can focus on the parts of the OS that best meet its needs. Linux is clearly becoming the Unix server replacement over time. IBM's AIX product space will be replaced. It can either

---

9   The first thesis in the Cluetrain Manifesto is "Markets are conversations." Indeed, most of the 95 theses are highly relevant to the discussion (*http://www.cluetrain.com/#manifesto*, Dec. 15, 2004)

actively participate and position itself on the leading edge of the curve, or wait until its product space is consumed.

SAP released a complete modern relational database for free in August 2002 to drive its core business into the mid-tier customer space where the customer may not already have an enterprise-class database and may not be willing to pay the "Oracle/IBM/Microsoft" tax to get SAP R3. It was released under the GNU GPL after a two-year, 100-person investment in updating the acquired Adabas technology. SAP then partnered with MySQL AB in Sweden to "manage" the database community.

Sun Microsystems worked in the GNOME desktop community to develop, acquire, and contribute the accessibility features it needed to meet U.S. government procurement policies to complement its Linux workstation offerings. For a relatively modest investment in the tedious and difficult accessibility technology, it is getting an entire full-featured desktop environment.

In each case, the corporation is getting more than it gives, developing a complement rapidly around core offering(s). They gain time-to-market for the complement at a reduced investment. While initially met with skepticism when a large company joins an existing community, as long as that company plays by the community's rules with respect to engagement and quality, it can become as accepted as any other active participant. Depending upon the nature of the product relationship to the core and company commitment, the company may make best efforts to hire key community developers. This is not altruistic, but neither does the company expect the developers to change their community engagement. It gives the company deeper insight into the community it is looking toward for support as it develops the complement.

There is a competitive edge to OSS community development as well. Often the company takes advantage of the reciprocal aspect of the licensing to salt the intellectual property fields around it by aggressively publishing prior art, holding the complement costs down, and preventing competitors from directly monetizing their original investment in the community project software. For example, SAP is not in the database business and so may feel comfortable publishing the investment in SAPDB (now MaxDB), but it probably doesn't want Oracle, Microsoft, or IBM directly making use of that investment in their respective database products. In this case, the reciprocal license is the most business-conservative license SAP could choose. As well as driving a complement directly, the community engagement also allows the vendor to work closely with partners, customers, and potential customers to build the relationships they will need to sustain the business over time.

The other competitive aspect happens when you consider two competing vendors' product-centric networks, and how they appear to the customer. The customer is looking at things as a "whole product solution" and does not really think (or care) about what is core or complement from the vendors' perspectives. A vendor can develop a complement community directly in the path of a competitor's core value proposition to a mutual customer. It need not be a deliberate move and the sole purpose of a community; it is the icing on the cake of the multifaceted approach of a business in using OSS development and engaging with its customers.

Small companies can also easily use the OSS buckets to bootstrap product complements. Clayton Christensen's original research[10] around disruptive business models shows how small companies assemble off-the-shelf parts into underperforming products compared to the industry norm, offering those products in their own niches with different business models. As the sustained innovation around the new disruptive product develops, it eventually becomes mature against the yardstick used to judge the incumbent but at a better price for the performance, and the incumbent's business is disrupted. Consider the development of the Linux operating system— from its inception in 1991, delivered by a university student, its growth in educational use, to simple infrastructure servers, to the point in history where it is presently challenging the traditional Unix vendors' products (though it has, in some cases, become too complex to teach anymore[11]).

There is also a situation, as we shall shortly see, where a product market hits the point when customers start to be overserved, and there is a call for standardization. This means that OSS components that already represent a package with well-defined interfaces may be a rapid way to bootstrap a "good-enough" product into that market.

One thing to note in this discussion using the network of core and complements together is that there is no "stack" of technology per se. Think back to the earlier discussion of customer-centric solution networks and vendor-centric product networks. Vendors may see their world as a stack with their valuable core at the top and all the commoditized complements below, but in reality, it is simply their view through their own product stack and its relationship to the customer and their partners. A chip manufacturer views the stack very differently from an operating system company or from a middleware company (hardware design in silicon is where the value is, with operating systems and middleware and apps being less and less interesting to the chip manufacturer). The terminology of eating up the stack may have more to do with the position in which the vendor perceives itself.

Open Standards Complements

Clayton Christensen further observed in his research[12] that as companies begin to overdeliver functionality in their product lines faster than customers are able to use the new functionality—and therefore faster than customers are willing to pay for it—

---

10 Clayton Christensen, *The Innovator's Dilemma* (New York: Harper Collins, 1997).
11 In interviews in February 2003, a number of university OS professors made reference to the current revision, with the addition of symmetric multiprocessor suppor, becoming too complex to teach. As a result, they were basing their course work on earlier versions of Linux.
12 Clayton Christensen, *The Innovator's Solution* (New York: Harvard Business School Press, 2003).

the market begins to call for standardization. Indeed, prior to the point where they begin to overdeliver, the market leader is often offering the technology in a tightly integrated fashion and best delivers to consumer needs in this space where the solutions typically are not yet good enough. This is the time when tight integration, not standards-based components, is the path to success. Standards develop once the marketplace reaches a point where the market leader begins to overdeliver. These are the circumstances in which a market-dominant de facto technology is at a critical point and the call for de jure standardization is possible.

The signal to standardize a technology is somewhat unclear, but there is likely a collection of factors:

- Competitors with standards experience and similar product offering networks but different core drivers likely use the opportunity to "call for standards,"[13] hoping to reduce their own complement costs while causing a competitor grief in a core revenue stream.

- Customers managing substantial procurement budgets will support and call for standards in the hopes of prolonging investments and attempting to reduce costs from vendors that are overdelivering. For example, the U.S. government as the largest IT buyer on the planet at the time, led the charge around the POSIX and C-language standards, quickly followed by the large companies in the petroleum and automotive industries.

If you are the one true implementer, and the market (i.e., partners, customers, and competitors) is calling for standardization in your core technology space, you have a problem. They're calling for the benefits of standards (expanding market and price competition) because they want the ability to replace you. Some segment of your customers wants the choice of multiple implementations. Your competitors are happy to support the call, as this is the thin edge of the wedge to break open your value proposition to your customer, all in the name of open systems. Your partners may be happy to support the call for standardization because they want price pressure as their margins diminish and perhaps your percentage of their Cost-of-Goods-Sold is increasing.

---

13 Geoff Moore argues that the first response in the market from competitors when they see a "gorilla" forming is to cry for "open systems" (Geoffrey Moore, *Living on the Fault Line* [New York: HarperBusiness 2002], 119). This might be more of a cause for standardization too early with all the attendant problems that ensue as has been observed by James Gosling of Sun Microsystems (James Gosling, "Phase Relationships in the Standardization Process", circa 1990). Gosling's observations are more closely in line with Christensen's, arguing that there is an optimal time in a technology's development for standardization. Some of us have always suspected that it is best to standardize existing practice and experience, instead of trying to standardize ahead of the market curve. Indeed, it would be interesting to do a survey of successful and unsuccessful standardization efforts to determine whether the unsuccessful efforts were undertaken too early in a marketplace, when vendors are still trying to define the marketplace itself and stake out claims with products and patents. First, of course, one would need to define the measure of a successful standard. Christensen's observations are likely more in line with standards forming at the optimal market time.

It is important to note that one needs to get the view of the market "right" for this sort of discussion, and hindsight is always 20/20. It is not necessarily the dominant vendor's product that is to be standardized, but the product market space. For example, one can argue that the POSIX standards (and the C-language standards, for that matter) were not about standardizing Unix systems, but rather, were an effort to standardize an OS interface for minicomputers. Digital Equipment Corp. was the dominant player in minicomputers (which became departmental servers and workstations). DEC was driving customers up the hardware upgrade cycle to support its market growth faster than customers were willing or able to absorb the change. Unix systems of the early and mid-1980s represented the best opportunity around which the market could form a minicomputer application programming standard to support customers' applications portability. While the Unix systems of the day were often less scalable, less robust, and less secure than VAX/VMS systems, the Unix operating system had been ported to most vendors' hardware (including DEC VAXen), so competing vendors could see the market opportunity.

At the same time, the PC arrived on the scene. Many have argued that the PC won against Unix systems by taking over the desktop, largely due to the inability of the Unix vendors to set a desktop "standard" fast enough. The PC certainly took the desktop by storm, but it was actually competing against nonconsumption. In a Christensen view of the world, it was put together from inexpensive parts, and when compared to minicomputers it was certainly underperforming, but it became the de facto business appliance in a document-centric world, enabling a whole new class of electronic document-centric applications. (Word processing systems companies vanished almost as fast as the minicomputer companies.) The PC was competing with nonconsumption, giving business users computing resources on their desktop instead of being stuck waiting for their business data processing applications to be developed by corporate IT, with its ever-growing systems development backlog. The Unix systems (driven by standards and an "open systems" message) were data processing-centric rather than document-centric, and caused DEC grief in a completely different space.

Christensen observed that as an area of technology is standardized, the value moves to adjacent spaces in the network.[14] The trick then becomes to ensure that one is building one's business efforts in the product network around the space being standardized. This would lead us to believe that the richer a product offering network a vendor has, among different software, hardware, and service components and products, the more opportunity that vendor has to move with the value or to define new components that the old components complement.

---

14   This was originally referred to as "the Law of Conservation of Attractive Profits," but is now referred to as "the Law of Conservation of Modularity."

This core-complement product network view allows one to very rapidly see how the vendor politics in a standards working group play out. A vendor with a de facto product technology that is being dragged by the marketplace into a de jure standards working group is likely a little less than enthusiastic about participating in its own commoditization. The vendor alliances within the working groups are participants in the complement space. The game is one of technology diplomacy, where the goal as a vendor representative is to expand your area of economic influence while defending sovereign territory. This holds true regardless of whether one is participating in a vendor-centric organization such as Ecma International, as an "expert" to a national delegation to the ISO (on behalf of her employer), or as an individual contributor to an organization like the IEEE (again, funded by her employer to participate). Vendor consortia offer a similar view. Which vendors formed the consortia and which vendors quickly and noisily joined shortly afterward says a lot about who the incumbent in a product space is and who the competitors are.

Conclusion

Businesses are often much more than simply hardware companies or a software companies or a service providers, offering breadth of product and service in overall value proposition to their customers. Successful companies use a collection of strategies to deliver a "whole product offering" for their customers, driving their core revenue generator with a host of complementing products and services.

Standards have traditionally been one tactic or tool for driving additional complement value to a customer by developing a complement space in a maturing market with a lot of implementations at a reduced price.

Open source software can also be used as a tool to develop a complement space that supports a core revenue product or service. The open source project can act as a quick and convenient bucket of technology around which other product offerings are wrapped, or plugged into an existing product offering network.

A number of models were presented on how to think about customer-centric solution networks and vendor-centric product networks (and for thinking about the product network from a core-complement point of view), alongside Moore's traditional Technology Adoption Life Cycle and Christensen's models for how product markets behave. Open source software projects and standardization efforts can be viewed as tools to be used to attain competitive advantage. A number of large corporations are now participating in OSS communities to the benefit of the corporation and the communities, just as corporations have historically driven voluntary standards engagements. The model-based view certainly doesn't take away from the excitement inherent in different OSS projects or the overall economic value of a successful standard. It merely provides context to businesses that want to understand how to adopt and participate in either.

CHAPTER 9

Russ Nelson

Open Source and the Small Entrepreneur

I've been giving away my software since 1983, full time since 1991. I don't do it for fun, although I enjoy it. I do it because it's a way for a small business to earn money *and* it's fun. Each of my software interests started as a hobby, and some have turned into a profession. Not every hobby of mine has turned professional, and I hope to explain why some have and some have not.

Three of my hobby projects, which I'll talk about in depth after I introduce myself, have turned profitable. They are Freemacs, Packet Drivers, and qmail.[1] Freemacs is an MS-DOS text editor, styled after Emacs. It's still used today as the official editor of the FreeDOS project. Packet Drivers hide the difference between Ethernet cards in an MS-DOS system. If you've ever eaten at a McDonald's restaurant, your order was communicated through Crynwr Packet Drivers. Qmail is a mail transfer agent (MTA) for sending and receiving Internet mail. Qmail is the engine behind Rediffmail's 30-million-user, multiterabyte, 100-node email cluster, and many smaller sites.

Introduction

I did hardware hacking long before college. Digital electronics was too expensive for me: $1 per TTL quad nand gate at a time when vinyl records cost only $5. So, I fiddled around with analog electronics. I invented a trigger sweep for my dual-beam oscilloscope, and an analog computer throttle for my model railroad.

---

1 Qmail is an all-lowercase name, and will be capitalized here only at the beginning of a sentence.

My high school was a member of LIRICS: Long Island Regional Instructional Computer System, which had a PDP-10 students could use to learn to program, via teletypes operating over modems. I was at Baldwin Senior High School from 1972 to 1975, but took advantage of the program only in my last year, from '74 to '75. I learned BASIC and wrote a four-banger calculator program. I also wrote a word processor in BASIC, for which I had to do all sorts of horrific string manipulation. It took hours to format a two-page social studies paper. Partway through I got an Instant Message (IM) from an operator who asked me what I was running, and if it was looping.

During this period I learned PDP-10 Assembly language. JRST, HRRLZ, and SKIPNE are all familiar friends to me. Unfortunately, none of my candidate colleges had a PDP-10. MIT almost certainly did, but I was a poor scholar who was more interested in getting an education than in proving that I had one. Of course, everything on the PDP-10 was what we'd now call "open source." Nobody thought of holding back the source code in those days.

Many colleges will teach you how to become a businessman. I didn't have that desire upon entering college, and so I sought a degree in electrical engineering. Only later did I decide to run my own business, but how to learn? I started slowly, learned through experience, and didn't take too many risks.

I learned "on the job," and discovered new ways to profit from open source software. Everyone in the business had to teach themselves. There is no master's degree of open source business administration—not yet anyway.

Hewlett-Packard recognized my "genius" and hired me and my wife to work in its calculator division doing integrated circuit design. I missed programming, so I bought a RadioShack Color Computer (CoCo). This led to my first freelance income associated with programming.

I wrote programs for fun and sold them to *CoCo Magazine* for distribution. They were just little cute things, but they were in Assembly language, so they were fast, small, and easy to distribute. The standard distribution was on audio cassette through a paid subscription. It was nice to receive money for writing a program for fun. Without a local user community, it was also the only way I could distribute my software.

Freemacs and Open Source

After I returned to graduate school, I started writing a programmable editor for MS-DOS. Instead of writing it *de novo*, I thought it should be compatible with Emacs. I had used Emacs while working at Hewlett-Packard, where I had done some hacking on an editor. That experience convinced me I wanted an editor without any distinction between editing and typing. I purchased a copy of the MIT AI Lab memo describing Emacs, written by Richard M. Stallman (RMS). I recognized his name because of his GNU Manifesto, a new document then. I sent email to RMS to get permission to sell a copy of the memo along with my editor. The document was in the public domain, but I thought that asking permission was only polite.

I got a call from RMS (this was back when his wrists hurt so badly that he couldn't type). He persuaded me to give away my version of Emacs rather than sell it. He appealed to my sense of fairness. He asked why I should profit from a manual that I had not written. I was impressed that so stellar a personage as RMS would take the time to call me. I decided that it would be best to give it away. I didn't really have any idea whom I might sell it to. All my previous software sales had been for the RadioShack Color Computer, and had been sold to *CoCo Magazine*. My editor was for MS-DOS, so this was a completely new situation for me. Once I decided to give it away, I gave it the catchy name Freemacs.

In graduate school, I had access to worldwide networks, and it was actually possible to distribute software "for free." Nobody was kidding themselves; the Department of Defense was paying the bills for the ARPANET, universities for BITNET, and companies for CSNet. Nothing was free to the institution, but the users perceived the networks to have no incremental cost. This led to allocation by congestion, but I'm not talking about economics yet. Regardless, a software author could give away software for the mere cost of uploading it to a distribution point.

Freemacs was distributed from SIMTEL-20.ARPA, an FTP site with copies of most useful MS-DOS software. It was run by the Army at the White Sands Missile Base, but nobody cared about where it was physically. The point was that they had good stuff, and they were sharing it. This made me a contributing member of the open source community.

Freemacs and Business

Freemacs was a hobby, and I had no intention to turn it into a business. My computers (even my home computers) were paid for by my employer, so I had no expenses to cover. As the program gained users, they told other MS-DOS users about it. Those users wanted updates, and none were available on any of the worldwide networks of the day. Having no other recourse, they asked me for copies. I knew that they were gaining from these copies, so I asked for a portion of those benefits in the form of a copying fee. Between 1985 and 1991, most of the activity of Crynwr Software consisted of putting software on floppy disks and mailing it to customers.

Two interesting stories about mailing floppies: one is about a customer in Ireland who had two floppies go bad on him. Guessing that his email was going through some kind of antiterrorist scanner (as the Irish Republican Army was quite active at the time), I sent him a third floppy wrapped in 1mm-thick lead foil. That floppy got through OK. Another is of a customer who, although a part of the defense department, had no Internet access, or even have a modem—and this was after almost everybody had gotten on the Internet, so I was surprised that they were even allowed to telephone out, but I sent them a floppy with software on it and they were happy with that.

Staying in Touch

It's crucially important to stay in touch with your users. The biggest advantage an open source developer has is close contact with users. If you're the primary user for the software, of course you know what users want—you just sit and cogitate. Quite a bit of open source software is written to "scratch your own itch."

This is easy and rewarding because very little communication is needed. Programmers are not typically great communicators (most programmers fall into one of the four NT classes on the standard Myers-Briggs personality test). A programmer who can listen and talk is worth her weight in chips (and chips are worth more per ounce than gold).

With Freemacs, I started with a single mailing list, which proved to be a mistake: some people don't need any help and just want to know when new releases come out; other people want to get or give help but don't want to code; and still other people are interested in every miniscule detail of the program. One list cannot serve everyone. I found that I needed three lists.

One list carried only announcements of new releases. You really want to have an announcements list, and you need to remember to use it. You might send only one or two pieces of email a year, but those are crucial. First, you need to remind people that they've given permission for you to send them email. Second, people need to know that you're in business even if they don't currently need your business. Any user might suddenly find himself needing to become a customer. You need to be the proprietor of the relationship between the software and the user, as you'll use that relationship to make money.

A second list was for user-level help. Some programs are exceptionally powerful (I'm not thinking of the Unix "cat" or "tail" here, but something more like sendmail or qmail) and in-depth knowledge to properly exploit all that power. Some users want to acquire that depth. Others do not, but will dip their toes into the depths by asking a question on the user mailing list. If there are sufficient users of the program, you will have other businesses competing with you. One of the ways they will compete is by offering to help other people. No need to worry, though! By virtue of your proprietary interest in the program, you will have a built-in advantage over these other businesses. In any case, customers like competition because they perceive it as ensuring fair prices.

The third list is for developers. I am of two minds here. You could have the developers list open to all comers, regardless of their to contribute to the project. Or you could have the developers list be open only to those who actually have contributed. The main tradeoff is protecting the time and attention of your contributors. You don't want them signing off the developer mailing list because it has too many user-level questions being asked on it. You really need their attention to help you make decisions that will affect them. For example, if you change an API, you need to clear that change with your developers—first, because it keeps them involved in the process, and second, because they may be relying on something you're doing.

FreeDOS has adopted Freemacs as its standard text editor, so it still has a user base. I only rarely do any MS-DOS work, and when I do, I'm happy with the state of Freemacs, so it's now frozen in time. There were never any commercial users, so apart from selling copies on floppies, Freemacs managed only to buy me my own computer for home.

Packet Drivers

Crynwr Software came into its own with packet drivers. A packet driver allowed the sharing of an Ethernet card between two protocol stacks. For about a year, the only possible way to get Novell network clients on the Internet was by using a packet driver. Also, a packet driver would hide the differences among Ethernet cards.

Unlike video boards, which are at least compatible at the VGA level, Ethernet boards have never been compatible.

Back before packet drivers existed, there were network clients and servers—largely Novell NetWare. The manufacturer-written network driver was linked to Novell's code in a single executable. The resultant program had no API for an external program to send or receive an Ethernet packet, which was very bad for any competition to NetWare. Maybe Novell planned it that way, but I doubt the company was that Machiavellian.

Anybody wanting to send packets other than NetWare's had a problem.

The 3Com 3C501 was the market leader, but it was a very insufficient card. It had one buffer shared between transmit and receive, so a packet could be lost if it arrived when the buffer had not been emptied, cleaned, and turned around. However, everybody had drivers for it. Novell, in an attempt to improve the state of the art, took National's 8390 demo board and put it into production as the NE1000 (and later as the NE2000). This board had sufficient memory with separate transmit and receive buffers.

Just about then, other Ethernet controllers were coming on the market. People were using the Intel 82586, the AMD LANCE, and the National 8390 (in non-NE2000-compatible ways). Only NetWare included a device driver development solution. It had a driver development kit (DDK), and a certification house (Novell Labs).

Other protocol stack vendors were doing the same thing—producing drivers linked into their own products. No vendor had a driver that could be shared, however. While Novell and Microsoft pondered, little FTP Software (now owned by NetManage) had the same problem as everyone else: too much hardware and too few drivers. It came up with its own specification for a shared Ethernet driver and, unlike other vendors, published it as an open standard.

I was working for Clarkson University at the time, and we had the same problem as everyone else: how to support multiple pieces of software and hardware at the same time. I was using Phil Karn (KA9Q)'s NOS, and he had packet driver support. So, I wrote packet drivers for the two Ethernet adapters in use at Clarkson (the 3c501 and Racal-Interlan NI5210) and published them as open source software.

A number of fellow Internet users contributed drivers, and before long, we had covered a considerable portion of the industry. This led to more support from TCP/IP vendors, and a group at Brigham Young University wrote a NetWare driver that could use a packet driver. We really got the ball rolling then, because anyone with a NetWare network could put it on the Internet.

There were some holdouts, notably Microsoft and Novell, both of whom started promoting their own standards: NDIS and ODI, respectively. The NDIS document was published from the start, but there were no sample drivers, and no base of code from which to build. ODI documentation was available only with an expensive DDK purchase. A packet driver distinguishes itself by coming with source code, by having a simple, approachable API, and by being small in size. The typical driver was 5K of executable, compared to 20K for ODI, and 40K for NDIS.

By the mid-1991, I realized there was money to be made providing packet driver programming and certification services—the latter for drivers not written by Crynwr. So I left Clarkson and, after a five-month placeholder stint at a local PBX company, I started Crynwr Software. Up until 1998, Crynwr's main source of income was from packet drivers.

Packet Driver Income

Most likely, everybody has heard that the way to profit from open source software is to sell services. That's true, but there are many different types of services. I'll list some of them in the following paragraphs.

The first, and most profitable, is contract programming. Various people need packet drivers written, or features added, or bugs fixed. I contract with them to fix it, either for a fixed price (if I understand the problem), or at an hourly rate (if discovery is needed). Buyers don't like cost uncertainty—they really like to know what something will cost up front—but whenever you bid a fixed price, you are taking on the risk that the project will be much harder than you thought.

I have actually been successful doing what appears, at first sight, to be the worst of both worlds: charging per hour with a minimum and maximum price. If you set the minimum and maximum to reasonably sane values, the risk is reasonably shared between the two parties.

Business Tutorial

Here's a quick tutorial, which I wish I had had when I started, on how people do business. First, customers expect to do business first, and pay you for it later. The customer accomplishes this by issuing a private currency called a purchase order (PO), with a face value and a serial number called the "PO number." Purchase orders owed to you are Accounts Receivable. Purchase orders you have issued and will pay are called Accounts Payable. I call a PO a currency because you can get a loan against good receivables, and you can sell bad receivables (customers who don't pay, or pay very late) at a discount.

Never do work on a promise to pay you. If someone is really going to pay you, they'll be able to cut you a PO. People change jobs and companies go into bankruptcy. If you have a PO number, that's as good as gold, because a company's ability to purchase things depends on its reputation for paying on terms. If it loses that ability, nobody will accept a PO from the company, and then it has to pay cash for everything. Get that PO!

Some companies have intricate purchasing systems, where you have to be a qualified vendor, you have to sign a W-9, and you have to sign a nondisclosure agreement just to work with them. Other companies just whip out the credit card, and you're good to go. For any company larger than 50 people, though, you'll be dealing with a buyer. Most buyers are used to purchasing software as a product.

Although they are starting to understand that software can be a service, you might still might run into a confused buyer, because sometimes they're told to "buy this software" and they don't understand that they're purchasing a CD and a support contract. Take the time to educate them about the difference, and you'll have an easier time working with them.

I've also sold proprietary packet drivers, although this was a special circumstance (and one that was very profitable to me). I had a customer who wanted a new packet driver, but who didn't want to pay the entire price for it. He wanted to pay only half. He persuaded the vendor (SMC Semiconductors, now SMSC) to pay the other half, since the packet driver would be useful for all the vendor's customers. That seemed fair to me. I had his purchase order and SMSC's promise. Unfortunately, he paid up and SMSC didn't, so I had a packet driver that the vendor hadn't paid for. If I made it freely copyable, no vendor would ever bother to pay me, so I decided to license it to SMSC's customers until SMSC paid me. The company never paid, so I sold it with a clear conscience.

I've also dual-licensed packet drivers. A vendor that was going to embed an Ethernet chip into its product and use an embedded processor wanted to freely copy code from the packet driver without taking a chance that its driver would become a derived work under the provisions of the GPL. So the vendor purchased a copy of the code from me, licensed for any use except resale.

I've also sold compatibility certification. Digital Equipment Corporation had written its own proprietary packet drivers. DEC wanted me to certify that the company was compatible with the open source packet drivers. I had written a test program that would exercise the edges of a packet driver to try to break it. If that program ran, it meant DEC had made no stupid mistakes reading the specification. I also ran a stress test for several days; if that didn't run into problems, it meant DEC had made no stupid coding mistakes.

I've also done pure consulting. Contracting is different from consulting. A contractor is someone who sells his work output, and a consultant is someone who sells his ideas. A customer wanted me to describe how my packet driver worked on their hardware, so they sent an engineer to my site for a day to get a debriefing. He took my family out to lunch, and I got paid handsomely for the day—not as much as I would have been paid to write the improvements myself, but you can't make *all* the money *all* the time.

Qmail

Qmail[2] is an MTA written by Daniel J. Bernstein, a University of Chicago professor. He needed a fast MTA to run his mailing lists. While I was working on Freemacs and Packet Drivers, I was also running my own mailing lists.

---

2   Qmail is not open source. It's freely copyable, but you can't redistribute modified executables. Although open source MTAs do exist, qmail is close enough to open source that I have insufficient reason to switch.

I experienced the same problems he did, so when he published qmail, I was on it right away. First, I ran it on a test machine, because I hate to lose email. Later, I ran it on my email hub after I learned to trust it. Slowly, the qmail community grew.

Working with qmail has been a different experience from Freemacs and Packet Drivers, as I did not write the qmail code. It's just as well, since that gives me something different to write about. It's a truism that you cannot sell something you don't own. A number of the techniques that I used for Packet Drivers do not work for qmail, as I don't own the copyright on it. Nonetheless, I have done quite well with qmail.

Open source software serves to promote the author. By writing quality software (code and documentation), the author shows everyone the quality of his work. The software pushes his reputation out into the world. But what if you haven't written the software? I have found the best technique is to spend time reading user forums and answering questions. You can advance your reputation in this manner.

There are other ways to establish your expertise with a particular program. You can contribute improvements to the code or documentation. You can write your own code that enhances the software, but that otherwise stands alone. You can also write extra documentation or maintain a web site about the software.

I did all these things for qmail. Dr. Bernstein was not interested in registering the qmail.org domain, so I did it for him and maintained the web site, and I wrote a POP3 daemon for qmail and gave it back to him. I answered questions on the qmail mailing list. I wrote add-on packages and patches that people have found useful.

Using all these techniques, I found paying qmail employment. Most often this came from people who needed qmail installed on their machines and had extra requirements that needed custom coding. Sometimes the customer wanted on-site qmail training—which I have provided in Stockholm, Mumbai, London, New York City, Oslo, and Istanbul. All these trips were, of course, paid for by my customers, who also paid me.

Open Source Economics

It was my reputation, or "brand name" if you wish, that got me involved in the Open Source Initiative (OSI). I had been running the Free Software Business (FSB) mailing list since 1993, and had some success and reputation. When I heard about the Open Source name, I immediately adopted it in describing my software. It's so much easier to explain to customers why they should pay for software when it isn't "Free Software." Sometime after that, I heard that Eric Raymond was seeking to create an organization to promote OSI. I had been corresponding with Eric about open source, and we had discussed it on the FSB mailing list, so I volunteered to be on the board of OSI.

I started the FSB mailing list so I could be more aware: I wanted to know what other people were doing to make money from open source. It seems that adding value to things others created is a revolutionary way to make money; even Shakespeare might agree, as he routinely "recycled" plots and storylines written by other people.

Certainly in the software business, an FSB is a new thing. With proprietary software, about the only way to add value is to sell the software for a higher price, or bundle products and sell them together for a lower price in toto.

Running an FSB interested me in economics. When you give up a payroll check and start paying yourself, you also have to pay the business's share of Social Security taxes and start paying estimated income taxes quarterly. The government imposes this dead load on businesses. Why do they do this? Who benefits? Who loses? Economics helps you answer these questions.

Think carefully about how you price your services. There is an economic concept called "price differentiation." It means that you charge different parties different prices for slightly different services. If you charge a single price to everyone, and that price is too high, you miss out on helping some people who cannot afford your services. If you charge a single price that is too low, you create value for some people beyond the amount they paid. It's not exactly fair that they should gain a lot more than you. To maximize your gain, sell things to people based on the value they receive rather than the cost to you. For example, everyone has seen software sold for less in a third-world country or to a school.

Of course, pricing is also related to costs. Think about both transaction costs and so-called "sunk costs." Every transaction consumes a small amount of value in the transaction itself. The buyer needs to evaluate whether the cost spent is worth the value received. The seller needs to take the money and provide the value. This cost is not received as value by either the buyer or the seller; it is simply wasted. One way you can sell support to a customer is on a self-renewing yearly support subscription. One month before the subscription expires, you send the customer an invoice. The subscription portion avoids having to bill the customer for each support request, and the self-renewing portion avoids having to ask the customer to renew the support contract.

A sunk cost is one you cannot recover by selling the thing you bought—for example a railroad or a run of fiber optics cable. You can pull up the rails and sell them, but you won't get back anywhere near the cost of building the railroad. Likewise, spending money to create or improve open source software is a sunk cost. Once you've spent that money, you have no way to sell the software (except by exiting the open source system by using a dual license). Just as a railroad needs to recover its sunk cost by selling transportation services, you need to recover your investment in the software by selling related services.

One of the ways to manage costs is by making use of "public goods." A public good is nonrivalrous (meaning that my use doesn't affect your use) and nonexcludable (I can't stop you from using it). Absent copyright or patent protection, information is a public good. Open source is typically copyrighted software, but is licensed under terms that make it effectively a public good. There is currently great debate over how much excludability is necessary to produce the optimal amount of software.

Previously, economics students were taught that public goods were always underproduced, with lighthouses as the canonical example.

Someone dug up the history of lighthouses, only to find that the early ones were built by voluntary organizations. In a similar manner, the Free Software Foundation (FSF) wrote the GNU tool set as a public good. Economists can no longer assume that public goods are underproduced.

People don't particularly care about products. People only buy things and own things for the services those things render to them. People don't want to own a washing machine; they just want clean clothes. Any desire to own a washing machine is secondary to having the clean clothes. The same thing goes for computers, only computers can provide many different services. The same services that someone can get from a software product can also come from open source software and a support contract.

Economists have discovered many principles helpful to the proprietor of an open source business. Nevermind the joke about 10 economists having 11 opinions. This chapter can but touch on the principles. (See the "For Further Reading" section at the end of this chapter for more information.)

Where Do We Go from Here?

I've brought you up to my present life. Well, not quite. I received an inheritance from my mother, which, when invested in the stock market, generates sufficient income that I no longer need to work for a living. I still take interesting jobs as they come up, and I support long-term customers because they've become friends, not just customers. In everything else, I just write open source and distribute it as I wish. My advice to you is to always pay your retirement fund first: put 10% of every project's check into a brokerage account and invest it in an Exchange Traded Fund.

Happy hacking!

For Further Reading

For a basic (and enjoyable) introduction to price theory, read *Hidden Order*, by David D. Friedman (Collins, 1997). In fact, read anything written by any member of the Friedman dynasty: Milton, Rose, David, or Patri.

For a basic introduction to economics, read *Economics in One Lesson*, by Henry Hazlitt (Three Rivers Press, 1998).

For an explanation of the proper function of the law, read *The Law*, by Frederic Bastiat (Foundation for Economic Education, 1987). It's in the public domain, so you can find it on the Web, in print, and as a free audio book.

For a very deep exposition on economics, read *Human Action*, by Ludwig von Mises (Fox & Wilkes, 1996). It's available at *http://www.mises.org/humanaction.asp*.

The FSB mailing list is at *http://crynwr.com/fsb.html*.

Freemacs is at *http://www.freedos.org/jhall/freemacs*.

Packet Drivers are at *http://crynwr.com*.

qmail is at *http://qmail.org*.

CHAPTER 10

Wendy Seltzer

Why Open Source Needs Copyright Politics

Some programmers and businesspeople draw a distinction between "Free Software" and "open source." Free Software is political, they say, and open source is pragmatic. Free Software developers want to recode the world; open sourcers just want to write good code. This distinction is, of course, exaggerated. Many people adopt these labels for their own reasons; some switch between them depending on audience or context. But even the most apolitical of open source developers and users should be concerned by the copyright battles waged right now. The copyright law being made and enforced today will impact the software we can develop and use for decades, and its impact reaches far beyond commercial media.

Imagine, for example, that you'd like to build an open source home multimedia server. Nothing fancy yet, just a place to play music, watch the occasional DVD, and record television programs—one machine to replace the menagerie of devices nesting in your media center. Easier developed than cleared legally. Technically, you (or others willing to share with you) will be able to meet the challenges with Moore's Law–fast processors, ever-cheaper massive storage capacities, and clever user interfaces. The legal obstacles are harder to hack. Start with the music. If you have standard CDs, you're all set: plenty of Free programs let you play them from the CD drive, rip them to Ogg, FLAC, or MP3 (with a nod, perhaps, to the patent licensors at Fraunhoffer). Try to connect to a streaming service or purchase music online, and things get tricky. Apple's iTunes, the "new Napster," and Rhapsody all lack open source clients, and none would be happy with reverse engineering to write one that plays the music they sell encrypted or by subscription.

Yet music is the easy part. Want to write a player for DVDs you've purchased or Netflixed? Because only closed source implementations have been licensed to decrypt the DVD's files, any DVD player you write is liable to be deemed a "circumvention" by the movie studios and courts, even if the only features you write match those of WinDVD or the standalone player under the TV. Television, then; recording over-the-air broadcasts shouldn't be too hard, since those are unencrypted. Watch out for the digital television transition and the broadcast flag, though. Unless public interest groups' challenge succeeds, the FCC's broadcast flag rule will ban open digital TV tuners that can be used with open source software. The only ones who will be able to play will be those making closed hardware or proprietary software decoders. You'll encounter these obstacles before you even try to take any of your media off the server to exchange with friends or family.

OK, but say for a moment that you have no interest in multimedia. Leave that minefield for another day and move on to business networks or productivity software. Even there, copyright law intrudes. Your security analysis of a system's encryption might be limited by what media companies have preemptively claimed as "technological protection measures;" your selection of replacement parts or add-on modules could be dictated by copyright-based tying more than fitness for use; your ability to interoperate depends on whether reasonable interpretations of the law prevail over some vendors' extreme copyright claims.

Like a group of once-healthy cells grown out of control, copyright law has metastasized to threaten the system of creativity it was once helping to support. No longer a "limited monopoly" to encourage creativity and dissemination of creative works to the public, copyright has become a blunt tool of exclusion, chilling development of software, among other creative endeavors. And so the fight to restore balance to copyright law cannot be dismissed as mere politics. Unless users and developers of open source software join the copyfight, they will find the new reality of copyright law restricting not just their freedom to play blockbuster movies, but also their core freedoms to write and run independent and interoperable software.

From Movable Type to MovableType

A balanced copyright law is enshrined in the U.S. Constitution: "to promote the progress of science and useful arts," Congress was empowered to grant authors exclusive rights "for limited times." The monopoly created was limited in time and scope. The first copyright law gave authors a 14-year term, renewable once, to publish and vend maps, charts, and books. Copyright protected original expression for a short time, while leaving others free to build around that expression (translations and dramatizations, for example), and then to recycle works entirely from the public domain once copyright expired.

Copyright law has changed with the introduction of new technologies. New means of reproduction often first challenge the copyright framework, then establish themselves as new creative tools for authors and their public audiences alike. At the turn of the last century, printers bought single copies of sheet music and punched holes into rolled paper to program "piano rolls" for then-new player pianos. Composers and music publishers sued, seeking to rein in this appropriation. When the courts held that punched paper didn't "copy" inked notes, Congress updated the law with a compromise—not to ban player pianos or mechanical reproduction, but to permit anyone to produce piano rolls if they paid a "mechanical license" royalty for every roll sold. As the player piano market grew, more music reached more people, and more composers got paid for creating it.

The pattern has repeated itself many times since. Songwriters and performers denounced radio until both found that it could promote sales. Movie studios deplored the videocassette recorder, saying Sony's Betamax would be the "Boston Strangler" to their industry. When they failed to shut Sony down, however, the industry converted its peril into a profit center, finding that viewers with home recorders were potential customers for rentals and sales of appropriately priced videotapes. Meanwhile, the Supreme Court's ruling that technology makers would not be liable for users' copyright infringements so long as their devices were "capable of substantial noninfringing use" fueled a technology boom. The public and the creators shared the benefits of new technology—the public could record movies from television to videotape; studios could sell or rent videocassettes more easily than reel-to-reel.

Despite making it through these earlier transitions, the entertainment companies haven't stopped fighting technological change and the competitive threats it represents. The MP3 player is a slightly more convenient cassette deck, and the weblog is just the next step forward from the typewriter and mimeograph. This time, however, the entertainment industry has swayed many in Congress and the courts to the view that "digital is different," and induced them to change the law in ways that are different and dangerous.

This expansion of copyright's control interferes with open source development. The changes manifest themselves in layers, most notably overassertion of protection for code itself; excessive protection of other copyrighted content that code is dealing with; and misuse of copyright to control markets and maintain cartels in technologies of distribution or manipulation of code and content. Together, the copyright layers build a shell around not only proprietary code, but also around culture and innovation.

Copyright in Code

Copyright protects original expression in code, as it protects any other "original works of authorship fixed in any tangible medium of expression." Some developers use that protection to enforce the openness of their code, as with the GNU General

Public License (GPL); others use it to reinforce a proprietary distribution model. But while copyright protects code's creative expression, it does not protect the functions, methods, or procedures that expression implements. Thus, even for closed code, a programmer remains free to study interfaces and functionality, free to interoperate or replace that code with his own.

Situated as it is in an environment filled with proprietary software and poorly documented interfaces, open source development frequently relies on reverse engineering to fit in. Anyone who has used Samba to bridge Windows and non-Windows networks has benefited from Andrew Tridgell's reverse engineering of the Windows protocols for network services; anyone who exchanges files to read and write them in OpenOffice.org rather than Microsoft Word appreciates the reverse-engineering-derived ability to edit files in Microsoft Word format.

Courts have long held that reverse engineering, the practice of examining something and taking it apart to figure out how it works, is a fair use, not an infringement of copyright. Even when programs are released only as closed, binary code, programmers can often discover a great deal about them by watching their operation or the file formats they use, or by disassembly. Reimplementing those discoveries in new code comports with copyright too. So, companies have been protected in taking apart a video game console to build a console emulator (*Sony v. Connectix*) or disassembling a game to build a new one compatible with the console (*Sega v. Accolade*). If someone builds a better mousetrap after examining those that exist, copyright law will not stop her from deploying it.

At least that's the black-letter law. In practice, though, many copyright holders try to extend copyright's limited monopoly to block reverse engineering, through a combination of copyright, contract, and anticircumvention. For example, Blizzard, maker of the popular *Starcraft* and *Warcraft* video games, claims that all players have "agreed," through click-wrap licenses they encounter before the programs run, to a contract that prohibits reverse engineering the Blizzard games. This contract plus copyright, Blizzard asserts, prevents anyone from interoperating without permission.

The bnetd project began when a group of programmers became frustrated by the poor performance of Blizzard's battle.net server for multiuser play of the games they had bought. Instead of putting up with the frequent downtime and rampant cheating on Blizzard's server, they started work on their own open source game server. By watching the communications between game and server, the bnetd programmers were able to reverse engineer their own compatible server, which they set up to give owners of Blizzard games an alternate place to meet, and made the source available for others to join the effort. The public got another option for playing *Starcraft*, and another reason to buy Blizzard games.

Blizzard rewarded the bnetd team's creativity with a lawsuit claiming, among other things, copyright infringement and breach of the click-wrap contract. The team had not seen any of Blizzard's source code, much less copied it, as they reimplemented uncopyrightable functionality, but that didn't stop Blizzard from pulling out the copyright sword. Copyright's lack of protection for functionality is deliberate and sound innovation policy—the public benefits from being able to choose among competing implementations of functionality, be they game servers or network services—yet Blizzard follows in the steps of many trying to get around copyright's limits with contract claims.

Blizzard was trying to limit the code others could produce, to extend the copyright protection on its own code. But many of those pulling out copyright's swords aren't trying to protect code, but other copyrighted content touched by—or that they're afraid will be touched by—code. These efforts, assertions of secondary liability, anti-circumvention regimes, and attempts to impose technology mandates, all limit open source developers' ability to produce new code, to learn from old code, and to compete in the market with proprietary code.

Secondary Liability

Copyright law has not been strictly limited to the direct infringers, but may be extended to those who "contribute" to the infringement in an ongoing relationship with the infringer, or with special-purpose equipment suited only for infringement. Thus, the proprietor of a hall, who looks the other way while paying guests listen to unlicensed music, can be held vicariously liable for the infringing public performance, and the seller of tapes of a length precisely timed for copying particular copyrighted albums could be held contributorily liable for the subsequent infringing use. Extended too far, however, secondary liability chokes off innovation.

Twenty years ago, the Supreme Court rejected Universal and Disney's "unprecedented attempt to impose copyright liability upon the distributors of copying equipment" with the ruling that manufacturers of devices "capable of substantial non-infringing use" could not be held secondarily liable. The *MGM v. Grokster* lawsuit, an attack by all the major record labels and movie studios against Grokster and Streamcast, maker of the Morpheus filesharing software, is nothing short of an all-out assault on the Sony standard.

The studios argue that Grokster and Streamcast should be liable because many users of the peer-to-peer software infringe copyrights—notwithstanding that many others transfer public domain works from Project Gutenberg or the Internet Archive; freely licensed works including open source software and Creative Commons–licensed media; or government works. They argue that the producers of software should be held liable for its "predominant" or "principal" use. Their standard is unworkable both to an entrepreneur financing an untested product and to an open source developer releasing software, any of whose users could adapt it to an unintended, infringing purpose.

Under the Betamax standard, makers of multiuse devices such as the VCR could thus offer them to the public without fearing that they might be held liable if customers misused them to infringe copyrights. With this assurance, hardware makers built components with open interfaces, including CD and DVD burners and massive hard drives, without fearing that someone might put the Plextor on the copyright hook by using that CD burner for large-scale copyright infringement. They built copying devices to transfer content. Software makers, too, have safely offered highly configurable and open source software with relative confidence that their users' configurations won't land them in hot water. The studios' proposed redefinition threatens that freedom to innovate.

Grokster is thus much bigger than peer-to-peer. An expansion of copyright liability, with a "predominant use" test, would make it safer to produce limited-purpose, non-user-modifiable devices and software than open hardware and open source software, regardless of the intent of the developer.

If secondary liability for those who "contribute" to infringement in some ill-defined way weren't enough, the entertainment industry is likely to return to Congress with pleas of renewed urgency to pass the INDUCE Act, which stalled last term. That proposed bill would add yet another level of indirect liability: "inducing" infringement of copyright would extend beyond those who made the tools, to those who explained how to make them work. Watch out that your documentation isn't too thorough!

Anticircumvention

The next stage of copyright expansion beyond direct infringement is the anticircumvention and antitools provisions of the Digital Millenium Copyright Act (DMCA). In the real world, these provisions do little to stop hard-core piracy, but they present a serious barrier to open source compatibility.

Section 1201 of the DMCA prohibits "circumvention" of technological protection measures controlling access to copyrighted works, and it bans manufacture, distribution, and trafficking in devices, products, components, or services that are promoted for, primarily useful for, or designed for circumvention of technological protection measures. Now, a copyright holder who employs a technological lock, such as simple encryption of content, gains the ability to control who and what programs or devices can unlock it, as part of the new right to control "access" to a protected work.

This is an important functional change from the world of printed books or even CDs, where anyone who purchased or borrowed the physical item had the right to use it as she chose—read the last chapter first or play the CD in the car or the computer. Someone who wants to develop a new shuffle mode for CD playback, or a new ripper with better compression for transfer to a portable music player, can do so without seeking permission from the recording companies. It's the misuse of those tools—say, to copy CDs and sell the copies without authorization, that can be pursued as an infringement of traditional copyright.

Under the new anticircumvention regime, however, those who control copyrights can take their control much further. Thus, copyright holders say, and courts have agreed, anyone who develops a decoder for an encrypted format without a license is producing circumvention tools in violation of the DMCA. So, groups of copyright holders, who couldn't individually control markets, join together in licensing cartels backed by the magic of the DMCA, by which they control not only copyrights but also the surrounding player technologies.

And so it is with digital video discs. Movie studios and consumer electronics and technology companies developed the content scrambling system, a.k.a. CSS, applied it to DVDs, and declared it to be a technological protection measure. By forming the DVD Copy Control Association to license the CSS specification as a trade secret on restrictive terms, they sealed themselves a nice cartel, simultaneously protecting themselves against disruptive innovation in video players from outside of the establishment and walling off their copyrighted works against fair use copying.

The CSS encryption was trivially easy to break, once Jon Johansen and some German programmers set their minds to it. But that's beside the point, since the law protects even weak technological protection measures with the full panoply of anticircumvention and antidevice prohibitions. Even weak measures set up the law's sharp division between licensed access and unlicensed circumvention. Thus, with DeCSS and its successor code, anyone can play or rip a DVD on any platform, but with DMCA, no one can lawfully do so in the United States, or even develop code for DVD playback, without a license from the DVD-CCA.

Kaleidescape is a small company with a rich clientèle—owners of hundreds of DVDs willing to pay nearly $30,000 for a DVD jukebox to organize and store them all. Kaleidescape built this machine—a computer filled with massive hard drives onto which customers could rip their DVD movie collections. For its efforts to help the movie industry's best customers get more out of their purchases, Kaleidescape earned itself a lawsuit from the DVD-CCA, claiming Kaleidescape had breached its contract licensing the not-so-secret DVD trade secrets.

Kaleidescape's system, which allows the creation of persistent digital copies of the content of DVDs and allows copying of the CSS copy protection data, is not designed in a manner and does not include features clearly designed to effectively frustrate efforts to defeat the copy protection functions....For these reasons, Kaleidescape has breached the CSS license.

Without that license, even if similar information could be derived from reverse engineering and publicly available information, using it to enable DVD "access" and playback would be labeled circumvention.

As the movie studios have done with DVDs, record labels and software companies have done with multiple incompatible formats for streaming and downloadable music: Windows Media, Janus, Apple's FairPlay. Many of these have succumbed to reverse engineering of varying degrees of sophistication, from "burn it to CD and rerip," to "run it through a simulated sound-card driver," through cryptanalysis (much of it from Jon Johansen, again).

Yet, like CSS, none of the music protection measures is licensed on terms that permit open source development, nor could they be. Open source is incompatible with both their stated aim, to prevent "piracy" of content, and the unstated underlying goal of technology control. Since the essence of open source is user modification, users could easily modify in or out any particular features—and the first to go would likely be the restrictions of digital rights management (DRM) and barriers to interoperability. Even if open source version 1.0 incorporated all the restrictions of proprietary clients, numerous versions 1.0.1 would likely disable them. Thus, anticircumvention regimes lock open source out of mainstream development of entire classes of applications to interact with these copyright cartels' media.

Of course, it's not just open source developers who need reverse engineering. Consider RealNetworks' attempt to sell music that would play on the popular iPod. Real could have transcoded downloads into standard MP3, which play on the iPod, but Real (or its record-label-relations department) wanted to include DRM on the files. So, it reverse engineered Apple's FairPlay format, in a move Real called Harmony, to encode Real files for the iPod. Apple fought back with legal and technical threats. Along with intimations that it might use the DMCA, Apple issued a statement that said, "We strongly caution Real and their customers that when we update our iPod software from time to time it is highly likely that Real's Harmony technology will cease to work with current and future iPods." Real wasn't threatening to infringe copyrights, but to give customers a way to interchange their iPods for other devices. Once again, DRM is market protection, not copy protection.

The Threat to Research

Again in the DMCA, copyright bleeds beyond entertainment media and content. Here, it also chills research into encryption that may be used to secure copyrighted works, regardless of whether that research touches entertainment content directly.

When the Secure Digital Music Initiative (SDMI) invited programmers to "attack" security technologies they were promoting to control digital music, Princeton computer science professor Ed Felten and his team stepped up to the challenge. They broke the security and prepared a scientific paper analyzing the weaknesses of digital audio watermarking. By scrutinizing these implementations, they could help the public, and particularly those considering watermarks to protect their own materials, to evaluate the security of watermark technologies.

But when the Felten paper was accepted for presentation at a computer security conference, its authors were threatened with a lawsuit from the Recording Industry Association of America (RIAA) and SDMI's technology providers. RIAA and the technology companies claimed that even the scientific analysis of flaws in security technology "would subject [the] research team to enforcement actions under the DMCA and possibly other federal laws." The RIAA suggested that the paper and conference presentation fell under the DMCA's antidevice prohibition, which bans offering to the public "any technology, product, service, device, component, or part thereof," designed or marketed for circumvention or having only limited commercially significant other purposes. Facing legal pressure on conference organizers and researchers, the team withdrew the paper.

While Professor Felten and his team ultimately published a version of the paper, "Reading Between the Lines: Lessons from the SDMI Challenge," they were forced to strip out technical detail. With the vague anticircumvention threat hanging, the authors felt they had to omit code samples that could be construed as aiding circumvention—even though that code would have helped other researchers and developers to understand the watermarks' weaknesses better and to learn from SDMI's errors in building their own security systems.

The work of the Felten team never infringed copyright. The researchers did not copy a single piece of music without authorization, nor even produce tools that enabled others to do so directly. And yet, they were caught up by the DMCA's vagaries because their paper—intended to educate other researchers and developers of security systems—might also have provided "part" of a tool for circumventing a copyright control. Will the same hurdles rise before someone who builds a tool to strip accidental copy protection from fonts he himself has created; security researchers who find vulnerabilities in network software they analyze; someone describing hardware modifications to make the Xbox a more general-purpose device? So far, the answer has been "yes" for Tom Murphy, SNOsoft, and Bunnie Hwang.

For those who are attracted to open source development because of its opportunities to learn from and share with others, this antiresearch aspect of the DMCA should be particularly troubling: research is being chilled precisely because it teaches too much, because its teaching might be misused. Of course, closing systems doesn't stop them from having security flaws, or even prevent those flaws from being discovered, and it does block some of the most effective information sharing around better security. Both the teaching that open source developers depend on to improve their programs and the teaching of open source code itself are at risk.

Further, the DMCA has been used in attempts to block competitive interoperability of devices including printer toner cartridges and remote control garage door openers, as manufacturers add little scraps of code to devices and hope to leverage its "protection" into market control. Though those arguments have been rejected so far, it's unlikely we've seen the last of them.

Technology Mandates

The final layer of copyright's expansion, so far, is to technology mandates, where an entire technology is redesigned by government fiat in the name of copyright protection. Senator Ernest "Fritz" Hollings proposed one of these in 2002 that would have required every "digital media device," including the personal computer, to be redesigned to protect copyrighted content. While the "Fritz chip" never came to be, a smaller version has been foisted upon us in the form of the Broadcast Flag, an FCC rule set to take effect July 1, 2005.

With the government eager to get broadcasters off the valuable analog spectrum and onto digital transmissions, movies studios threatened that they wouldn't allow their content to be broadcast digitally in the clear, and warned that there would be no transition without their "high-value" content. The FCC didn't want to abandon the notion of unencrypted over-the-air television broadcasts, but it did want to give the studios their "protection," so it put the restrictions into the hardware. At the studios' recommendation, the FCC adopted a rule that adds a "flag" to these unencrypted broadcasts and then requires every receiver to watch for the flag and output flagged content only to "compliant" devices or in low resolution. Only devices that can implement DRM in a manner "robust against user modification" will be deemed compliant.

The Broadcast Flag rule enforces copyright on communications through the devices that receive them:

> We conclude that in order for a flag-based content protection system to be effective, demodulators integrated within, or produced for use in, DTV reception devices ("Demodulator Products") must recognize and give effect to the ATSC flag pursuant to the compliance and robustness rules....This necessarily includes PC and IT products that are used for off-air DTV reception.

Instead of focusing on infringing uses of TV broadcasts (taping a show and selling copies, for example), this new kind of regulation puts the government in the business of redesigning products that *might* be used to infringe. In the process, it locks out many noninfringing uses, innovative technologies, competitive products, and open source developers. Building a device for time-shifting, pausing live TV, remotely scheduling recordings, and watching shows at double speed doesn't infringe copyright, but because the hardware/software to enable those capabilities isn't "robust," it is sacrificed to illusory copyright protection. Because these collateral harms are unavoidable, technology mandates should be a last resort, not a predictive strike against hypothetical danger.

The result of this rule is restriction on open source even greater than encryption would have been. Open source can implement encryption, but it can't offer "robust[ness] against user modification." Pre-flag, you could get an HDTV tuner card for a PC, pair it with open source software such as MythTV, and build your own digital video recorder to compete with TiVo. Post-flag, TiVo must use government-approved "robust" technologies to lock down its hardware and software, and open source will be shut out from access to the high-definition signals entirely.

Under the Broadcast Flag regime, market participants, bound up in the welter of licensing and preapproval requirements, can't offer the products users want. Where the market fails to provide fair-use-enabling technologies, the robustness rules prevent end users from correcting the problem. Absent technology mandates, users dissatisfied with commercial options can and do write their own software alternatives and often share them in open source. In a world of restricted, robust hardware, users are limited to the options the commercial market provides: the fully capable hardware HD tuner card can't be manufactured. Consumer-driven innovation is cut off when users can't tinker with existing technologies or develop new ones that challenge market leaders.

What About That Media Server?

Copyright in its historical form benefits open source developers. Along with the general public, they benefit from the incentive to creativity and the support copyright gives to open source distribution models. Copyright as special-interest law, however, hurts open source development, because the special interests are those of closed markets and closed content. DRM can't stop piracy, but it can prevent anyone from Betamaxing another industry, commercializing disruptive technology development without content-industry sanction. Where the entertainment industry can't stop infringement, it attacks openness instead, and the "honest person" loses.

Whether you want to build a media server or an embedded network device, you'll likely run across the snares of copyright law. It's time to peel back the layers of copyright protectionism and return copyright to its original purpose: "to promote the progress of science and useful arts."

CHAPTER 11

Jesus M. Gonzalez-Barahona
Gregorio Robles

Libre Software in Europe

The libre (free, open source) software[1] community is probably one of the more global and internationalized. Therefore, it may be a little artificial to try to separate the European share of it in the hope of finding peculiar characteristics. But at the same time, Europe is so diverse, so full of national, cultural, and linguistic boundaries, that it may be difficult to find common patterns in this already diverse libre software world. However, our feeling is that in between these two facts, there is plenty of room for writing about what is happening in the European libre software scene. While preparing the material for this chapter, we have come to the idea that, in fact, this is not such a global world, nor does its European fraction lack common patterns despite its diversity. With this focus in mind, we have looked for both the peculiarities and the commonalities. We have walked through the enormous amount of data concerning what is happening in the vibrant European libre software scene with the aim of offering the reader the more relevant and revealing trends and facts, providing a vision of a complex and diverse, but also uniform, landscape. And, for sure, a very personal one.

---

1   In this chapter, we will use the term *libre software* to refer both to *free software* (according to the
     Free Software Foundation definition) and *open source software* (according to the Open Source
     Initiative), except where making distinctions makes sense. *Libre* is a term well understood in
     romance languages, such as Spanish, French, Portuguese, and Italian, and is understandable for
     speakers of many others. It lacks the ambiguity of "free" in English ("libre" means only "free" as
     in "free speech") and is used by some people especially in Europe (although the term is rooted
     in the early U.S. free software community; see *http://sinetgy.org/jgb/articulos/libre-software-origin/
     libre-software-origin.html* for details). In this respect, it is important to notice that although the
     communities, the motivations, and the rationales behind "free" and "open source" are different,
     the software to which they refer is basically (although not exactly) the same.

In fact, although libre software can be considered to come from the U.S., the spread of the Internet (and before that, the Usenet) in Europe made it possible to develop a fragmented European libre software community as early as the late 1980s. With time, common trends that could be called "European" seem to be emerging from this framentation. However, even without strong pan-European relationships, and maybe due to the common sociocultural background, some trends are found now and again in different parts of the continent, producing a collage with many common patterns.

In this chapter, we provide some snapshots of that collage. Instead of focusing only on issues that can be truly called European (because they involve participants from many areas of the continent), we have tried to show the diversity of initiatives and environments, so the reader can have at least an idea of the details of a very complex landscape. We have also devoted some efforts to identifying common patterns and Europe-wide initiatives, especially when we find they are a signal of an emerging common trend. Intentionally, many examples which are European by nature or birth, but have evolved into global projects, are not included, or are mentioned briefly, since they are now more global than European and therefore make little sense here. There is, however, a clear intent to show the main European contributions to the libre software world, in terms of development, use, and promotion.

All in all, the set of case examples, and the issues presented on these pages, are just a (hopefully representative) showcase of what is happening in the European world of libre software, obviously filtered by our personal biases and backgrounds. For sure, the selection by any other observer of libre software in Europe would be different. We just hope that the reader will find our selection illustrative enough.

Brief Summary of an Already Long History

Before showing the current landscape of libre software in Europe, it seems necessary to provide some historical background. From issues like the European involvement and impetus in projects such as Linux (the kernel) or the KDE project, which influenced greatly the shape of the currently available libre software, to very specific use cases in European companies, which are mainly a consequence of the global importance of the phenomenon, there is a whole rainbow of milestones which will contribute to the understanding of the present situation.

The evolution of libre software in Europe during the early days was parallel (as it was in other parts of the world) to the penetration of the Internet (and before it, the Usenet). Therefore, it is not strange that areas which had an early and deep exposure to these Nets—such as the Nordic countries, the Netherlands, and the United Kingdom—also had the first cases of involvement in global libre software activities. However, it is important to also consider some linguistic issues. For instance, those countries where the English language (which is clearly the lingua franca in the global libre software community) has more penetration (either as the mother tongue or the second language) commonly had an earlier involvement in libre software. With the passage of time, maybe since the mid-1990s, specific dynamics started to show strength in other regions, with more or less of a relationship to the global evolution. Most European countries (Germany, France, and Spain are clear examples) have grown their own communities and libre software fabric in partial isolation, to the point that many initiatives that are very well known in one country are almost unknown in the others and outside Europe, despite their valuable contributions. In many cases, the flux of information among Europeans of different areas is still carried through global (or, for that matter, American) events and initiatives. Just as a case example, we still find out about relevant news in other European countries through the American web news service, Slashdot, despite being reasonably well linked to the libre software communities in several European countries.

Considering this situation, the contributions of Europe to the history of libre software are extensive. We can consider, for example, events such as the birth of many applications and projects (consider Linux, the kernel, by Linus Torvalds, Finnish; Python, by Guido van Rossum, Dutch; MySQL, by Michael [Monty] Widenius, Swedish; PHP, by Rasmus Lerdorf, Danish; KDE, by Matthias Ettrich, German; and many more); or the foundation and development of some of the first companies with a business model based on developing or distributing libre software (such as MySQL AB, Trolltech, or SuSE); some of the first studies and initiatives denoting attention by public administrations to the libre software phenomenon (such as those by the European Commission); and some of the first research projects considering libre software as a matter of study (such as those performed by the FLOSS project).

Most of those contributions will be mentioned and presented in some detail within this chapter. Instead of following a timeline approach, we have preferred to group matters according to the different topics involved, each one in its own section: developers, community, companies, public administrations, legal initiatives, licenses, education, and research. Of course, this implies a certain degree of artificial delimiter, since many issues in fact belong to more than one of those sections. But we hope that what is lost with respect to that precision is gained in readability and comprehension of the situation as a whole.

The Development Community

Since the emergence of libre software as a concept, European developers have contributed considerably to its growth. Only recently have we had evidence of how large this contribution is: the WIDI survey in 2001,[2] the FLOSS study in 2002,[3] and the

---

2   Gregorio Robles, Hendrik Scheider, Ingo Tretkowski, and Niels Weber, "Who Is Doing It?"; *http://widi.berlios.de/paper/study.html.*

3   Rishab Aiyer Ghosh, Ruediger Glott, Bernhard Krieger, and Gregorio Robles, "FLOSS Final Report, Survey of Developers"; *http://www.infonomics.nl/FLOSS/report/index.htm.*

FLOSS-US study in 2003[4] showed that large numbers of developers declare themselves as nationals of a European country. In the case of FLOSS-US, from a sample of almost 1,500 developers, more than 900 (or about 60%) declared themselves to be living in Europe (including Russia), compared with about 405 in North America. In the FLOSS study, more than 70% of the developers were living in Europe (14% in North America). WIDI reported developers were 54% European and about 35% North American. Of course, all these studies could be biased (and at least they are with respect to language, since they all were done in English), since the respondents were self-selected in nature and the studies were not focused on geographic characterization, which was more like a side product of other characterizations. But they were conclusive about the European contribution to libre software development being quite high and, as an aggregate, incomparable to any other region worldwide.

It is also interesting to notice the distribution of libre software developers within Europe, since not all countries are equally represented. In this respect, results from studies are variable, although some common patterns can be inferred. For instance, France and Germany seem to be the countries with a greater population of developers in absolute numbers, closely followed by the United Kingdom and Italy. According to the FLOSS study, France has the highest, with 15% of the total of respondents, followed by Germany (12%), Italy (8%), the UK, the Netherlands, and Spain (all at about 6.5%), and according to the FLOSS-US study the first country is Germany (25%), followed by the UK, France, and Russia (all close to 4%), Spain, the Netherlands, and Italy (about 3% each). WIDI results give 20% for Germany, close to 5% for France and the UK, and about 3% for the Netherlands.

These absolute figures are hardly surprising, since they match (with some exceptions) the countries with a greater population and GDP (which in general shows a certain correlation with the number of developers all around the world). But some countries are clearly overrepresented. Among them, the Netherlands is a clear case. Sweden, Spain, Poland, Switzerland, Finland, the Czech Republic, Denmark, and Austria also seem to have more developers than their population or GDP would suggest. In fact, the Czech Republic and Austria are two relatively small countries that have some of the highest figures for libre software developers per capita.

A much more specific case study is the Debian project, where developers have the option of indicating their residence. Of those who registered, the figures are in line with the previously described results. Seven of the top eleven countries in November 2004 were European: Germany (151), the United Kingdom (80), France (57), Spain (37), Italy (34), the Netherlands (30), and Sweden (28). The total number of European developers is well over 400, which compares to 364 in North America (U.S. and Canada), the second-highest region in this survey.

---

4   "Paul A. David, Andrew Waterman, and Seema Arora, "FLOSS-US The Free/Libre/Open Source Software Survey for 2003"; http://www.stanford.edu/group/floss-us/.

In summary, Europe clearly has a high concentration of developers, possibly higher than any other region in the world. Within Europe, the Western countries have higher concentrations (though the Eastern countries of the Czech Republic and, to some extent, Poland, buck this trend). An area extending from the UK to Italy seems to include most of the European developers (with countries such as France, the Benelux, Germany, Austria, Switzerland, and the Czech Republic). The Nordic countries also have high concentrations (with respect to population). In Southern Europe, Spain seems to be the country with the most developers.

Although some successful libre software projects could be considered European by origin, or (in some rare cases) because almost all developers are European, this identification is tricky: libre software projects are global by nature (any developer from any part of the world can, in principle, join in). Therefore, we will not try to identify strictly European projects.

On the contrary, infrastructure to host libre software projects can easily be assigned a nationality (although usually it is open to the world). In this respect, Europe is well behind the U.S., since SourceForge and other well-known hosting sites are physically located there. But some hosting facilities do also exist in Europe, usually focused on a given national or linguistic community. As an example, we can cite BerlioOS , the largest one (*http://berlios.de*; in Germany, with more than 2,400 projects and almost 12,000 developers registered), Software-libre.org (*http://software-libre.org*; Spain, 66 projects and 234 developers), and Gna! (*http://gna.org*; France, 360 projects and almost 2,000 developers). All of these figures (checked during March 2005) turn pale in comparison to SourceForge. This is one of several cases where, despite the importance of the European developer community, its infrastructure is not at the same level.

The Organization of the Community

The libre software community in Europe, despite being healthy and full of life, is also fragmented, reflecting the cultural, national, and linguistic diversity we have in the continent, but also lacking the strength and power it could have if it were more coordinated. However, more and more links (usually at the personal level) are being established, and something similar to a really European libre software community seems to be emerging. Although it is still some distance from speaking as loudly as it could—for instance, behind the European Union institutions, we are seeing more and more transnational initiatives—as with any other community formed around the Internet, it has its own virtual and real-world meeting points, such as news sites, conferences, and associations. However, as we will show in the next paragraphs, most are closely linked to relatively small geographic areas, and can hardly be identified as being "for Europeans."

Many news sites are strongly related to libre software. Maybe the most transnational one is The Register (*http://theregister.co.uk*) (probably because it runs in English), which of course, is the most popular in the UK. Although it carries news on many IT-related subjects, libre software receives more than reasonable coverage. Similar things can be said for Heise Online (*http://www.heise.de*), but applied to the German-speaking community. Other sites worth mentioning are Barrapunto (*http://barrapunto.com*; Spanish), LinuxFr (*http://linuxfr.org*; French), Linux.pl (*http://linux.pl*; Polish), gildot (*http://gildot.org*; Portuguese), and Svenska Linuxföreningen and Gnuheter (*http://gnuheter.com*; Swedish, both read also by Norwegian and Danish speakers). Some of them are built as Slashdot-like sites, others are not, but each has its own flavor. Most of them are helping to foster a sense of community and are even assisting in the foundation or consolidation of more formal organizations related to libre software. There have been some attempts to establish a "European Slashdot," but without success (for now). Maybe that is simply not possible where there is such cultural and linguistic diversity.

If we consider real-world meeting places, we can find some that come close to being truly European. Probably the clearer examples are FOSDEM[5] (Free Open Source Developers Meeting), which is especially oriented to libre software developers and has for several years been pulling together a good quantity of European hackers (also attracting some from other regions), LinuxTag,[6] which is more oriented toward users and companies, although also attractive to developers, and LSM/RMLL[7] (Libre Software Meeting/Rencontres Mondiales du Logiciel Libre), for developers specifically devoted to libre software. Several projects also have European meetings: GUADEC (GNOME User and Developer European Conference; *http://guadec.org/*), YAPC::Europe (*http://yapceurope.org*; for the Perl community), EuroBSDCon (*http://eurobsdcon.org*), ApacheCon Europe,[8] OOoCOn (*http://marketing.openoffice.org/conference*) OpenOffice.org Conference, not specifically European, but always held in Europe), and many others. These are usually held every year in a different location.

In addition, we have the national events: the aforementioned LinuxTag in Germany, LSM/RMLL and Solutions Linux (*http://solutionslinux.fr*; a more commercial event) in France, Linux Forum (*http://linuxforum.dk*) in Denmark, Congreso Hispalinux (*http://congreso.hispalinux.es*) in Spain, Linuxwochen (*http://linuxwochen.at*) in Austria, and so on. In some countries, there are no national events, but smaller meetings are usually organized by Linux User Groups (LUGs) or libre software associations.

With respect to associations, diversity is also the rule. In addition to more or less local LUGs (spread through all of Europe), there are some organizations which either represent groups of local associations, or have a wider membership, usually at the national level. Some of them are: APRIL (Association pour la Promotion et la Recherche en Informatique Libre; *http://april.org*) and AFUL (Association Francophone des Utilisateurs de Linux et des Logiciels Libres; *http://aful.org*) in France, NUUG (Norwegian

---

5   *http://fosdem.org*, which takes place in Brussels, Belgium, usually in late February.
6   *http://linuxtag.org*, which takes place in Karlsruhe, Germany, usually in June or July.
7   *http://rencontresmondiales.org*, which takes place somewhere in France, usually in July.
8   *http://apachecon.com/2005/EU* for the 2005 edition.

Unix User Group; *http://nuug.no*) in Norway, DKUUG (*http://www.dkuug.dk*) in Denmark, Atviras kodas Lietuvai (Open Source for Lithuania; *http://akl.lt*) in Lithuania, SSLUG (Skåne Sjælland Linux User Group; *http://sslug.dk*) in Sweden and Denmark, ANSOL (Associação Nacional para o Software Livre; *http://ansol.org*) in Portugal, PLUTO (*http://pluto.it*) in Italy, AFFS (Association for Free Software; *http://affs.org.uk*) in the UK, PLUG (Polish Linux User Group; *http://www.linux.org.pl*) in Poland, and Hispalinux (*http://hispalinux.es*) in Spain. At the European level, there is also the FSFE (Free Software Foundation Europe; *http://fsfeurope.org*), and the corresponding FSF-related associations in some countries, such as FSF France (*http://fsffrance.org*), Verein zur Förderung Freier Software (*http://ffs.or.at*) in Austria, and Associazione Software Libero (*http://softwarelibero.it*) in Italy.[9] Each of these associations has its own history. Some of them were formed as Unix associations (now more than 20 years old) and have evolved into Linux and libre software associations with time. Some others were formed specifically as LUGS, mostly in the mid-1990s. Still others are devoted to libre software in general, again mostly founded in the late 1990s. Membership, activities, involvement of companies, etc., vary a lot from one to another.

On many occasions, there have been discussions suggesting the convenience of a European umbrella association devoted to libre software, which could speak on behalf (and be some kind of federation) all of these organizations. But so far, none has crystallized.

In addition to these "community" associations, there are also some others representing corporate interests. Among them, probably the most notable is Open Forum Europe (*http://www.openforumeurope.org*), which has backed some actions related to libre software promotion in Brussels, but performs most of its activity in the United Kingdom.[10]

Libre Software in the Private Sector

Since the early 1990s, it became clear to some European entrepreneurs that libre software was interesting for business. Approximately at the same time in other parts of the world, companies started to use libre software, and a new market niche for companies providing services based upon libre software emerged. During the mid- 1990s, many small companies started to offer services based on the then-new Linux-based operating system (MandrakeSoft in France and SuSE in Germany are well- known cases that will be discussed later in more detail, but they were not unique). Some other companies

---

9   A more complete list (which also includes some organizations that are not associations of the libre software community, but includes information from almost any country in Europe) is available at the Open Source Observatory, *http://europa.eu.int/idabc/en/document/1631/471*.

10 Open Forum Europe was widely criticized by libre software advocates when its president signed a public statement in favor of software patents, in 2003 (see *http://swpat.ffii.org/letters/ofeu034* and *http://www.kuro5hin.org/story/2003/5/6/8355/78133*).

focused on the libre software infrastructure of the Internet or on specific products (such as Trolltech AS in Norway and MySQL AB in Sweden). By the late 1990s, interest in libre software was strong enough to maintain several mid-size companies devoted (completely or in part) to generic consultancy, support, and development of libre software, such as Alcove in France, Andago in Spain, and ID Pro AG in Germany.[11] Later, starting in the early 2000s, large companies began to show interest in libre software, especially in the secondary sector (these were intensive users of software, although that was not their main line of business—such as telecommunications, automotive, aerospace, and banking). In 2005, the list of large European companies with a significant use of libre software is too long to include here.

A good exponent of the interest of European industry in libre software is the ITEA Report on Open Source Software,[12] published in January 2004, which is aimed at elucidating the libre software world (from legal and economic, to development and quality issues), and uncovering business opportunities and issues to be resolved in relation to it. The following sentences, taken verbatim from the report, may provide some insight on the opinion of those drafting it (and maybe on the view of the companies for which they work):

> Depending on the business they are in, companies are likely to have different reasons for using OSS. In some cases, OSS can help them lower the cost of the service, system or product that they offer. In others, their contribution to OSS can help to establish new standards worldwide. By carefully applying license conditions it is certainly possible to derive considerable benefits from OSS, while minimizing the risks. Open Source Software is not the "magic bullet" to solve Europe's software development competitiveness. However, OSS is an important new development and an interesting option for software-intensive systems.

Different sources provide different figures for the market share of libre software in Europe, but they are consistent in indicating its continuous growth. The market for libre software in Germany in 2003 was estimated, according to a study by Soreon Research,[13] to be in the region of EUR 131 million (with a projection of more than EUR 300 million for 2007). The manufacturing industry was the one with the highest penetration, with 18% of the companies making significant use of libre software. The structure of the market was heavily based on support services (EUR 81 million) and training (EUR 27 million), and direct sales of software accounted for only EUR

---

11 Some of these companies stopped business during the early 2000s. This was the case of ID Pro AG and Alcove (which still maintains a web site in early 2005, http://alcove.com. Andago, http://andago.com, is an example of those still remaining in the market.

12 ITEA, http://www.itea-office.org, is a joint effort by many European companies to stimulate precompetitive research and development, specially in the field of information technologies, including in its partnership companies like Alcatel, Bosch, Bull, Daimler-Chrysler, Italtel, Nokia, and Siemens. See http://www.itea-office.org/newsroom/publications/Open_Source_Software.htm.

13 Soreon Research "The Market for Open Source Software in Germany" (July 2003).

10 million. However, a year later IDC[14] estimated that USD 98 million would be spent on IT services for libre software in the whole of Western Europe, and predicted that USD 228 million would be spent by 2008 (for the same region).

To provide an overview of the landscape of European libre software companies, we have selected a short list of them, probably the better-known ones. This list is, of course, not exhaustive, but will hopefully be illustrative enough to show the European contribution to the world of libre software business models:

- SuSE[15] was one of the first companies providing services for Linux-based distributions, almost since it was founded in 1992 in Germany, as S.u.S.E., by Hubert Mantel, Burchard Steinbild, Roland Dyroff, and Thomas Fehr. They started by selling floppies of Slackware Linux partially translated into German. Later, they decided to build their own distribution, which was released in 1996 and quickly becoming the most-used GNU/Linux distribution in Germany. Based on the success of this distribution, the company grew, with a business model that also included support, training, and consultancy for libre software. It had a workforce of more than 500 and had one of the more well-known GNU/Linux distributions in the world when it was acquired by Novell in 2003, for USD 210 million. SuSE has also been known for its contributions to the libre software community, supporting or directly contributing to many projects, from KDE to ALSA to the Linux kernel itself.

- MandrakeSoft (*http://mandrakesoft.com*) is a French company with major operations in the United States and other countries. It was funded in 1998 by Gaël Duval, Frédéric Bastok, and Jacques Le Marois, after the success of the first release of Mandrakelinux some months before. One of its innovations was to let people download the entire distribution from the Net, which assisted its spread around the world. Agreements with the Pearson Technology Group (then Macmillan Software) and other distributors helped too and, by 2000, had made it one of the companies with a larger market share in the libre software segment in the United States and France, having gone from three to about a hundred employees in just two years. In 2004, the company had revenues of more than EUR 5 million. Currently, its business model seems to be mainly linked to its distribution, although it also provides consultancy and support services for libre software in general. MandrakeSoft has recently acquired Conectiva, one of the other companies producing a major GNU/Linux distribution, and has changed its name to Mandriva (*http://www.mandriva.com*).

---

14 IDC, "Services around Linux, Open Source, and Free Software—Western European Market Forecast" (June 2004).
15 SuSE was bought by Novell, but still maintains its original url: *http://suse.com*.

- Open CASCADE S.A. (*http://opencascade.com*), now in the AREVA group, is a French company providing services around the Open CASCADE system (a set of 3D modeling components and libraries) and SALOMÉ (a platform for the integration of numerical simulations), both distributed as libre software. The company is rooted in Matra Datavision, also French, which was a major player in the CAD/CAM market from the early 1980s to the mid-1990s. One of its lead products was CAS.CADE (Computer Aided Software for Computer Aided Design and Engineering), released in 1993. In 1999, after the decision to change the focus of the company from products to services, Matra Datavision decided to release CAS.CADE as libre software, under the name Open CASCADE. One year later, the company of the same name was segregated and acquired by Principia (in the AREVA group). Open CASCADE S.A. is now focused on providing customized development, training, consulting, and other services, with a team of about 80 developers. The company is also fostering the building of communities around its products, channeled through a specific site (*http://opencascade.org*).

- Trolltech AS (*http://trolltech.com*) is a Norwegian company well known for producing Qt, an essential component for KDE and many other systems. It was founded in 1994 by Haavard Nord and Eirik Chambe-Eng, with the aim of building cross-platform C++ GUI tools. Now a company with more than 90 employees, it pioneered a dual-licensing model for Qt (and other products). In the beginning, this software was gratis but not free. However, after the launch of the GNOME project (promoted by the FSF, among others, and backed by some companies worried about the dependence of KDE on the proprietary Qt), Trolltech decided to distribute it under QPL (a libre software license). Finally, Trolltech moved to the GPL (for the X11-based version of Qt), and gave those not willing to comply with the GPL the option of purchasing proprietary licenses.

- MySQL AB (*http://mysql.com*) is the company that owns the code for the MySQL database server. It was founded in Sweden in 1995 by David Axmark, Allan Larsson, and Michael Widenius (the founders of the MySQL project) and has since opened offices in many countries. Its business model is based on selling support and services for MySQL, and selling licenses to those unwilling to fulfill the conditions of the GPL (that is, dual licensing MySQL). This is why my SQL AB has been careful to maintain the ownership of the code, by having all the developers of MySQL as employees of the company. MySQL AB has run on venture capital since 2001, had revenues of about EUR 15 million in 2004, and about 160 employees.

To complete the vision of this landscape, let us introduce some examples of large European companies involved in libre software. The examples were chosen at random from our experience and therefore are not representative, but again, hopefully illustrative of what is happening out there:

- ObjectWeb (*http://objectweb.org*) is a consortium created in 1999 in France by Bull (*http://bull.com*), France Telecom R&D (*http://rd.francetelecom.com*), and INRIA (*http://inria.com*) to develop libre software middleware, ranging from specific software frameworks and protocol implementations (such as CORBA) to integrated platforms. In 2002, it evolved into an international and independent nonprofit organization open to companies, institutions, and individuals. The software developed by ObjectWeb includes more than 40 products in the application platforms, workflow engines, IDE plug-ins, and software engineering domains. ObjectWeb uses GPL and LGPL licenses, and has managed to create a large community of companies providing services around those products.

- Based in Spain, Telefonica Investigacion y Desarrollo (TID (*http://www.tid.es*) a subsidiary of Telefonica (*http://telefonica.com*), one of the largest telecommunication companies in the world), launched in late 2004 the Morfeo Project (*http://morfeo-project.org*), in collaboration with some other Spanish companies, universities, and public administrations. Morfeo is a framework for distributing and developing as libre software some products that TID either has produced or needs, mainly in the field of platform software (middleware, workflow, communications, etc.). It has already released products such as CORBA systems, and is trying to build a community of developers. In the long term, this could be a first step toward a strategy based on libre software for some of the activities of TID.

- Ericsson (*http://ericsson.com*), based in Sweden and not especially well known for its contributions to libre software, has distributed several products as libre software. Ericsson's implementation of the Erlang programming language[16] was released in 2000, and has an active developer community. Erlang provides facilities for concurrency, distribution, robustness, and soft real-time processing. Another contribution is TIPC (*http://tipc.sourceforge.net*), a protocol for intracluster communication implemented as a loadable module for Linux.

- Nokia (*http://nokia.com*), based in Finland, and also not normally known as a libre software producer, has distributed some software under NOKOS (the Nokia Open Source License, an OSI-recognized open source license). But recently another event related to the use of libre software by Nokia hit the news: the availability of a Python environment for some of its products (*http://www.forum.nokia.com/python*) using the Symbian OS, in what could be a strategy of letting libre software developers in the Python community build applications for Nokia devices.

- Symbian (*http://symbian.com*), with headquarters in the UK and owned by Nokia, Siemens, Ericsson, and others, is releasing one of its products for Symbian OS, the Open Programming Language (OPL; *http://opl-dev.sourceforge.net*), as libre software, using the LGPL. OPL is a BASIC-like language used in Symbian OS phones for rapid prototype development.

---

16 Open-source Erlang, *http://erlang.org*.

These examples show how, despite their general strategy of being more or less oriented toward proprietary software, many European companies are experimenting with libre software models and, in some cases, are considering new lines of business based upon them. In fact, similar cases of exploration of the libre software world can be found in almost any medium to large-size company heavily involved in the software business.

Public Administrations and Libre Software

Before the year 2000, libre software was almost completely off the political radar in Europe. But since then, and with widely varying intensities, it has entered the political agenda in many European cities, regions, and countries. In some cases, it is considered as one possible choice for public administrations in their role as intensive users of services based in software. In some others, it is deeply linked to efforts to promote the information society. In this respect, some public administrations are actively proposing libre software as a viable alternative for citizens and companies in their area of influence. Finally, some legislative bodies have also considered law proposals which deal specifically with libre software.

Initiatives are happening at all levels: European Union institutions, national governments, regional administrations, and municipalities. However, the situation varies a lot from country to country and from region to region. All in all, our feeling is that we have reached a point in Europe where it is strange to find institutions that have not at least considered the use of libre software. In fact, several studies signal the public sector as one of the driving forces behind libre software in Europe for the coming years. In the rest of this section, we will take a look at some of these initiatives.

Actions by the European Commission

The European Commission (the institution in the European Union most similar to an "executive branch") has promoted several actions related to libre software. By 1999 an informal group of experts, the European Working Group on Libre Software,[17] was meeting in Brussels at the request of some officials of the DG-INFO (Directorate General on Information Society). The main interest of these meetings was to explore specific opportunities for Europe in the field of libre software, to provide the Commission with some input about its impact in the European IT sector. The most widely known output of the group was the report titled "Free Software/Open Source: Information Society Opportunities for Europe?" (*http://eu.conecta.it/paper.pdf*) which was presented in Brussels in March 2000. This was probably the first public activity of the Commission in relation to libre software and represented a kind of turning point which led to many other actions by European institutions. The report identified some signals that

---

17 Although the group is no longer active, *http://eu.conecta.it* hosts some information about it. There is also an open mailing list, freesw, that is still used for announcements and discussions related to libre software in Europe.

evolved later into trends. It also recommended both the consideration of libre software solutions in public administrations, and activities to inform the European industry about new possibilities. In addition, the report presented a complete landscape of the libre software world, from technical, legal, and economic points of view.

Since the days of that group, libre software has been present in several actions funded by the European Commission, which has had a policy of researching and publishing information about it, without promoting it explicitly. To describe just a few of those actions, we will concentrate on some initiatives promoted by the IDA (now IDABC) program and by the IST program (in the context of the fifth and sixth R&D Framework Programs).

IDA (Interchange of Data between Administrations) was a program of the European Union, started in 1999 and aimed at the funding, development, and coordination of pan-European services for public administrations. Since 2004, it has been continued by the IDABC program. IDA and IDABC have performed many activities related to libre software, which are referred to in the IDA Open Source Observatory (*http://europa.eu.int/idabc/en/chapter/452*). Among them, the following can be highlighted:

- "European Interoperability Framework for pan-European eGovernment Services" (*http://europa.eu.int/idabc/en/document/3761;* 2004). This reference document on interoperability was written after an extensive consultation process. IDABC considers it the highest-ranking module for the implementation of e-government in Europe. It includes several references to libre software, among which the following can be highlighted:

  > Open Source Software (OSS) tends to use and help define open standards and publicly available specifications. OSS products are, by their nature, publicly available specifications, and the availability of their source code promotes open, democratic debate around the specifications, making them both more robust and interoperable. As such, OSS corresponds to the objectives of this Framework and should be assessed and considered favourably alongside proprietary alternatives.

- "IDA OSS Migration Guidelines," (*http://europa.eu.int/idabc/en/document/2623#migration;* November 2003). One of the best guides for the migration to libre software. Specially targeted at public administrations, many of its analyses and recommendations are, however, valuable for any party considering moving from proprietary to libre solutions. It includes a detailed methodology for estimating the convenience of the migration, and for putting it into practice. It also provides complete descriptions of some typical scenarios, and configurations for the usual cases (email, desktop, server, etc.).

- "Pooling Open Source Software, Feasibility Study," (*http://europa.eu.int/idabc/en/document/2623#feasibility*) June 2002. A study on the opportunities for sharing libre software among public administrations, from technical, legal, functional, and financial points of view.

- "Study into the Use of OSS in the Public Sector," (*http://europa.eu.int/idabc/en/document/2623#study*) June 2001. One of the first reports on the use of libre software in public administrations. Includes some general information on libre software, and details on libre software solutions (about 100 examples showing specific systems that could be useful in the public context). It also analyzes the deployment of libre software in Europe at the time of the report, and presents some interesting conclusions.

- Organization of meetings and symposiums for sharing experiences on the use of libre software in public administrations in different countries. These have been helpful for coordinating actions and establishing links among the promoters of different initiatives.

- The IDA Open Source Observatory. This is worth mentioning, as it provides a good compilation of information about libre software, its situation in Europe, and many issues especially relevant to public administrations.

Within the IST (Information Society Technologies; *http://www.cordis.lu/ist*) research program, the Commission has funded (and is funding) several projects related to libre software. Many of them are aimed at the production of libre software in a given domain, such as AGNULA (*http://www.agnula.org*; libre software distributions specialized in audio and video). A detailed listing of those projects is available in the area devoted to libre software in the Information Society Thematic Portal (*http://europa.eu. int/information_society/activities/opensource/european_activities*). Some others are devoted to researching libre software as a matter of study, with the aim of improving general knowledge about it:

- FLOSS (*http://www.infonomics.nl/FLOSS*) was the first academic research of the libre software phenomenon as a whole, looking at it from many different points of view. It included studies on the developers themselves, based on a survey and on the analysis of author information in source code, focusing on sociological data about them. It was the first to provide some insight about why developers participate in libre software projects, what professional profile they have, what amount of time they devote to libre software, and where they come from. The study was also successful in the introduction of the name *FLOSS* (an acronym for "free, libre, open source software"), which has since been used in many other cases, especially within the research community.

- AMOS (*http://www.clip.dia.fi.upm.es/~amos/AMOS*) was a project to research the feasibility of building a system capable of categorizing and allowing searches among libre software package descriptions. This is especially useful for developers looking for code to reuse in their systems.

- COSPA (Consortium for Open Source in the Public Administration; *http://cospa-project.org*), started in 2003, aimed to analyze the effects of the introduction of libre software and open standards in European public administrations.

- FLOSS-POLS (Free/Libre/Open Source Software: Policy Support; *http://www.floss-pols.org*) is a project started in 2004 as a follow-up to FLOSS, and includes research tracks on government policy toward libre software, gender issues, and the efficiency of libre software development methods for collaborative problem solving. FLOSS-POLS will also deal with libre software in e-government, and will look for feedback from governments in relation to policies about libre software.

- CALIBRE is an action to coordinate the research on libre software in Europe, and to help transfer its results to industry. It started in mid-2004, and has already organized several conferences with a special focus on showing the results of libre software engineering, or on the use of libre software within European industry.

In the following years, it is expected that more and more projects related to the study of libre software will be approved in future calls, in what seems to be a growing interest by the European Commission research work programs to understand how libre software works from several points of view.

National Initiatives

At the national level, the situation is different from country to country. And even among those who have started some kind of action related to libre software, approaches are diverse. However, some common patterns can be identified. It is unusual to find a national government that has not issued studies and recommendations for the use of libre software in public administrations. There is also a certain consensus on some matters that have been proposed again and again, such as consideration of libre software for public acquisitions, adherence to open standards and interoperability, the need for inspection, and the importance of retaining proprietary rights on software. In any case, the following brief descriptions of the state of affairs in several countries should show both consensus and diversity of approaches:[18]

France

ADAE (Agence pour le Développement de l'Administration Électronique, Agency for the Development of the Electronic Administration; *http://www.adae.gouv.fr/index.php3*), formerly ATICA, maintains a good deal of information related to libre software in public administrations, and organizes activities related to that topic. It has also published several reports of special interest. Among them, it is worth mentioning the "Guide de choix et d'usage des licences de logiciels libres pour les administrations"[19] ("Guide to Choosing and Using Free Software Licenses for Government and Public Sector Entities), a complete guide to the legal implications of using libre software licenses, either for external software (obtained with or without

---

18 Some of this information was obtained from the European Information Society Thematic Portal, *http://europa.eu.int/information_society/activities/opensource/cases*, and from the Open Source Observatory, *http://europa.eu.int/idabc/en/document/1677/471*.

19 *http://www.adae.gouv.fr/upload/documents/free_software_guide.pdf*, linked from *http://www.adae.gouv.fr/article.php3?id_article=172*, which includes a translation into English.

cost) or for software produced by the administration itself (recommending the GPL in this case). There are also some recommendations for the promotion of libre software in the framework of e-administration programs, and cases of large-scale deployment, mainly of OpenOffice.org, in French public administrations.

Also in France, two of the first proposals of laws related to the use of libre software in public administrations were produced. In 1999, Laffitte, Trégouet, and Cabanel drafted in the French Senate the 2000-117 law project, aimed at enforcing the use of libre software in public administrations in those domains where technical solutions were already available (considering a whole set of exceptions and temporary measures to facilitate the transition period). It also considered the creation of a Libre Software Agency, funded by the government, which would help public administrations in the deployment of libre software technologies. In 2000, another law was proposed by Jean-Yves Le Déaut, Christian Paul, and Pierre Cohen. It was similar in objectives and rationale, but was not compulsory about the use of libre software in public administrations, but more focused on the availability of source code for applications and on the principle of "right to compatibility of software," which aims to guarantee the interoperability principle, common in European legislation. Although neither of these projects was approved, both have influenced later law initiatives in many other countries.

United Kingdom

There have been several studies and pilot experiences, which led in 2002 to the publication by the OGC[20] (the Office of Government Commerce) of a formal policy on the use of libre software, "Open Source Software: Use within UK Government" (*http://www.ogc.gov.uk/oss/OSS-policy.pdf*) which mandates not only the consideration of libre software in procurements, but also that decisions must be made considering "value for money" (a policy that has been widely copied in many other countries, especially in the developing world). At the same time, it establishes a policy of avoiding lock-in by proprietary software providers, supporting open standards and specifications, and exploring the use of libre software licenses for dissemination of research and development funded with public money. A document on implementing this policy was also published. "Guidance on Implementing OSS" (*http://www.ogc.gov.uk/embedded_object.asp?docid=2498*) provides details on how and when to consider libre software, and includes a detailed study of proprietary software lock-in practices and how to avoid them.

Some trials have also been conducted, with an interesting final report about them, "Open Source Software Trials in Government: Final Report" (*http://www. ogc.gov.uk/embedded_object.asp?docid=1002367*; published in late 2004), which states the viability of libre software solutions, with different perceived levels of maturity depending on the area of implementation.

---

20 *http://www.ogc.gov.uk/*. Information related to libre software is at *http://www.ogc.gov.uk/index. asp?docid=2190.*

In March 2005, the Open Source Academy was announced (*http://www.egovmonitor.com/node/319*). Funded by the UK government, supported by several municipalities and other institutions (including OpenForum Europe and Open Source Consortium), and with the help of the private sector, it is aimed at the promotion of the use of libre software in local government.

Germany

The German federal government produced in 2002 one of the first official documents dealing with the use of libre software in public administrations, "Open Source Software in der Bundesverwaltung"[21] ("Open Source Software in the Federal Administration"), by the KBSt (Coordination and Advisory Agency of the Federal Government for Information Technology). Later, in 2003, it produced the "Migration Guidec" (*http://www.kbst.bund.de/Anlage303777/pdf_datei.pdf*), one of the more complete documents about how to migrate to libre software, including detailed technical information about possible paths for migration in several domains. The German government has also funded some libre software developments, or improvements to existing systems, that were critical for its IT strategy. In 2005, it announced the Open Source Software Competence Centre,[22] a web site aimed at spreading best practices regarding the use of libre software in the public sector.

Italy

In February 2004, the Italian government issued rules regarding the use and acquisition of libre software in public administrations, "L'Open source nella pubblica amministrazione."[23] hey were formally made public after the release of an official report, "Indagine conoscitiva sul software a codice sorgente aperto nella Pubblica Amministrazione,"[24] which presents some rather interesting conclusions on the characteristics of the use of software by public administrations. The rules set the criteria to consider when acquiring software (which include interoperability, nondependence on a single provider or on proprietary technologies, availability of code for inspection, etc.), and specifically included libre software as a possible choice.

Finland

A working group promoted by the Finnish Ministry of Finance produced in 2003 the report "Recommendation on the Openness of the Code and Interfaces of State Information Systems,"[25] which (among other interesting recommendations) proposes the consideration of libre software for the custom developments funded by the public administration and for the acquisition of software.

---

21 *http://www.kbst.bund.de/dokumente/Publikation/,-300432/dok.htm*. A long summary in English is also available at *http://www.kbst.bund.de/Anlage302856/KBSt-Brief+-+English+Version.pdf*.
22 OSS-Kompetenzzentrum, *http://www.kbst.bund.de/oss-cc*.
23 *http://www.governo.it/GovernoInforma/Dossier/open_source/index.html*.
24 *http://www.governo.it/GovernoInforma/Dossier/open_source/open%20software%20PA.pdf*.
25 *http://www.vm.fi/tiedostot/pdf/en/65051.pdf*, referred in *http://www.vm.fi/vm/liston/page.lsp?r=65052&l=en&menu=2678*.

Denmark

The strategy of the Danish government with respect to libre software is exposed in the report "Danish Software Strategy,"[26] officially adopted in 2003. In summary, the approach is based on the principle of obtaining the maximum value for money, irrespective of the type of software (which is also the reason several detailed studies on total cost of ownership in libre and proprietary software scenarios are being carried out), but also not forgetting the importance of promoting competition, interoperability, and flexibility. Some preliminary studies performed prior to this report showed potential major savings could be made through the use of libre software.[27]

Sweden

In 2003, Statskontoret, the Swedish Agency for Public Management, published "Free and Open Source Software, a Feasibility Study,"[28] a complete and detailed study of libre software, including cases in Swedish public administrations, with very positive conclusions.

Spain

Many of the interesting developments related to libre software in Spanish public administrations have been achieved at the regional level. However, there are also some interesting actions by the national government. One of the most revealing is the inclusion of measures related to libre software in the document "Criterios de seguridad, normalización y conservación de las aplicaciones utilizadas para el ejercicio de potestades"[29] ("Criteria of Security, Standardization and Conservation for Applications Used in the Exercise of Authority"), edited by the Consejo Superior de Informática (Higher Advisory Board on Informatics), an interministry body of the Spanish administration. It details the issues to consider for all the applications used by the public administration and recommends specifically the use of libre software whenever technically feasible. It also recommends requiring the availability of source code for programs acquired by the administration, the use of open formats, and the use of libre software applications to access some kinds of data.

The Netherlands

The OSOSS program (http://www.ososs.nl) is aimed at encouraging the use of libre software and open standards in public administrations. OSOSS is a program of ICTU, the national organization for IT in the public sector, founded by the Ministry of the Interior. In the context of this program, libre software is promoted as a

---

26 http://www.oio.dk/files/Softwarestrategi_-_Engelsk.pdf (available as a part of the Offentlig Information Online, http://www.oio.dk/software).

27 http://www.tekno.dk/subpage.php3?article=969&survey=14&language=uk&front=1.

28 http://www.statskontoret.se/upload/Publikationer/2003/200308A.pdf (in English).

29 http://www.csi.map.es/csi/criterios/pdf/criterios.pdf. A summary available at http://www.csi.map.es/csi/pg5c10.htm.

full-fledged option. OSOSS is basically an informative advisory body, supporting policy makers in exploring the relationship between libre software and public administrations.

Norway

The issue of libre software in the public administration has been dealt with by the Norwegian Board of Technology (a public, independent think tank on technology) in its report, "Software Policy for the Future"[30] (November 2004). It recognizes the potential interest of libre software, and recommends a policy similar to that of Denmark: pilot programs and careful case-by-case studies.

Other Initiatives in the Public Sector

There are many other initiatives in the public sector. Among them, we have selected a short list which we have found especially meaningful as illustrative examples of the whole landscape:

- Extremadura (Spain) is a small, with a population of about a million, and relatively cash-poor region that has defined a strategy based on libre software to catch up on information society issues. The main principles of this strategy are connectivity and IT literacy for all citizens. One of the key projects for implementing it is gnuLinEx (*http://linex.org*), a GNU/Linux distribution originally targeted for primary and secondary education (deployed in tens of thousands of computers in all public schools), but which is now also used in the public administration and offered to SMEs and individuals. One of the latest initiatives announced by Extremadura (jointly with Lambdaux (*http://lambdaux.com*), a Spanish libre software company) is the CompatibleLinux (*http://compatiblelinux.org*) catalog, an analysis of the hardware available in the market with respect to its compatibility with GNU/Linux distributions. This initiative has also led to the AENOR[31] compatibility certificate, which can be specified by public administrations and companies seeking to purchase hardware for use with GNU/Linux distributions.

- The French police (Gendarmerie Nationale) started a plan in 2004 to switch to OpenOffice.org in all its desktop machines[32] (about 80,000 PCs). They expect to complete the switch by the end of 2005. They estimate savings at about EUR 2 million.

- The Kolab Groupware Project (*http://kolab.org*) was initiated in 2003 as a spinoff of the Kroupware contract (*http://kroupware.org*), which was funded by the German Bundesamt für Sicherheit in der Informationstechnik (BSI, Federal Agency

---

30 *http://www.teknologiradet.no/files/endelig_rapport_programvarepolitikk_0066_20041109.pdf* (full report, in Norwegian); *http://www.teknologiradet.no/files/english_summary_041223_copy.pdf* (executive summary, in English).

31 AENOR is the Spanish standardization organization; *http://www.aenor.es*.

32 *http://www.solutionslinux.fr/document_conferencier/420c7d6295f27.pdf*.

for IT Security) and won by a consortium of three companies: erfrakon, Inteva-tion, and Klarälvdalens Datakonsult. Kolab is today a libre software system that allows for the interaction among mixed groupware environments: KDE, Out-look, and web-based tools. This is one clear case of the promotion of a new libre software project by a public administration (in this case, because it was inter-ested in overcoming this missing functionality in the libre software world).

- The city of Munich (Germany) started in May 2003 a plan to migrate to GNU/ Linux and libre software (LiMux)[33] most of its desktop machines (some 14,000 PCs). This initiative, which started as a political one (including in the process a delay to get attention to the proposed European directive on software patents, dur-ing the summer of 2004), is backed by detailed studies and has had a lot of media attention. Despite this attention, the project had not been completed at the time of this writing. However, it seems to have started a trend followed by some other European cities (although there are also earlier cases, such as the city of Florence, which passed a law in 2001 mandating the use of libre software when feasible[34]).

- Rijkswaterstaat[35] (Directorate for Public Works and Water Management, the Netherlands) has been using the Geoservices system, heavily based on libre soft-ware, since 2003. Rijkswaterstaat has the responsibility of maintaining dikes, roads, bridges, and canals, and uses Geoservices for web-based access to geo-information obtained from many different sources.

- The Junta de Andalucia (regional government of Andalucia, the most populated Spanish region) has instigated two of the few laws related to libre software that are actually in force. The first one was the "Decree of measures to push the knowledge society,"[36] which deals with (among other issues) the use of libre software in education. It fosters the use of libre software in public schools (not mandating it exclusively) and mandates that computers purchased for that use be compatible with libre operating systems. The second law is an order approved on February 21, 2005,[37] which mandates the distribution as libre software of any program owned by the Junta de Andalucía. This order basically amounts to releasing a large quantity of code to the libre software community, doing the same for new programs built on behalf of the Junta. To the knowledge of the authors of this chapter, this approach is completely novel and marks the begin-ning of a new path in the promotion of libre software by public administrations.

---

33   *http://www.muenchen.de/Rathaus/referate/dir/limux/89256/* (note by the city of Munich) *http:// news.zdnet.co.uk/software/applications/0,39020384,39171380,00.htm* (note in ZDNet).

34   *http://www.softwarelibero.it/portale/legislazione/mozione_comune_firenze2.shtml.*

35   *http://www.rijkswaterstaat.nl*; report on the experience available at *http://europa.eu.int/idabc/en/ document/3934/470>.*

36   Decree 72/2003, March 18, BOJA of March 21; *http://andaluciajunta.es/SP/AJ/CDA/Secciones/ Boja/AJ-BojaPagina/2004/10/AJ-verPagina-2004-10/0,20748,bi%253D696836605883,00.html.*

37   Published in the BOJA of March 10, 2005; *http://www.andaluciajunta.es/SP/AJ/CDA/Secciones/ Boja/AJ-BojaPagina/2005/03/AJ-verPagina-2005-03/0,22557,bi%253D699234368885,00.html.*

- Bergen, the second largest city in Norway reported a strategy, already deployed in large part, of using Linux on servers[38] (including the servers of the network for schools). The experience seems to allow for cost cuts both in hardware and in software, and includes the wide use of libre software instead of proprietary solutions for many services.

- The city of Vienna (Austria) announced in early 2005 a plan offering its departments migration to OpenOffice and GNU/Linux on the desktop.[39] The migration plan is voluntary, linked to lower costs charged by the city's IT department, and is currently targeted for about 7,500 desktops. For this solution, a Debian-based distribution (Wienux) has been created.

Legal Issues

Legal issues are still largely undecided for libre software worldwide. However, some of these issues are specific to Europe. Among them, we have selected two cases: the European Union directives (affecting most of Europe) that have (or may have, depending on approval) a negative impact on libre software; and the concerns about the validity of libre software licenses within European jurisdictions.

EU Directives with Negative Impact

For sure, not all legislative initiatives in Europe in the field of software have a neutral impact on libre software. In some other cases, important laws have been passed (or are in the process of approval) that cause serious problems by producing an environment hostile to libre software. Two of the most relevant cases are:

Directive on software patents[40]

Although software patents may affect any kind of software, the libre software community is especially concerned about the problems it poses for the freedom of innovation. The directive on software patents (actually, "Directive About Patentability of Computer Implemented Inventions") was proposed by the European Commission in February 2002. If approved as such, it would mean the introduction of software patents in Europe very similar to those in the United States.

Early in the process, groups all over Europe started to explain the problems this directive would cause to European software developers (be they individuals or

---

38 http://www.linuxforum.dk/2005/program/slides/LinuxIBergen/Linux_i_Bergen_-_Tuftedal.ppt (slides, in Norwegian), http://europa.eu.int/idabc/en/document/3471/469 (note in IDABC), http://news.zdnet.co.uk/software/linuxunix/0,39020390,39173557,00.htm (note in ZDNet).

39 http://news.zdnet.co.uk/software/linuxunix/0,39020390,39185440,00.htm (note in ZDNet).

40 There are many web sites with information on the directive on software patents. Probably the most complete about software patents in general (including information about the directive itself), and the European campaigns against them, is the site of the FFII group, http://swpat.ffii.org.

companies) and users. In part thanks to the awareness caused by this campaign, led to a great extent by libre software activists, the European Parliament passed in September 2003 a set of amendments which, together, would amount to the invalidation of software patents in Europe. Meanwhile, the European Council of Ministers (representatives of EU national governments) has approved a text even more radical than the proposal of the Commission, allowing more clearly for software patents, in a rather strange meeting in March 2005 (*http://kwiki.ffii.org/Cons050307En*). Things are in quite a mess at the time of this writing. The Parliament asked the Commission to withdraw its proposal, what was refused, entering into the "second reading" stage, which will lead to a new vote in Parliament, probably during the summer of 2005. All in all, this directive proposal is having one of the more complex, strange, and time-consuming paths ever seen in Brussels. There is a strong perception in the libre software community in Europe that the introduction of software patents would be a strong barrier to the development of libre software. On the contrary, if Europe were to remain free of software patents, libre software development would benefit from much more legal certainty, a friendlier environment, and a more level playing field.

The European Union Copyright Directive (EUCD)[41]

This was approved in 2001. It is in many respects similar to the Digital Millenium Copyright Act (passed in the U.S.). It poses risks for libre software: the impossibility of distributing programs for handling certain file formats (for contents subject to the EUCD), and of interoperation with certain systems. This directive can make it illegal to produce libre software programs for handling DVDs, for instance. The libre software community is concerned about this problem, but is not mobilized to the same extent as in the case of software patents.

Although there are other legal initiatives in Europe hostile to libre software, these two are the more well-known ones. In particular, there is now discussion about DRM systems and the legislation surrounding them, which could develop into very dangerous laws making it impossible for a libre software system to handle content such as e-books, movies, and music. Only the future will say whether the legal environment that was, until the late 1990s, basically neutral to libre software will evolve into a fairly hostile one.

Libre Software Licenses in Europe

Most (in fact, almost all) libre software licenses were formed in the United States, in a jurisdiction alien to European countries. Therefore, for many years now, there have been concerns about the validity of those licenses in European jurisdictions. This has spawned many efforts in different directions: either to assess the validity of, or to trans-

---

41   There is information about how EUCD affects libre software in the FSF Europe web site, *http://www.fsfeurope.org/projects/eucd/eucd.en.html.*

late and localize such licenses, having versions valid in every European country. Until now, the former approach has had more impetus, as is shown, for example, in the aforementioned "Pooling Open Source Software" study, which includes a detailed review of the validity of the GPL and other libre software licenses, concluding that it is valid for practical purposes. The latter approach is tricky, since it could contribute to the fragmentation of the libre software world and to endless problems in cases of international collaboration (so common in libre software projects). That is why many people from the legal community are considering the proposal of international regulations, which would complement intellectual property treaties with the consideration of libre software, given clearer international support for these licenses.[42] However, two of the clearer cases of the validity of libre software licenses in Europe happened in Germany. On April 14, 2004, a German court granted a preliminary injunction to stop distribution (by a company called Sitecom; *http://www.sitecom.com*) of a router that included code (Net-filter/IPtables) licensed under the GPL, yet failed to comply with its provisions, because Sitecom did not distribute the source code. This preliminary injunction was confirmed on July 23, 2004,[43] along with a significant judgment, after which Sitecom started to provide the source code on its web site. The second case was also a preliminary injuction (*http://yro.slashdot.org/article.pl?sid=05/04/14/2024258*), also for the use of Net-filter/IPtables code in some firewall products distributed by Fortinet.(*http://www.fortinet.com*). Both cases have lent weight to the GPL worldwide, but particularly in the German jurisdiction (and in other European jurisdictions of similar tradition).

Libre Software in Education

One of the fields where libre software has entered with most impetus in Europe is education. This does not mean that libre software is mainstream in European educational institutions, but that there are several very clear examples that seem to have been successful and that are currently being considered in many other realms. For several reasons—the specific advantages of libre software in the education field, the importance of localization, the lack of suitable tools for many educational tasks, the funding problem so ubiquitous in education, and the readiness of large parts of the educational community to accept and embrace its assumptions and philosophy—this field seems to be especially receptive to libre software.

For illustrating this rich landscape of experiences, we have selected four examples that have come to our attention:

- SkoleLinux (*http://skolelinux.org*) is a successful case of a grass-roots effort to bring libre software to the education world. It was formed in 2001 as a project

---

42 One of such proposed regulations is the Free Software Act, *http://www.fsc.cc/node/view/69*.
43 *http://www.heise.de/newsticker/meldung/49377* and *http://yro.slashdot.org/article.pl?sid=04/07/23/1558219*; see also *http://www.oii.ox.ac.uk/resources/feedback/OIIFB_GPL2_20040903.pdf* for a translation into English of the Court decision.

for developing software systems for schools in Norway. It was originally aimed at the localization of a GNU/Linux distribution for that country (mainly by translating it into Norwegian written languages), to improve the installation and maintenance so that it would be suitable for the needs of schools (including distributed administration), and to promote the introduction of the product in Norwegian schools. In this respect, it has been successful, being used in many schools in Norway and other countries, with a healthy community of developers and users around it. The project has been funded by a loan from the SLX Debian Lab Foundation, which pays for three employees and has strong relationships with the Debian project.

- gnuLinEX (*http://gnulinex.org*) is promoted by Junta de Extremadura, the regional government of Extremadura, Spain. It is a part of a larger project (already mentioned in the section about public administrations). gnuLinEX is a Debian-based distribution, completely localized, which is currently deployed in the whole public education system of the region (about 66,000 computers in 2004, mainly in schools) and is now being considered for other kinds of environments. Teaching materials and specific applications for education are also being developed (usually under libre software or libre documentation licenses). A complete strategy encompassing training, support, development, and dissemination within the society of Extremadura is also being put into practice. The project started in 2002 and is evolving into a complete strategy for the promotion of the use of information technologies based on libre software. gnuLinEX was the first of a series of education-oriented libre software distributions that have been deployed in many other Spanish regions.

- AbulEdu (groupe Éducation de l'ABUL; *http://abuledu.org*) is a French project oriented toward the use of libre software in schools. Its best known product is a GNU/Linux distribution, completely in French, developed mainly by volunteers. It includes many educational software products, and is designed to be simple to maintain in the environment usually found in classrooms. It is currently in use in many schools all over France.

- SIGOSSEE (Special Interest Group on Open Source Software for Education in Europe; *http://ossite.org*), co-funded by the European Commission, has been established to investigate, inform, and advise about libre software in education. It is a kind of umbrella project providing a common space for many working groups, organizing many conferences, workshops, and seminars, and acting as a framework for relationships with other projects (such as JOIN, devoted more specifically to libre software learning management systems). This is a good case of a mixture of grass-roots and government-promoted efforts and has been successful in disseminating the advantages of libre software for educational organizations all around Europe.

Another interesting development related to education has been observed over the last two years: the appearance of studies specifically oriented toward explaining the libre software phenomenon, usually from many different angles, including technical, economic, legal, and sociological. We are not referring here to technical courses about software systems which happen to be libre, but to studies about libre software itself, which are usually aimed at developing an understanding of the complex interactions between technology, development processes, business models, licensing schemas, volunteer motivation, etc., which are inherent in libre software. Those would be needed, for example, to drive the libre software strategy of a company. At the time of this writing, we know about some master's-level programs which point in that direction: those delivered by Universidad Oberta de Catalunya[44] (Spain, started in 2003), and Universitá di Bologna[45] (Italy, started in 2004). More programs are due to start in 2005. And an informal group of universities, the MoLOS group (Master on Libre, Open Source Software), is designing a curriculum suitable for being taught as a master's study in the context of the new European Higher Education Space.

Research on Libre Software

An active research community is concerned with libre software in Europe. From sociologists and economists to software engineers, the interest in studying and understanding this phenomenon is on the increase.

One of the first projects specifically devoted to analyzing the libre software world was the aforementioned FLOSS Survey and Study (led by Rishab Ghosh, University of Maastrich, Netherlands, and finished in 2002). It opened several lines of research, from authorship of libre software code to motivations of libre software developers. Some other pioneering works were performed by Stefan Koch (*http://wwwai.wu-wien.ac.at/~koch/uni.html*) in Wirtschaftsuniversität Wien (Austria), who in 2000 was already studying the GNOME project from a quantitative point of view, and by the group to which the authors of this chapter belong, at Universidad Rey Juan Carlos (Spain), who were studying Debian at the same time.

Also in 2001, one of the first research workshops on libre software engineering took place, the Workshop on Open Source Software Engineering (*http://opensource.ucc.ie/icse2001*), organized by Joseph Feller and Brian Fitzgerald (both then at University College Cork, which hosts an active group on libre software engineering (*http://opensource.ucc.ie*); Brian is now at the University of Limerick, also in Ireland) and Andre van der Hoek, and continued every year since. It is interesting to note that in that many of the papers presented at the workshop were by European groups, even though it was held in Toronto.

---

44 *http://www.uoc.edu/masters/softwarelibre/esp/index.html.*
45 *http://www.unibo.it/Portale/Master/Master+Universitari/2004-2005/Tecnologia+del+Software.htm.*

Since those early day's many research groups have joined this field in Europe. Just to name a few of those researching libre software as a matter of study, we can mention (in no particular order):

- The FLOSS group at MERIT, University of Maastrich (*http://www.infonomics.nl/FLOSS*; the Netherlands; focus on the economics of libre software and the motivations of developers)

- The Software Engineering Group at Aristotle University of Thessaloniki (*http://sweng.csd.auth.gr*; Greece; strong emphasis on studying the development processes)

- The Open Source group at University College Cork (*http://opensource.ucc.ie*; Ireland, research on libre software processes)

- The Software Engineering team at the Department of CSIS, University of Limerick (*http://www.csis.ul.ie*; Ireland; focus on processes, organization, and coordination)

- The Libre Software Engineering group at the University Rey Juan Carlos (*http://libresoft.dat.escet.urjc.es*; Spain; focused on the quantitative and qualitative analysis of projects)

- The Distributed Software Engineering Group at University of Lincoln (*http://facs.lincoln.ac.uk/Research/Distributed*; the UK; focused on the relationship of libre software development and agile methods, and on its distributed component)

- The Software Engineering Group at Politecnico di Torino (Italy; focus on evolution and maintenance)

- The Center for Applied Software Engineering at Free University of Bolzano-Bozen (Italy; research on metrics applied to libre software and its relationship to agile methods)

- The team at the Institute of Computing Science at Poznan University of Technology (*http://www.cs.put.poznan.pl*; Poland; focus on data mining of publicly available information)

- The Open Source Group at the University of Szeged (*http://www.inf.u-szeged.hu/opensource*; Hungary; publishing on quality and complexity metrics)

- The Science and Technology Policy Research team at SPRU, University of Sussex (*http://www.sussex.ac.uk/spru/1-4-9.html*; the UK; focus on the economics of libre software development)

- The Open Source research team at Technical University of Berlin (*http://ig.cs.tu-berlin.de/forschung/OpenSource*; Germany; research on economics and politics of libre software)

There are, of course, many more research groups, and not finding one here implies nothing but my poor knowledge (please, forgive me if you are one of those not named). In particular, note that only groups, and not individuals, have been mentioned.

Some of these groups are partners in the CALIBRE (*http://calibre.ie*) coordinated action (already mentioned in the section about public administrations), funded by the European Commission and aimed at coordinating some of the research on libre software being performed in Europe and transmitting its results to industry.

Although it is difficult to tell, we think that European research on the libre software phenomenon is at a very high level, and when compared to similar efforts in other parts of the world (mainly in the United States), it may be more focused on understanding how libre software projects work (whereas in other cases, the understanding is more a side effect of analyzing software development in general).

The Future Is Hard to Read....

We have tried to show how libre software is flourishing all over Europe. Of course, there are many differences throughout an area where diversity is the rule, but also many coincidences. For now, Europe is an important pillar of the libre software world, and is maintaining an equal leadership of it. We have a large share of developers; there are companies producing, maintaining, using, and providing services; and our public administrations seem to be aware of the libre software phenomenon.

However, experiences are fragmented. Few companies are based on libre software in Europe as a whole, although an increasing number are working at the national and regional levels. The developer community, and the libre software community in general, is in fact a collection of loosely linked national or linguistic communities, with very little coordination among them. We do not have common news sites, and there are very few umbrella organizations, or even meeting events, recognized throughout Europe. The initiatives of the public administrations may be a bit more coordinated, but even those are wildly different from country to country. Maybe all this is just a consequence of the fragmentation of Europe—or maybe it is a first step toward a real European space of libre software. Whatever the reason, for now the real impact of European initiatives in the libre software world is far lower than the relative importance of libre software in Europe. In a few specific cases (such as the campaign against the directive on software patents, which is not carried only by libre software activists), we are starting to see coordinated movements that show the real strength of libre software in Europe.

In this context, we still have to wait to see whether Europe will capitalize on its current leadership in libre software penetration or, on the contrary, will lose this position in favor of other regions with a clearer and more active policies of promotion. The coming years will tell but, for now, we have the potential to be the first economic area to experiment with the benefits of large-scale deployment of libre software, creating a whole new industry around it, and promoting not only companies, but also the individual developers who are making this a real possibility.

In case libre software provides real advantages in terms of innovation, competence, and social benefits, Europe is well placed for advancing in that direction. Are those opportunities not worth exploring? Can we risk losing our advantageous position in what could be the next revolution in the information society?

CHAPTER 12

Alolita Sharma and
Robert Adkins

OSS in India

In modern times, India has accomplished miracles through the power of collaboration. Free and Open Source Software (FOSS) has the potential to accomplish yet another set of miracles in automating government and industry, and producing affordable education for all.

Three earlier revolutions using collaboration have dramatically improved the basic infrastructure within the country. The first revolution was called the *Green Revolution,* which started in the 1970s and took India from being a grain deficit to a grain surplus country. The second revolution, in the 1980s, was the *White Revolution,* which used the power of dairy cooperatives to enable large-scale milk production. Not only could India's own population be satisfied, India also became an exporter of dairy products. The third revolution, in the 1990s, was the *Gray Revolution,* which used India's plethora of English-speaking engineers and scientists to capture a significant share of the world's outsourcing business in software and pharmaceuticals.

Open Source Software (OSS) is poised to become the next revolution—perhaps named the *Gold Revolution.* OSS promises to build India's local infrastructure and create new wealth based on information services.

Business

OSS is a boon for the Indian export market. However, automation of any sort is only beginning in the domestic market. Hence the local market languishes in the adoption of all automation tools, whether proprietary or open.

Domestic Market

The localization and adaptation of computer-based solutions to move the local economy in India from pre-automation to automation continues to be a very slow march. Business processes remain predominantly manual. For example, it is reported that most doctors in India are practicing the same way they did 75 years ago—with pencils and pieces of paper. Ideally, OSS can promote the cost-effective adoption of automation, especially when legacy constraints are minimal. However, ground realities often discourage adoption of OSS.

Developing economies such as India's tend to foster low wages for services and support while permitting low prices for proprietary products because of piracy. This has led to a proliferation of proprietary technologies with affordable support structures and, at the same time, a resistance to OSS.

The resistance to OSS is driven by three main factors:

- The parity of "purchase price" when equalized by piracy
- The perception of a lack of maturity of OSS solutions
- A higher cost of support due to relative scarcity of Linux-trained labor

However, a recent trend is the emergence of local businesses that provide support for point solutions important to small to medium-size enterprises (SMEs) in India. For example, companies have sprouted up in metropolitan areas to support the migration of email from large-scale proprietary server environments to the equivalent OSS solutions. This is partly because companies which have significant server-based infrastructure have been recent targets of licensing enforcement campaigns by proprietary vendors and government enforcement agencies. The perception of increased risk in using unlicensed software has provided the impetus for OSS adoption.

In contrast, automation for the export economy in India, with its highly skilled, English-speaking workforce, is beginning to exploit some of the new business opportunities offered by open source, especially in providing services for migrations from Unix to Linux and from Microsoft platforms to Linux.

Outsourcing and OSS

Meeting the needs of the outsourcing market, low wages in India have fueled a substantial generic services economy with a global reach. Now, the outsourcing industry is beginning to take a serious look at using OSS for Information and Communication

Technology (ICT) solution development and implementation, for migration services, and for complex systems integration. As early as 1998, Dr. Ajay Shah, a consultant to the Indian Ministry of Finance, realized the importance of exploiting the inherent characteristics of OSS to build a services industry which could amplify the traditional Indian outsourcing services business. Today Indian companies like Tata Consultancy Services (TCS) have translated Dr. Shah's realization into sophisticated methods of services provision—for example, employing the state-of-the-art "ongoing cost reduction formulas" for client companies using the year-over-year economic advantages of OSS. The emerging market for OSS-based development and services has created high-value jobs in India for developers and, in addition, for business process analysts and service providers.

A principal portion of OSS outsourcing requirements centers around migration from older software platforms to Linux. For example, interest is rapidly increasing in retooling software from earlier, proprietary Unix platforms to Linux. Much of the conversion of traditional Unix applications and tools, like the earlier bonanza of Y2K work, is being done in India.

A second set of OSS outsourcing requirements involves building custom business applications using the new open source environments.

Companies taking advantage of these new outsourcing opportunities include HP India, Cognizant, Infosys, Wipro, Mindtree, IBM India, and many others.

Infosys, a top Indian IT company, is building a Linux migration practice as part of its multidimensional systems services and integration business strategy. Recent projects at Infosys illustrate both migration services and custom application development. For example, to meet the needs of a large petroleum industry client, Infosys ported applications for visualizing oil exploration data to Red Hat Linux from Solaris and IRIX. For another client, Infosys migrated a multinode high-availability application cluster from Solaris to Linux. In a project to help a leading peripheral manufacturer in Japan develop a new cost-effective product line, Infosys built Linux-based POS terminals using Java POS international standards.

Well-publicized projects at Wipro in the financial services and messaging services markets also illustrate the harnessing of OSS to drive cost-effective outsourcing services.

Government

OSS is providing the first steps toward an information society in India and thereby helping to close the digital divide. Examples include the Open Source Simputer project, CoIL-NET & TDIL localization projects, e-governance projects, and others.

Maharashtra and Kerala state land record systems have separately demonstrated the cost effectiveness of applying OSS, including database technologies, to what traditionally has been a slow and manual process.

Government-sponsored software technology parks of India (STPIs) are often used by companies to demonstrate their solutions to large government customers. One example is a 2003 MoU between IBM and the state of Karnataka to build an OSS center of excellence in the government-built Hubli software technology park. IBM has also set up a similar Linux Center of Competency in Bangalore, which provides development and testing services for Linux applications. Other OSS resource centers have been built or are being planned.

Two 2004 MoUs between IBM and the state of Uttaranchal have initiated a statewide university education program and an eGovernance program for an OSS-based framework addressing both legacy and new applications. These applications cover municipal services for record keeping, taxation, and social and health programs.

Also, IBM signed an MoU with the union territory of Chandigarh to set up an eGovernance Solutions Center for Linux for the local government. The center will help Chandigarh develop eGovernment applications using open standards and IBM's open source–based development framework.

Oracle's eGovernance Center of Excellence, set up in partnership with HP in the state of Haryana, helps government agencies develop better ICT policies and deploy improved ICT systems using OSS and Linux.

Challenges in Local Adoption of OSS

The trend toward OSS adoption faces a number of significant challenges.

Support

The perceived lack of support available for OSS in India is largely due to support services being readily available for legacy platforms. The legacy support industry has been built over many years. However, today there are a growing number of channels of support for OSS. Because of its collaborative nature, a great deal of high-quality yet inexpensive or free help for OSS is provided online. Furthermore, if a user has money, the same level of support is available for OSS as for proprietary solutions, and at the same prices. It should be noted that, while readily available, legacy support is often of poor quality. Furthermore, users or organizations that already have Unix skills find few difficulties in supporting OSS applications or systems. As more students trained in OSS enter the workforce, increasing support services options will emerge. In addition, as more OSS services revenues are derived by the software industry, the OSS services infrastructure will mature, and greater fulfillment of support requirements will be possible.

Piracy

The high seas are unfriendly to both OSS and proprietary products.

According to industry sources, more than 70% of proprietary software is pirated in India. Rampant piracy equalizes the price between "free" software and proprietary software. Since there is little legitimate market value for proprietary desktop packages, there is little financial incentive to develop a local software product market. IT growth is consequently stunted. OSS is seen, by some, as an antidote to these effects because it has the potential to transform the technology consumer of proprietary products into a technology collaborator of open solutions.

India's commitment to maintain compliance with World Trade Organization (WTO) and WIPO standards in the protection of intellectual property will encourage the proliferation of OSS packages. In particular, antipiracy drives and subscription licensing models are already improving the attractiveness of functionally equivalent OSS packages in India.

Localization

Proprietary as well as OSS vendors have committed considerable resources to localizing software in India. Microsoft has pledged millions of dollars to the localization of its proprietary software in Hindi, Marathi, and other major Indian languages. Red Hat has announced a plan to build a U.S.$250 million center to support localization and other software development. IBM has initiated various multimillion-dollar localization projects. Government agencies such as the Center for Development of Advanced Computing (C-DAC) and the Department of Information Technology (D-IT) also have active programs to develop localization solutions. And OSS community resources have initiated projects to localize various components of the OSS suite. Examples include the Indic-Computing Project, and IndLinux.

Localizing the killer applications of open source, such as OpenOffice.org, the KDE and GNOME suites, and the Mozilla browser and email applications, has increased adoption and usage by the non-English-speaking majority of Indians.

In fact, localization efforts are the main channel by which Indian contributions are being made to OSS projects. Desktop localization projects are the most active collaborative efforts in India. Projects include BharateeyaOO, IndLinux, JanaBhaaratii, and AnkurBangla.

The BharateeyaOO project represents the "Indianization" of OpenOffice.org. It is a cross-platform project for translating a rich office productivity suite into languages appropriate for non-English-speaking Indians. Availability of major computational tools in local languages is already helping to bridge the digital divide and spread computer usage and learning in the rural areas of India.

The IndLinux project has created a Linux distribution that supports major Indian languages, including Hindi, Punjabi, Oriya, Telegu, Bengali, Gujarati, Kannada, Malayalam, and Tamil. Like BharateeyaOO, this project tries to bridge the digital divide by bringing the benefits of computer and information technology to non-English-speaking Indians.

The AnkurBangla project has created a Bangla-language Linux distribution as well as Bangla support for some major applications such as office suites, databases, development tools, and desktop environments like GNOME and KDE. The project's objectives include developing and maintaining open source software targeted toward Bangla-speaking users.

JanaBhaaratii is an Indian government project run by the C-DAC and is funded by the D-IT. This project uses OSS to promote localized computing applications. The project will develop and deploy technology in Indian languages for a broad range of areas such as home use, mass applications, education, rural areas, info-kiosks, cyber-cafes, and e-governance.

Other projects include localized regional voter registration applications, such as the Voter List project, which uses a bootable CD distribution called GNUBhaaratii that is based on Morphix.

Culture

India's work culture embodies a complex mix of both rigid hierarchy and elastic opportunism.

Opportunism drives the consideration of special favors at all levels of economic activity in India. It is common to hear senior Indian administrators say that there is "no money" to be made in procuring OSS. Today, the Indian form of guanxi in the OSS world is a trickle at best. But some form of guanxi may always be needed to successfully conduct business in India. Guanxi "reciprocity" may ultimately be based on the growing wealth from the IT services economy. However, if some measure of transparency and containment of corruption is to be achieved, cost-effective and pervasive automation is key to reducing discretion in the conduct of government and in the application of governance. It's icing on the cake that automation can be achieved using legitimate, nonpirated OSS tools.

Organizational rigidity is the other face of India's work culture. India's tradition of social hierarchy in the workplace tends to reduce the value of collaboration. While innovation is a strong Indian value, collaboration is not. Traditional Indian business culture is strictly hierarchical. Collaboration with peers is less valued than performing a prescribed duty according to one's place in the organizational structure. This is slowly changing, as more relaxed and flexible Western business practices are adopted. Collaboration inhibitions are reflected in the lack of contributions to collaborative OSS projects.

Software Patents

Intellectual property concerns affect all software, whether proprietary or open source. Globally, the status of software patents is unclear, with a number of initiatives in various stages of contest. Software patents are allowed in both the U.S. and Japan. The European Union is examining its options. In India, unconstrained software patents are not yet allowed. However, recently, the government of India amended the Indian Patents Act to support patenting of embedded software to conform to WTO/WIPO agreements. There is strong pressure from industry bodies such as NASSCOM to extend IPR protection to all forms of software as a way of strengthening the Indian software industry. The counterview—that patents serve to inhibit innovation in software—is not widely recognized and, unfortunately, the FOSS community appears to be having minimal influence on keeping software free from patents. The new patent protections for embedded software came into effect January 1, 2005.

OSS in Education

Despite its IT prowess, India lags behind in contributing to large-scale OSS projects. It also lags behind in the use of OSS in the educational curriculum at all levels. Nonetheless, OSS is viewed by many in Indian industry and government as a key for improving the quality of education.

The enthusiasm for OSS in education is still in its infancy, since India's secondary school curriculum is today oriented toward proprietary products such as Microsoft Office, Windows applications, and development environments like Visual Basic and database applications like SQL Server and Oracle. In the curriculum, there is little support for generic computing concepts or platform-neutral software applications.

But there are exceptions, where FOSS advocates have worked with local school administrations to teach computing concepts using open tools and development environments.

More significantly, industry players such as Red Hat, Novell, IBM, and Intel have initiated open source resource centers and internship programs to grow the talent pool of open source engineering in India. Intel, in conjunction with the Department of Information Technology, has established an Open Source Resource Center to promote ICT education and curriculum development. IBM and the C-DAC have created an Open Source Software Resource Center (OSSRC) in Mumbai to foster OSS development, to increase understanding of OSS models, and to develop courseware which promotes OSS skills and builds a national OSS talent pool. Red Hat has launched a scholarship program with IIT Bombay to encourage OSS development skills. Novell has started an OSS internship program to boost student participation and contributions from India in OSS projects such as Mozilla, GNOME, and OpenOffice.

Conclusion

OSS is still early in its influence in India. While the outsourcing business of Indian IT industry is profiting from the new services and integration market provided by OSS, the domestic market is still immature. The domestic adoption of OSS is also handicapped by the ready availability of inexpensive, pirated, proprietary software products. However, there are signs of increasing use of OSS in government information processing and provision of services, as well as in educational curricula and in tools used to create educational content within many knowledge areas.

However, the long-term potential of OSS is recognized by some of the leadership of India—for example, by the current president of India, Dr. APJ Kalam. With continued advocacy, OSS can become a Gold Revolution that powers export as well as domestic industries across all economic segments and realizes the promise of a shared knowledge and collaboration-based information society.

CHAPTER 13

Boon-Lock Yeo, Louisa Liu,
and Sunil Saxena

When China Dances with OSS

One of the key challenges for China's Information and Communication Technology (ICT) industry is to ensure that China has the right software solutions to support usage models and value requirements. The advent of open source software (OSS), along with its business model with respect to intellectual property and value proposition, brings this business force to the fast-growing software industry in China. That, in turn, will provide an opportunity for the OSS community to promote and popularize this model throughout the People's Republic of China (PRC). This chapter analyzes how OSS has developed in China and where it is heading.

What OSS Was and Is in China

With the Gross Domestic Product (GDP) growing at an average rate of 9.8% year over year (YoY) from 1979 to 1997,[1] China has been pegged as a fast-growing economy. The high-tech industry had become the number-one pillar industry as of 2002,[2] and is expected to contribute 7% to the country's overall growth in 2005.[3] All of this is part of China's tenth Five-Year Plan, with the vision that "information drives industrialization" and ICT will continue to be a national focus onward.

---

1   *http://www.stats.gov.cn/tjfx/ztfx/xzgwsnxlfxbg/t20020605_21437.htm.*
2   *http://www.ccw.com.cn/htm/news1/dt/inland/02_9_16_10.asp* and *http://www.chinabyte.com/20020327/1603813.shtml.*
3   *http://www.c114.net/policy/policyread.asp?articleid=249.*

There are many growth opportunities within China's ICT, particularly in software and services, hardware, and telecommunications. Currently China is experiencing significant growth and success in hardware and telecommunications, notably with companies like Lenovo[4] and Huawei Technologies,[5] which are recognized as global industry players. The software industry, however, is slower to produce such success stories.

There are three key objectives to consider when examining the future of China's software industry:

Grow the local independent software vendors (ISVs) who can drive more PRC market success
> China currently has many small and medium-sized software companies, 90% of which employ fewer than 200 people. Today the revenue generated by the Chinese software companies competing in the global market is relatively small, with the most successful local vendors generating approximately $70 million in revenue, only 1% of that of their leading global software counterparts.

Grow infrastructure software to take advantage of the specific needs of the PRC market
> There is a perceived opportunity to develop specific infrastructure software for the Chinese market. Currently 85% of the local ISVs fall into the application software segment. China's software industry is largely reliant on multinational vendors, which are the predominant suppliers of infrastructure software such as operating systems.

Continue education on the importance of intellectual property (IP) rights to grow a healthy software industry
> It will be important to continue to educate Chinese enterprises that software and services are not complementary to purchased hardware. Piracy is perceived by many organizations as the norm, particularly for popular software and tools. Sharing of such IP not only negatively impacts the vendor's bottom line, but also hurts the local software industry, making it difficult for vendors to build capital with which to compete and innovate. Today more Chinese companies have begun to develop original content, which means they will think more about adequate safeguards for their IP. This is borne out by the increasing number of patents and trademarks that are being filed in the PRC.

Other factors worth considering to improve and grow the domestic software industry include:

Minimize piracy by encouraging fair-priced software and services
> China's World Trade Organization (WTO) entry means it will need to confront and contain software piracy. The government has adopted a multipronged approach targeting piracy to combat the issue of copyright infringement over time.

---

4   http://www.lenovogrp.com/cgi-bin/main.cgi?section=about&sub_section=chair_message
5   http://www.huawei.com/was/wps/portal/!ut/p/.cmd/cs/.ce/7_0_A/.s/7_0_1K4/_th/J_0_6A/_s.7_0_A/
    7_0_1JJ/_s.7_0_A/7_0_1K4

Enhance national security and decrease viruses and hacker attacks when running software

As China works to develop infrastructure software, it will continue to rely on what is available. To protect national security, China recognizes that it has to complement the core parts in the value chain of the software industry, as it sees the direct relationship between the software and hardware industries.

Grow a strong software industry in pursuit of IT industry growth

As software is where the profitability goes in the ICT industry, China needs to position itself to supply the innovation, new software, support, and usage models which the growing Chinese economy will need. By considering local needs of the various market sectors and using standards that would allow for international usage, Chinese software companies can build a home base of customers to support IT industry growth.

At this time, OSS has entered onto the PRC SW sage.

What OSS Means in China

Picture a generic software stack (Figure 13-1), with levels for the different types of software, starting from the firmware/basic input/output system (BIOS), all the way to the applications. OSS traditionally has focused on the operating system (OS) part and above; Linux has played a key role in the OS layer.

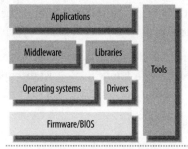

Figure 13-1. *A generic software stack*

The introduction of OSS in China can be attributed to a group of technical enthusiasts in the early 1990s. It is said that copies of Linux were brought back by Chinese visiting overseas from the University of Helsinki, UC Berkeley, and MIT, and quickly spread throughout universities and research institutions such as Tsinghua University and the University of Science and Technology of China.[6] At that time, computer science students and professors focused on becoming familiar with the system and localization.

---

6   http://www.linuxforum.net/forum/gshowflat.php?Cat=&Board=linuxtalk&Number=
     492466&page=0&view=collapsed&sb=5&o=all&fpart=all&vc=1.

In 1997, OSS was officially recognized by the government with the development of "Free Software Research and Application Development," a sub-branch of the China Software Industry Association, along with a wide range of OSS communities, such as a free software database—freesoft.cei.gov.cn—and some other bulletin board systems (BBSs), newsgroups, and Linux User Groups (LUGs).

In 1999, the first Chinese Linux company—Xteam—was founded and delivered to the industry the first commercial operation of Linux/OSS in China. Consequently a nascent market started on this initial engagement, which drew the attention of competition and keen interest from entrepreneurs.[7] Companies such as Red Flag, China Software Network Technology Co., Ltd., and BluePoint, as well as multinational ones such as TurboLinux have since begun activities in China.

Status of OSS in China

Since 1999, China has placed a stronger focus on OSS. There are many different components in OSS, from infrastructure software like Apache (web server), MySQL (database), and JBoss (application server), to tool and application software. In China, many of the efforts and activities have focused on Linux.

As seen in the market, the server side of Linux is relatively established in enterprise infrastructure, with a healthy growth rate. Increasingly Linux is being deployed for application servers and backend databases. This has been mainly seen in the financial services industry (FSI) and telecommunications. Vendors such as TurboLinux and Red Flag have taken an active part in these segments to grow their corporate revenues.

Since the Linux OS is the most well-used OSS in organizations, enterprises, and government bodies, a snapshot of China's Linux market is an appropriate way to study the market potential. Table 13-1 shows the key players in the market in China.

Table 13-1. *Linux distributors in China*

Type	Company	Background/description	Commonality
Local	Red Flag	Market leader in Linux client operating environment (COE) and second in server operating environment (SOE).[a] Co-founded by Software Research Institute of the Chinese Academy of Sciences and NewMargin Venture Capital in 2000. Got investment from CCID Capital incorporated under the Ministry of Information Industry in 2001. Offers a complete portfolio of Linux products for desktop, server, and embedded systems.	Strong government background. Fewer than 200 staff. Revenue mainly from government IT purchasing.

---

7   *http://www.linuxforum.net/forum/gshowflat.php?Cat=&Board=linuxtalk&Number=492466&page=0&view=collapsed&sb=5&o=all&fpart=all&vc=1.*

Table 13-1. *Linux distributors in China (continued)*

Type	Company	Background/description	Commonality
	China Standard Software Co., Ltd. (CS2C)	Ranks as third Linux distributor in COE and SOE markets, respectively.[b] Spun off from China National Software and Service Co., Ltd. (CS&S) in 2003. Focus on desktop, server, and office products.	
	Co-Create	Joint entity formed by tens of local IT companies in 2001. Received investment from CapInfo in 2003. Targets Linux desktop and office suite.	
	Xteam	First Linux distributor in China. Listed in Hong Kong Stock Exchange two years after the company was established. Transitioned to Linux server from its initial focus on Linux desktop. Focused on solutions for government and education after 2004 investment by Beijing Enterprises Holdings Ltd., which had a strong government background.	Government background.
	BluePoint	Transitioned from a pioneer Linux distributor to an embedded firewall provider.	No government background.
Foreign	TurboLinux China (TLC)	Existed as an American company, then acquired by a Japanese company. In 2004, local company Hinge Software became the No. 1 shareholder of TLC,[c] which made it a local entity. Strong in server side, especially in the telecommunications industry.	Presence in China.
	Novell SuSE	After the acquisition of SuSE Linux, Novell, with over a decade of operations experience in China, started its new Linux business locally. Focus on training with local institutions.[d]	
	Red Hat	Leading distributor globally. Commenced operation in China in November 2004. Plans to invest $1 billion for development in China.[e]	
	MandrakeSoft	Ranks as third distributor in the world.[f]	Plans to enter China.

[a] Nielse Jiang, "China Linux 2005–2009 Forecast and Analysis." IDC, Feb. 2005.
[b] Nielse Jiang, "China Linux 2005–2009 Forecast and Analysis." IDC, Feb. 2005.
[c] *http://www.smartpartner.com.cn/sp1/index/article.php?storyid=8334.*
[d] *http://tech.ccidnet.com/pub/article/c308_a184083_p1.html.*
[e] *http://www.ciweekly.com/ciweekly/inforcenter/A20041123364459.html.*
[f] *http://www.chinabyte.com/homepage/219015092686028800/20050110/1898897.shtml.*

In China, Linux distributors are broken into two camps: local and foreign.

In the local camp, there are five major distributors. The first three distributors, composed of the tier-one groups, make up more than 60% of the market share among local players. Although they have different characteristics and strengths, all of them possess certain commonalities—i.e., they have government backing, fewer than 200 employees, and revenue coming mainly from government IT purchasing.

As to the rest, the main difference is that they do not have government backing. That may change the fate of these companies. Some have transitioned to new business, such as BluePoint. Some have turned to getting government support, thanks to capital infusions, as seen in Xteam. Still, there are some distributors who continue to fight for a ticket into the tier-one group.

Although there are many indigenous distributors in China, the market potential has attracted the attention of foreign distributors also. This group of foreign distributors in China includes almost all the leading global distributors: TurboLinux, Novell/SuSE, and Red Hat. Some other distributors, such as MandrakeSoft, are also expressing interest in making inroads into China.

OSS Business Models in China

Many people may wonder how companies make money based on open source products. There are different approaches, but one that has a proven track record is related to the service model—i.e., making money mainly in services while selling products for a reasonable price premium.

Several companies have positioned themselves toward the services business model. They may compete with each other by providing additional value-add or niche products. Open source distributors like Red Hat, SuSE, and Red Flag are clear leaders of open source Linux and have positioned themselves as services companies. These companies tend to get service-level agreements for support for the Linux distributions they supply to their customers. They have made their distribution as rich as possible, validate them on many platforms themselves, and get the help of many platforms through their OEM partners.

The other successful business model is proprietary applications above the open source products that run on top of open source distributions and link only dynamically with user-level libraries. Oracle server products, and Office products from Kingsoft, are clear examples of products that deliver value above the open source stack.

Other examples include set-top boxes like TiVo,[8] and Linux-based cell phones. These products use Linux as an embedded operating system and provide dedicated proprietary services above the stack. These vendors can easily charge for their

---

8   http://www.tivo.com.

individual products, as they provide very visible value to end users. These companies clearly use the Linux operating system as the base for their solution stack, leveraging open source to bring cheaper systems into the marketplace.

In China, both business models are common and advantageous. Red Flag, known as a services company, distinguishes itself from its competitors by focusing on providing value-add via localization for the PRC market. Other additional value that it provides is management and security solutions not available from other local vendors.

Red Hat, on the other hand, distinguishes itself with a large volume of ISV support and validated stacks from thousands of ISVs. This value-add may fetch additional customers, however the core of its business remains as a services company. Another key factor in favor of Chinese companies is the strong manufacturing base of consumer electronics and cell phones—leading to opportunities for value-added software on top of the OSS infrastructure.

SWOT Analysis of OSS in China

We have come up with the strengths, weaknesses, opportunities, and threats (SWOT) and have done comparisons among China and other geographies with regard to OSS adoption. In Table 13-2, we summarized the SWOT analysis of OSS, to further separate factors common with other geographies from those unique to China. In the sections that follow, we examine strengths and opportunities for OSS in China.

Table 13-2. *SWOT analysis of OSS*

Category	Shared with other geographies	Unique to China
Strengths	Free source code. Strong multinational company (MNC) support.	Strong government support.
Weakness	Shortage of applications.	Lack of localized applications. Lack of Linux developer talent. Lack of understanding and participation in Commons. Young software industry. Entrepreneurship skill sets are at an early stage. Shortage of successful OSS businesses.
Opportunities	Opportunities to develop value-added software on top of OSS.	Opportunities in China beyond desktop/server, such as embedded, cell phones, set-top boxes, and telecommunications, all of which are China's strengths.
Threats	Competition with incumbent software and infrastructure.	Software IP [Editor's Note: I would argue this is NOT unique to China.].

Strengths

The future of OSS in China is bright because of the government's strong support, as the traits of OSS match well with what China is in pursuit of:

Availability of free source code
> This helps develop China's own software products with customized needs and requirements.

Availability of infrastructure software
> Among the OSS, there is a suite of infrastructure software, such as Linux as OS, Apache as web server, MySQL as database, and JBoss as application servers. These complement what China currently lacks.

The strategic value proposition of China's developing OSS capabilities includes:

Financial perspective—cost
> In 2003, government IT spending totaled U.S.$2.8 billion, or RMB 23.1 billion. Government IT spending is expected to maintain a fast growth rate of around 18% through 2008 in anticipation of the 2008 Beijing Olympics and Expo 2010 Shanghai. By then, total government IT spending is projected to reach U.S.$5.7 billion, or RMB 47.2 billion. The 2003–2007 CAGR will reach 15.3%.[9]

The packaged software market was calculated at U.S.$264.3 million or RMB 3.5 billion, in 2003 and represented 9.5% of overall government IT spending. Along with the increasing standardization of government application software and government emphasis on software, especially security and storage software, the packaged software market will maintain a fairly high CAGR of 21.7% in the next five years and will eventually reach U.S.$706.3 million or RMB 5.9 billion in 2007.[10] That spending may help jump-start the local software industry to be profitable and create a revenue stream for OSS.

Security consideration
> By allowing access to the source code, OSS-based applications allow organizations to help ensure that the software they use can protect against viruses and hackers, which are becoming more and more of a concern to a well-functioning IT capability. While cost is important, security can be more significant. Concerning information security, it is imperative to get transparency from the solution and software products, which is hard to realize from proprietary products. OSS shows all the source code, which greatly relieves government concern.

---

9   Enid Du, "China Government Industry Solution 2004–2008 Forecast and Analysis." IDC, Nov. 2004.

10 Enid Du, "China Government Industry Solution 2004–2008 Forecast and Analysis." IDC, Nov. 2004.

Growth opportunities to create a balanced software industry

   The infrastructure OSS applications are prominently featured in parts of China's software industry.

Developing a competitive edge in the software industry

   The long-term goal of Chinese government of boosting the domestic software industry is to realize the transition of China from IT consumer to IT provider. It wants to play a leading role in the Linux/OSS community, in the hope of being recognized as one of the global standard makers and enablers.

Policies such as File 18[11] and File 41[12] have been published to support growth of the local software industry. It is suggested that for IT spending, government agencies consider support to domestic products/services.[13] If there are no appropriate local products/services, foreign ones can be considered. As such, certain governmental IT purchasing projects have favored indigenous vendors.

For example, the Beijing Municipal Government's IT purchasing by the end of 2001 had approved all six local vendors plus Microsoft, the only foreign vendor in the bidding.[14] The local Linux vendors and ISVs that provide products running on the Linux platform therefore have been given chances to make inroads into the public sector. That kicked off the first round of government purchasing of local software vendors.

The execution will be run from the top down—from central government to tier-one government agencies in places like Beijing, Shanghai, and other provinces—to tier-two and tier-three ones.

Apart from the efforts of the government and Linux distributors, multinational vendors (MNVs) are another significant force to foster growth. The MNVs include Dell, HP, IBM, Intel, Sun, and SAP, to name just a few, many of which are the founding members of OSS communities, such as Open Source Development Labs (OSDL). Some of the endeavors that these MNVs have made include:

Dell

   Back in 1999, Dell started to sell Linux-based servers. It also invested in Red Hat.[15]

HP

   For details, please see *http://www.hp.com.cn/services/education/edm/itm/0409/17.asp*.

---

11   *http://www.istis.sh.cn/zlxx/zcdh/list.asp?id=1421*
12   *http://it.anhuinews.com/system/2005/01/07/001098472.shtml.*
13   *http://www.chinabyte.com/homepage/219001834121986048/20031115/1745026.shtml and http:// tech.tom.com/1121/1793/200485-115570.html.*
14   *http://tech.sina.com.cn/s/n/2001-12-31/97972.shtml.*
15   *http://www.blogchina.com/new/source/source.asp?bid=148.*

IBM

Since 2000, when IBM invested $1 billion in advertising and R&D on Linux, it has fully supported Linux with all the offerings.[16] Support has been conducted from the Linux server side to the desktop. February 2005 has seen IBM make further commitments to Linux. It plans to spend another $100 million to develop a Linux version of Lotus Workplace to deploy on a Linux client.[17]

Intel

Intel advocates the idea of "platform of choice"—i.e., meeting the natural demand of the market and customers. Through its worldwide programs, Intel has brought together multiple players with an interest in OSS, including government agencies, institutions, and vendors, to collaborate on key projects as beacon/reference sites, nurture the native demand for cost-effective and high-performance localized solutions, and drive toward a broader customer base in a variety of industries with China's self-sustaining OSS ecosystem.

Oracle

Linux is one of the key business focuses at Oracle. It has worked with Red Flag as well as Asianux.[18]

SAP

In addition to supporting Linux on the server side, SAP Labs China has worked with Red Flag since its establishment.[19]

Sun

Sun, as the innovator of Java™ and OpenOffice.org, has entered into a licensing and co-development deal with CS2C for its Java Desktop System. By releasing its StarOffice codebase into OSS together with specific guidance on localization to support emerging markets in 2000, Sun essentially provisioned the desktop Linux market. Such movements have helped the local distributors to directly gain the technological strength and methodologies, and bridge the gap between Chinese companies and foreign ones.

These efforts have helped contribute to the rapid development of OSS on computer systems in China, with more boxes sold with more services and hardware (mainly servers); and with more PCs for market segments such as education. But to China's software industry, their involvement escalates a healthy growth circle.

Last but not least, the impact of government's strong push has a ripple effect throughout the whole economy, at different levels.

---

16 *http://www-900.ibm.com/cn/servers/eserver/linux/news/summarize.shtml.*
17 *http://www.enet.com.cn/enews/inforcenter/A20050218391054.html.*
18 *http://www.linuxfans.org/nuke/modules.php?name=News&file=print&sid=2160.*
19 *http://www.sapchina.com/china/company/press/press.asp?pressID=2948.*

Momentum has built around the ecosystem of Linux/OSS. As previously stated, the Chinese government has taken the lead in adopting open source products in IT systems. This pattern continues and gradually carries over into hardware, business applications, and the IT services sectors. Some of the major Chinese hardware manufacturers, such as Lenovo, have preinstalled Linux in desktop PCs and ISVs, such as Kingsoft, have provided applications that can run on the Linux platform.

Educational institutions also appear to be rapidly adopting Linux, as do financial services and telecom companies. Communications and process manufacturing verticals will likely be the most aggressive adopters, with plans to deploy Linux in two to five years. As for SMEs, they have also shown signs of moving toward Linux, mainly due to consideration of total cost of ownership. Some mission-critical applications have been seen in these verticals.

Another recent trend of OSS effort in China is that of standardization. First is Asianux, initiated by Red Flag and Japanese Linux vendor Miracle Linux, and created to standardize Linux distribution in Asia. Asianux involved validation of major ISV applications for compatibility. Recently, Korea also joined in the Asianux effort.[20] At the same time, a China-level Linux standardization effort was initiated in 2004 by China Electronics Standardization Institute (CESI). This standardization effort was later aligned with the Linux Standard Base (LSB) effort of the Free Standards Group (FSG).[21]

All these efforts and contributions have greatly strengthened the local OSS community's development, as well as helped government to roll out its Linux/OSS initiatives.

Opportunities

Currently, many OSS efforts and activities are centered on the Linux OS and Linux applications. Much of the effort is based on driving compatibility with other alternative OS or proprietary software applications. Compatibilities come in various forms, ranging from file formats, GUIs, user experience, plug-in availability, and so on.

Rampant piracy in the China market means that OSS software is expensive compared to pirated incumbent software, which is effectively free and thus cheaper and better, as OSS is trying to catch up. To convince end users to switch to something inferior and more expensive can be a very difficult value proposition.

While such attempts will continue from current players leveraging OSS to their advantages and strong government support behind the scenes, we believe success can also come in a different form—that of treating OSS as a form of disruptive technology and leveraging China's current strength to build successes in the software industry.

---

20 *http://www.cnii.com.cn/20040423/ca253202.htm.*
21 *http://www.csip.cn/new/soft/2004/0916/1010.htm.*

The disruption cannot be based on chasing after the incumbents in a well-established market. Only when OSS-based software offers unique end-user values unavailable from alternatives will end-user adoption follow. This conclusion follows from the theories laid out in *The Innovator's Dilemma* (Christensen). There are at least two implications:

- Impact in niche markets without strong incumbent players. Examples include the cell phone handset market and the set-top box market. Both are Chinese strengths.

- Impact in existing markets via delivering unique and new end-user values not currently served by incumbent software solutions.

Let's look at these two sets of opportunities.

The market for embedded software outside the conventional desktop or server opportunities

China is the world's largest consumer of cell phones and an emerging force in cell phone handset design and manufacture. With the world transitioning to next-generation cellular and wireless technologies, there are tremendous opportunities for Chinese cell phone manufacturers to use OSS—whether operating systems or applications software or software tools—to build a credible and sustainable cell phone handset business.

Similarly, China is in the forefront of producing consumer electronics devices, ranging from DVD players to next-generation digital TVs to set-top boxes. As the world of entertainment goes digital, such consumer electronics devices will demand more software, from operating systems to media codecs to GUI software to supporting new usages such as Personal Video Recording (PVR) capabilities.

The availability of OSS software potentially allows these consumer electronics manufacturers to more easily create the entire software stack and to build the entire system with greater ease. The Chinese embedded market players, which are good at producing high-volume systems at competitive price points, may view OSS as the foundation on which to add innovative usages and to deliver new business models for this age of digital communication and entertainment. The creators and suppliers of such software stacks may be ISVs, rather than the system manufacturers themselves.

Delivering innovations and unique end-user values on top of available OSS—values not currently served by the presently available software.

For example, the difficulty entering Chinese characters using alphanumeric keyboards means opportunities to create value-added pen-based or other, better input methods for Chinese characters. This type of end-user value can be delivered on top of OSS and can help to offset some of the current shortcomings of the current OSS offerings.

Another example specific to China is the strong focus on education by Chinese families and society at large. The opportunities here would be in the form of better learning software for schools and homes and easier-to-manage e-classrooms, the values of which would be delivered and built on top of already available OSS software stacks.

The preceding two cases present the potential opportunities for OSS as a disruptive technology when combined with China's strengths and core competencies.

Another opportunity that is also unique to China lies in the intersection of the growing Original Design Manufacturing (ODM) base in China and Chinese Taiwan, and the recently announced open source implementation of a next-generation BIOS technology code-named Tiano,[22] formally known as Intel Platform Innovation Framework for Extensible Firmware Interface (EFI).

The open sourced Foundation of Tiano, combined with the standardizations of the firmware interface via EFI, opened up opportunities for ODM and its partners to deliver more platform-level values to end users, in particular in areas of manageability, security, serviceability, and administrative interface, which are otherwise more difficult to implement in the old BIOS environment. This also represents the first example of serious OSS effort in the firmware/BIOS layer in the software stack shown in Figure 13-1.

Where OSS Is Going for China and Beyond

Although the road ahead is full of challenges, the government and end users are helping to drive OSS adoption and enhancement in China to be a significant component of the software industry. More importantly, China recognizes the value of helping drive OSS to build a healthy and sustainable software industry. It is our belief that what is needed is to prove OSS can be used to create several successful businesses in China, and to do so soon.

The SWOT analysis of OSS in China presented in this chapter pointed to various scenarios, mostly happening in parallel. First, strong government support will be a key force for continued experiments and deployment of OSS in government-related usages. There will be many attempts to replicate the success of OSS in server environments and in client environments. The challenge of trying to replace clients served by strong incumbents with OSS alternatives will remain—innovations to deliver unique end-user values are critical here.

Software piracy will likely continue, but the effort to contain it will strengthen; we expect the results and impact to be felt over time. China's strong manufacturing base in computing, communication, and entertainment systems as the world becomes more digital further present unparalleled opportunities for China to leverage OSS as a strong base for innovations and to bring more value to end users.

---

22 *http://www.tianocore.org.*

With all this happening, the Linux/OSS talent pool will continue to grow, and many more developers will make attempts to leverage OSS to create meaningful business models.

One thing is certain. We expect many changes in the Chinese OSS landscape in the years ahead. China's policy choices with respect to OSS will have a profound impact on how China educates its talent pools, how it grows its software industry, and how it competes in the global arena. A SWOT analysis at the turn of the next decade—i.e., 2010—will most certainly have a very different outcome.

Bruno Souza

CHAPTER 14

How Much Freedom
Do You Want?

"Free as in freedom" has been used many times to express the objectives of Free and Open Source Software (FOSS). Although the Free Software Foundation (FSF) has created a very precise definition of what "free" is with respect to software, freedom itself is one of those difficult things to define and agree on, especially since freedom always assumes some form of compromise. "Your Freedom cannot be so broad as to negate someone else's Freedom" is a common saying. And when we talk about freedom in software, it is normal to have different views on what it is and how it is achieved.

The notion of free software was born in the United States, from inside the software development community. Long before the birth of free software, collaboration among developers from different companies and universities was the norm. Once software started to be seen as a company asset, barriers were built to protect these assets, and it became increasingly harder for developers to collaborate on and share code. Better ways to facilitate and guarantee the necessary collaboration were needed. Free software, and later open source software, allowed this to happen: across company boundaries, via the Internet,and between people that didn't even know each other.

Most of the software in the world today is (or was) developed in the United States, but software development itself is a borderless activity. It happens everywhere. And although many have tried to apply software development in a repetitive "factory-like" format, it fortunately is still largely a creative activity that favors the best and insightful developers no matter where they are.

The Internet has put these developers in touch, connecting people and cultures without the same background as the FOSS movement's original history. Also, other ways of limiting freedom are affecting the ability of governments to reach their sovereignty and users and companies to decide their own technology future. Now, developers that believe in freedom in every country are trying to adapt and expand the ideas of the FOSS movement, to map to their needs and realities, and even to apply them in their legal systems.

The diversity of collaboration that resulted from FOSS is one of the greatest achievements of developers in the last several years. Discussing how the ideas and philosophy are being applied by diverse governments, companies, and cultures will strengthen freedom for everybody.

Livre Versus Gratis

It is interesting to look at the terms that were used for freedom—*free* and *open*—both highly overloaded. Many corporations and developers are still confused because "free software" is not necessarily without cost, and "open source software" is not necessarily related to open standards or the simple ability to inspect the source code.

Developers in Brazil had an easier time coming up with a term. *Livre* is the Portuguese word for free as in freedom, and *gratis* (from the Latin *gratis*) is the word for "no cost." No one doubts that "software livre" is a way to achieve freedom, not savings. Maybe because livre is so obviously a good thing, the idea of software freedom spread throughout the country, and successfully reached companies, developers, and especially government policy makers. On the other hand, Portuguese for open source—*código aberto*—is much harder to grasp, because it can have different connotations. This is reflected in the fact that the vast majority of the Brazilian FOSS movement uses the expressions *software livre* and *free software*, and much less frequently, *open source*, although most people do consider these interchangeable.

But why discuss terms now? Wasn't this debated over and over when the term *open source* was coined back in 1998? Haven't we had enough time already to sort them out and be clear on what we mean? Maybe we have, but that discussion took place in the developer community, where the terms were defined. Now, we're not talking only to developers, we're discussing with governments. And companies. And judges, lawyers, and politicians....But more important than that, we are now discussing more than free software, we are really discussing livre, liberdade, freedom.

Background for Freedom: The Market

Like many countries, Brazil has a tradition of aiming for independence and sovereignty in strategic markets. This brought excellent results in many areas, from oil extraction (today Brazil produces almost 100% of the oil it consumes), pharmaceuticals, energy, and also technology. Software development is seen as very strategic; in fact, the country's domestic development market is larger than India's export market. Unfavorable currency conversion means that software from outside companies, priced in dollars or euros, is usually much more expensive than similar solutions developed in-house or acquired from local companies. Also, because of a unique, complex tax system, much of the existing commercial software is not suitable to run in the country. Add this complexity to the low buying capacity of small and medium-size companies', and you see why Brazil is not favored by international companies as one of the targets for software customization or translation. For commercial software, it is common to feature Portugal Portuguese translations over Brazilian Portuguese translations, even though Portugal has only about 5% of Brazil's population.

The Brazilian government develops large, countrywide solutions, mostly tailored to the needs of the poor, large, and dispersed population. The sheer size of those systems is enough to make the government the single largest buyer of software and the focus and target of most software companies, especially infrastructure software vendors. Moreover, extremely high inflation existed for many years, having been controlled only in the last decade—but not before it pushed the banking industry to invest heavily in technology, putting Brazilian banks among the most technologically advanced in the world. Because of specific legislation and finance standards, banks are also some of the largest investors in software development.

For historical reasons, Microsoft and IBM hold strong positions in this market, both in government and in private companies. Oracle is by far the largest database vendor. Custom software developed to run on those companies' platforms is a large legacy, which is used to justify most large expenditures in software; the result is an effective lock-in that is extremely hard to break. With the large amount of custom development, what mainly prevents vendor choice in Brazil are the government custom-developed, single-vendor solutions. Especially in the 1980s and early 1990s, when the country had rigid policies disallowing hardware imports, there was huge promotion of software product adoption, mainly in education. These included incentives—either explicit or unofficially hinted at—to use unlicensed copies, and donation of software to universities and schools at all levels. This created a large legacy of knowledge. Many use only Microsfot Windows and Office and don't consider any other options. The comparison of this strategy to that of drug dealers, who offer the first dose for free, was what recently made Microsoft sue a leading software livre activist.

Many older PCs still automate supermarkets, bank branches, and public offices and agencies. Many of those are doing their jobs perfectly running older versions of Windows, but Microsoft is pushing for software upgrades and more licensing. The lack of options has forced many companies to spend their budget on licenses for new OS versions, and on new computers able to run the new OSes—all just to be able to run the same preexisting applications.

Developing the Software Livre Movement

In this market, dominated by preexisting, international companies, the Brazilian-organized software livre movement started in the south of the country. Although there existed some prior activity, the initial work of Mário Teza and Ronaldo Lages joined users and system administrators interested in the GNU/Linux OS and started to organize the community. At that time, a discussion about the use of GNU/Linux was the main objective, but when this initial group was ready to launch its first event, it decided that it should be more than simply a GNU/Linux event and created the first FISL—Fórum Internacional software livre, the International Free Software Forum—effectively launching the software livre discussion in the country.

The moment was a turbulent one. On one side, the Internet shed a new light on the notion and importance of standards and the real possibility of avoiding lock-in. The desktop dominance of Microsoft was being heavily questioned—first because of the indisputable reality of security problems with Microsoft products, but also because of the possibility of multiplatform development raised by new technologies and products such as Apache, HTTP, CGI, Java, and HTML. In the middle of this, decisions in the United States on Internet governance and security technologies made clear, not only to Brazil but also to the whole world, how much the IT industry was relying on U.S.-controlled technologies. Microsoft being declared a monopoly and accused of power abuse, raised questions about how much trust Brazil (or any other country) should put in a single company, especially for such a strategic market.

Microsoft launched an astonishing campaign against software copying, or "piracy," that backfired by giving a clear demonstration of how important choice is. After many years of promoting the copying and usage of its software inside universities and companies, by sponsoring local "software protection" organizations, Microsoft initiated a fear campaign that included strong TV commercials comparing software copying with drug dealing and other major crimes. The company also encouraged police raids on large companies. The campaign was legal and was supported by Brazilian antipiracy legislation, but it was an obvious intimidation move. The message was clear: stop doing what we told you was OK when you were in the university. Revenue from licensing increased overnight, but the initiative backfired in an unexpected way. Many of the initial migrations to open source solutions were done for fear of investigation. Companies and, especially, universities, lacking the budget to suddenly buy hundreds or thousands of very expensive licenses (remember the currency exchange rates), and risking fines up to 3,000 times the license cost for each unlicensed copy, switched to GNU/Linux and OpenOffice in a matter of days.

These initial migrations were instantly successful. Part of the success stems from the Internet's inherent cross-platform nature. Furthermore, people were acutely aware of ongoing security problems with Microsoft products. Add the dependency that most companies and developers, and even the government, had on systems from these large corporations and the *U.S. v. Microsoft* lawsuit. All this came into play around the same time. The stage was set for the discussion of how dependency on proprietary software was affecting the Brazilian economy and, especially, its sovereignty.

At about the same time, state governments were using free software products to offer Internet access to the population. A project called Telecentros, led by Sergio Amadeu, the software activist recently sued by Microsoft and one of today's top software livre advocates in the country, proved that GNU/Linux, OpenOffice, Mozilla, and a full set of open source products could be easily mastered by the population, even by people who had no previous contact with computers. The reliability and security of these systems, the possibility of running on cheaper donated machines, full support for all users, and essential ease of use made the program a huge success. Three years later, the balance showed that what was spent implementing software livre in the Telecentros would have paid for only half as many systems if Microsoft products had been used. Although the cost savings were significant, the fact that users, the government, companies, and the country could now see a real choice instead of being locked in was the main achievement of the initiative.

Freedom is something that every politician understands, so many bought the idea early on. Walter Pinheiro and Simão Pedro were pioneers on the wave that swept almost all political parties and a large number of politicians. Software livre reached the legislation houses, first at state level and then at the federal level. Now the discussion has reached the courts, where the laws and decisions taken in favor of software livre are being questioned by those who oppose it; even open source licenses are being questioned.

In just a few years, this important discussion has basically surpassed the IT industry, reaching all levels of society.

We have gone from the simple use of FOSS operating systems to freedom in general, to patents and intellectual property, Internet governance, international standards, and relationships. By going further than the simple four liberties of free software, the discussion reached the music and cultural sectors, promoting the sharing of knowledge—all knowledge—and today, even the Minister of Culture, the famous singer Gilberto Gil supports the software livre movement. Not just for free software, but for freedom of choice—and of knowledge sharing.

Not About Price, but About Choice

Many people see the movement to FOSS as an ideological movement. It probably is. Every time freedom has been battled for, there has always been some ideology involved. Different people have always tried to get as much freedom as possible for their partisans, usually ignoring others' views of freedom. Eventually, we get to a middle ground, where freedom wins—not necessarily the myopic view held by the partisans on each side, but a more compromised freedom that can be held longer.

The main discussion is not about economics, but about choice. For too long there was no choice but to keep using the software provided by Microsoft. The Internet opened the possibility of another option, one that proved to be cheaper and more reliable, and one that guaranteed more flexibility and customization.

So, why not simply move to this new option? If software livre is all that it is claimed to be, why even in a government where so much uproar has been generated are there still purchases of new hardware, new licenses, and new versions of the old lock-in option? There are many reasons. A huge lobby has been set in motion to slow any adoption of FOSS. Microsoft is not the only one supporting this lobby, but is surely very active in it.

Microsoft started by giving licenses away—or making them a lot cheaper—to governments and schools. This is understandable. It reasoned that the issue driving the growing popularity of software livre was economics, so it tried to be even cheaper (by avoiding training and migration costs). Then there was some more intimidation. It is reported that companies which were promoting their plans to migrate to GNU/Linux and OpenOffice in magazines or events would receive visits or letters from antipiracy organizations and, eventually, a friendly visit from the police. Some companies do not allow employees and consultants to talk about what they are doing with FOSS, for fear of this antipiracy raid. There are reports from some inspected companies that officers concentrate only on Windows machines, ignoring GNU/Linux, Macs, Solaris, and so on. Either these systems and their companies are not entitled to receive protection, or maybe it is just that somebody else is asking for the inspection....

Microsoft surely did a lot of talking. Many Microsoft employees have been vocal about software livre, many times through half-true statements and fear, uncertainty, and doubt tactics. Microsoft's country manager for Brazil, Emilio Umeoka, even went so far as to directly criticize Brazil's president, saying that by supporting software livre he was taking the country in the wrong direction—an arrogant comment coming from the head representative of a company that was considered guilty in the U.S., Europe, and recently Brazil, of hurting competition and governments through its commercial practices. Recently, even Microsoft's Bill Gates. The intimidations may have gone too far. At the time the software livre movement in Brazil was promoting the fifth edition of one of the world's largest international FOSS advocate gatherings, the FISL, Microsoft decided to sue the highest authority of the Brazilian government's software livre initiative, the president of the National Institute of Technology (NITI) and one of the originators of the celebrated "Telecentros" project, Sergio Amadeu. An important representative of the software livre principles, Sergio has a deep understanding of the issues and knows that software livre is all about freedom. He has been promoting a carefully considered and reasonable strategy that aims to give to the Brazilian government the choice it needs to decide about the country's technology future. By personally suing Sergio Amadeu (or as Microsoft puts it, requesting explanations in court) for what he has said, Microsoft generated an uproar from people all over the world. Because of the weak basis of the suit, Sergio simply stood by what he said and ignored Microsoft's requests for an explanation in court or otherwise. Sergio said he will not be intimidated, because he's sure that he's doing the right thing for the freedom Brazil needs and deserves. Microsoft's actions show that it doesn't understand what is going on in Brazil. The country is reaching for an option. And the price of choice is not measured in the number of licenses you get for free. Software livre is a way to achieve choice.

The huge lobby against software livre does not come only from international vendors. Many software companies in Brazil don't understand how they will compete in the software livre market. They don't understand the business model and don't see the benefits of Yonchai Bentler's "commons-based peer production" model. It is not clear how they will compete nationally and internationally by developing free software. In fact, not only traditional software companies are at a loss: many in the software livre movement still see any ties with business as "evil" or at least as undesirable. Even those that understand that the only way for software livre to survive is to have a strong and viable software ecosystem around freedom are still struggling in how to create this ecosystem. This has to be addressed by the software livre movement, and it is refreshing to see that some of the most influential software livre entrepreneurs are already joining forces to expand their ability to compete and participate in this market.

Besides the strong lobby, there are other, more important reasons why a faster move to software livre solutions is hard, even when there's a will.

Choice Requires More Than Free Software

One of the main lock-in problems governments and companies have is the legacy custom software already in place. Although it is reasonably easy to move your Windows users to GNU/Linux and OpenOffice and your web site from IIS to Apache Web Server, it is a lot harder to move your internal applications from one platform to the other. And without applications, there's no user adoption, of course.

Every internal or custom system that is developed is tied to a specific infrastructure software product—be it an operating system, a database, a messaging or security system, a file format, or a runtime library, you're locked in. Unless you have this solution available in the system to which you are trying to move, you'll have a hard time porting or rewriting your software.

As Microsoft, IBM, and Oracle are the dominant players in Brazil, most applications that are developed are tied into their infrastructure products. Microsoft is the most obvious target of the software livre movement, since it is the king of the desktop and it powers many departmental servers where licensing issues are more visible. And Microsoft is the one that's pressuring for more licensing and more upgrades. It is really the main bully in the playground. But IBM mainframes, with thousands of legacy applications and huge Oracle databases, among all kinds of other systems, fall under the same reasoning that says choice is better. Being free to move away from those systems requires that applications are ported, and this takes time.

And once the decision to port the applications has been made, where do you port them? Is it to another specific (although software livre) OS? Does choosing any software livre solution promise freedom? Does limiting the decision to any solution that has a specific FOSS license simplify the question? Should we move to one of the great FOSS-licensed solutions (database, framework) that in fact has just a single company and no community behind it? Maybe we should put up some criteria, based on community side and adoption, that will help in choosing a high-quality FOSS product to tie our application to?

The fact is that once you choose any one product to port your system to, you're creating ties to it. If this is a software livre product, that can be a big help, but it is not enough to guarantee your freedom, your choice.

Governments and companies do not get locked into a vendor or a platform simply because they use closed software, but rather, because they develop their own applications tied to a specific product, be it a free product or a proprietary product. Once all applications are written to a product, and all data is saved into a product-specific format, to move to another offering (free or proprietary) is a big effort. And the longer the ties to that product endure, the more difficult it is to move. Vendors drive their customers to stay as much as they can. Although software livre makes you less dependent on a specific vendor (because you can make your own changes), it does not necessarily keep you from getting locked into a product....

That's why the best option is to guarantee that developed software be effectively free of product lock-in: custom applications are based on open standards, and all data is saved into open formats. That's why many of the most valuable software livre projects are not simply products. They are open source implementations of open, royalty-free standards. The powerful combination of FOSS implementations based on open standards is what gives us the choice we need.

The use of open standards that can be implemented as open source is a strong way of promoting software livre. Once applications are free from specific products and standards, it is possible to replace closed products for the FOSS implementations that have a real chance to compete and show their technical advantages. Defending software livre on technical merits is a much stronger argument to governments and companies. Keeping custom applications ready to benefit from FOSS implementations of open standards favors software livre even in the (quite common) case where the FOSS implementation is still in development.

How Java Technology Can Help

That's where Java technology enters the story. Yes, I know many of you complain that Java is not FOSS, and so how can it even be cited after so many arguments in favor of software livre?

Well, Java is as much software livre as HTTP is. Or as PDF is. Or, for that matter, as C is! That is, none of these technologies is software livre. They are standards. Some are more open or compatible than others, but all of them are books, pieces of paper, and web pages. Standards are not software livre. The fact that you have the GNOME PDF Viewer licensed as GPL does not make PDF software livre. It does not even make PDF an open standard (it is a royalty-free standard, but it is not open since it's controlled by, and only by, Adobe). Having GCC as a C compiler does not make C software livre. It means only that it contains software livre implementation of the ANSI C standard. In addition, just because they are standards, even standards having software livre implementations, does not mean they can be modified at will. Apache HTTP Server is the software livre implementation of the W3C HTTP standard, meaning that although Apache HTTP Server code is legally modifiable, you cannot modify it at will and still say it is compatible with the W3C-HTTP standard.

Then there's Java. As a standard, defined and controlled by the Java Community Process (the JCP), the Java standard can now also be implemented royalty free as software livre (under any FOSS license). Actually, Java is a set of standards. You have things like the Java 2 Enterprise Edition (J2EE), the Java Virtual Machine (JVM), and the Java 2 Standard Edition (J2SE)—several class libraries defined as separated standards. Even the class file format and the Java language are standards.

Until recently, the JCP rules did not really allow for a software livre implementation (although many initiatives were being developed). However, this has changed, thanks to the initial work of Jason Hunter on behalf of the Apache Software Foundation and supported by Sun and other JCP members. Since the release in 2002 of the JCP 2.5 rules, there have been no barriers for compatible software livre implementations of the many Java specifications. Because of the large number of Java specifications, "Java needs to be open source" and "Java is not free software" are usually meaningless statements, especially because of the fact that since JCP 2.5, many of these specs already have a FOSS implementation.

One of the important things that the JCP rules mandate is that every contribution made to a standard has to be licensed royalty free to anyone implementing the standard. This guarantees that the FOSS community will be able to implement all of Java. Many standards bodies (ISO and ECMA, for instance) do not mandate this, simply requiring the infamous Reasonable and Non Discriminatory (RAND) clause that is not necessarily reasonable for the FOSS community, and as such discriminates against it. Standards generated by those standards bodies can be, and usually are, encumbered by patents and RAND agreements, making them, for all practical purposes, impossible to implement legally by the FOSS community.

However, because of Java's very broad objectives, the Java standards are hard to implement. Different from other standards, Java tries to reach the binary compatibility promise: the ability to run your Java binary unmodified on any platform with a Java runtime. For this even to work, a compatibility test was created to guarantee that a Java implementation meets the standard's requirements. Technology Compatibility Kits (TCKs) are provided, requiring implementations to pass these tests to claim compatibility. Compatibility requirements and the sheer number of libraries and standards involved, coupled with Sun's quite restrictive licensing in its own proprietary implementations have delayed software livre implementations of Java.

This does not mean that initiatives don't exist or that we cannot have a software livre implementation. It is important to say that many Java standards already have a software livre implementation: servlets, JSP, JSF, EJB, J2EE, JMS, and JDBC are among the most important examples. What is missing is the very important underlying runtime: the complete JVM and the set of J2SE libraries. Many projects do exist to fill this gap and are under development right now. Several are quite capable of running much of the Java code out there, including some very complex applications. These implementations are far from perfect, but we're getting there.

Java Provides the Other Side of the Choice

But what does all this have to do with freedom and choice? We saw that software livre did wonders to provide freedom for developers and to allow developer collaboration across the Internet, pushing the whole practice of software development to a new level. Freedom was well served and well used.

It is understandable that developers need their freedom to expand, explore, and literally change the world. Now we're discussing freedom with not only developers, but also companies, users, and the government. We saw how these deployers are looking for freedom of choice. They want to be able to choose. They want—actually they need—options: the ability to choose different vendors, different implementations, different software houses, different licensing...in short, to not be locked in.

Going back to the many visions of freedom, the standards community sees freedom as the possibility to choose freely from multiple vendors, since usage of standards prevented lock-in. Standards are the way society normally defines its rules—we see this in very different markets, from water distribution to electricity, from TV to telecommunications—and this was applied to the software world. There were no explicit discussions about how developers would collaborate on code; on the contrary, standards usually hide the underlining implementation. And the focus was not on developer freedom: standards are actually restrictions on what a few developers can do, so a larger group of deployers and users can benefit from choice and option.

The free software movement was looking for freedom for the developer to collaborate and create, but more than this, freedom as an ultimate goal—*a social good*. Quoting the FSF, "for the Free Software movement, non-free software is a social problem and free software is the solution." The GPL puts all the responsibility in the hands of the developer, who has all the freedom, and only one very strict and strong restriction: the copyleft. Once-released code cannot ever be restricted, making it difficult (although not impossible) for developers to derive commercial benefit directly from the code they write. This is a strong compromise that free software developers accept for the benefit of all developers.

Then the Java community comes along, looking for another type of freedom, one that would free the developer to run his application on any platform. Actually, the main goal of Java was to free the user to run an application on whatever platform he chose, no matter what the platform the original developer was targeting. That was another type of freedom altogether, and it had to be based on standards to allow for multiple vendors. This view puts a restriction on a very narrow group of developers, the ones that deal with the creation of the runtime. They have to obey strict rules, for the benefit of the vast majority of developers, deployers, and users, from all platforms, who are then free to choose what they want to use and run.

At last the open source movement was formed, and promoted the discussion that although free software was good, having companies investing in and commercially benefiting from free software would be even better, proposed a similar but different view. Seeing freedom in a more practical way, which promotes developer collaboration, but they are willing to accept that not all software must be free, not considering proprietary software a social issue. Open source is referred to as a software development methodology, having dropped the political manifesto, and making the notion more usable and acceptable for the commercial market. The open source movement compromises the ideology and accepts that developers will benefit commercially from code they did not write, to achieve the benefit of allowing more companies and probably more resources to be applied to the evolution of software. This vision tries to achieve the much-needed critical mass that made FOSS the success it is today.

Are these freedoms incompatible? They can complement each other in very powerful ways, but freedom is always a compromise. We usually impose restrictions on a few to the benefit of a larger group.

Unfortunately, some people like to see things narrowly, creating unsolvable conflicts among these different freedom-promoting movements. As a Java developer, many times I have been excluded, ignored, and generally considered an "outsider" of the FOSS community. Many other Java developers feel the same way. All you need to do is mention Java in a FOSS discussion forum, and you'll be flamed. You see a lot of Java bashing at FOSS events. At a recent event, one of the main speakers claimed there are no good open source Java developers. How would developers from the Apache Jakarta project—who have implemented so many great FOSS projects, and so many FOSS implementations of the Java standards—feel about this? What about the developers who put a lot of effort on the Kaffe VM, the GNU Java Compiler, and the GNU Classpath? These are all great FOSS software projects highly respected implementations of the Java standards, done by top-level, committed Java developers.

Even FSF founder Richard Stallman has not proposed prohibiting the use of Java. In his essay, "Free But Shackled—The Java Trap," which seems to be misunderstood by many in the FOSS movement, Stallman actually recommends that developers use the already available free Java implementations and, more important, that they help improve them. "We do have free implementations of Java, such as the GNU Java Compiler and GNU Classpath, but they don't support all the features yet.[...]Fortunately, [the Java] specification license does permit releasing an implementation as free software[...]To keep your Java code safe[...], install a free Java development environment and use it," says Stallman.

Stallman is not only very clear on the importance of the free Java implementations, but he also gives us an important argument. He says, "In the early days of the Free Software Movement, it was impossible to avoid depending on non-free programs.[...]It was inevitable that our first programs would initially be hampered by these dependencies, but we accepted this because our plan included rescuing them subsequently." This was a reasonable compromise to make to reach a much more worthy goal. But Stallman continues: "The situation is different today. We now have powerful free operating systems and many free programming tools. Whatever job you want to do, you can do it on a free platform; there is no need to accept a non-free dependency even temporarily." Although Stallman uses this rhetoric to say that you should stick to the free Java implementations, the effect really depends on who he means by "you."

One of the powerful freedoms that Java brings is platform independence. As we have already seen, the main roadblock for companies and governments in Brazil to choose and migrate to the "powerful free operating systems" is the lock-in of existing legacy applications. This is a hard problem to crack, since we cannot move the systems until all needed code is ported. So, until we're able to expend enough effort to migrate all the existing applications, we have to keep adding to today's Windows system. Since a simultaneous migration is improbable, we then have to coexist both systems for awhile.

Here is where Java comes to the rescue. Applications can be developed to run on today's Windows systems, coexisting with any new software livre system that is introduced, thus making the eventual complete migration a lot easier. Certainly there are other technologies that allow you to develop cross-platform applications, but Java is one of the best options, not only technically but also in terms of available tools, information, and market penetration. Anecdotal evidence from the open source community on the number of non-Java applications that were developed on GNU/Linux but can run on Windows shows that platform independence is not that easily achieved. It is possible, but the large number of Windows-developed Java applications that can run on GNU/Linux is a clear statement that Java makes cross-platform deployment a much easier task.

There are many corporate and government developers who need to deal with a migration strategy, who need to start moving from Windows to GNU/Linux, or maybe to an open source Solaris. Many more are not planning to migrate now, but would like to have a number of choices in the future. So, if the "you" Stallman was referring to are those developers, they will be better off by accepting a nonfree dependency temporarily, to get the freedom of platform now provided by Java, and be able to move to a "powerful free operating system" when desired or needed. Moving Java software to the free operating systems should be easier, and promoting multiplatform development and standards for governments and companies is a powerful strategy to guarantee choice. Minimizing dependencies on software that cannot be implemented by the FOSS community is a clear road map to allow the migration.

In the meantime, we, the FOSS community, should stop pretending Java does not exist, and that freedom of platform is not important to users, and make the effort to accept, use, and finish the free Java implementation and tools, as Richard Stallman suggests.

The existing free Java standards implementations may lack functionality if compared to the full proprietary ones, but the huge amount of functionality provided is at least comparable, and in many cases is much better than that offered by other major open source implemented languages. There's no reason why lots of great software could not be developed and used on those free Java implementations, but as long as the FOSS leaders insist on downplaying the importance of Java, this important free software will be ignored both by the FOSS community and by the Java community at large.

When the FSF talks about usage of the term free software, it mentions that "to stop using the word *free* now would be a mistake; we need more, not less, talk about freedom." This is an interesting point. Although the FSF has its strong and idealistic view on what is software freedom, it is clear that its views of freedom are not the only important views, and that the freedoms we're talking about are not mutually exclusive. They may be different, and not everybody may want all of them, but they can be explored together. Platform independence is an important, easy-to-grasp freedom, that can go hand in hand with the notion of free software. The fact that we still don't have a compatible FOSS Java implementation should not be a reason to stop promoting the notion of freedom of platform.

Walking the Path

With such restrictions from the FOSS community, openly discussing Java and software livre seems to be easier said than done. However, we have a series of successful examples in Brazil that are helping put some light on the possibilities that Java and software livre can bring to developers and deployers.

We saw how Brazil has chosen the freedom path and some of the strong reasoning behind that. One of the pillars of that choice is, without a doubt, software livre: it has pushed the idea of technology independence to other levels, and has prompted the whole country to discuss freedom in other related areas, like the Internet, music, and information sharing. It has also captured all the attention, and many what are now discussed as a "software livre" initiative, although more related to freedom in general.

Because of this discussion of choice and freedom alongside software livre, other initiatives are strongly pursued. Standards, for example, receive special attention from the many government agencies. Serpro, the largest Brazilian federal IT agency, an organization with thousands of developers working in governmental software, has declared that one of its main responsibilities is to create, defend, and apply national standards. Standards are defined and then used to guarantee vendor independence, and then to allow the inclusion of FOSS solutions—if they exist and are stable—as a preferred choice for acquisition.

But even before software livre was ever discussed in Brazil, multiplatform applications became increasingly important in the quest for vendor independence. Microsoft lock-in was the main catalyst, because lock-in to any vendor is one of the main worries inside companies and the government. Add to that the almost mandatory requirement that every government's software must support GNU/Linux while still supporting the huge legacy of Windows systems, usually developed by Windows developers that had literally a few hours to start generating multiplatform code while still using their Windows machines....

Because of all this, from the start Java played an important role in guaranteeing freedom for both the Brazilian government and Brazilian companies. As early as 1996— long before software livre was actually considered a real option anywhere in the world—software companies and the many government development agencies saw in

Java the opportunity to gain freedom from Microsoft lock-in. This was an important step: for the first time, Brazilian developers were even considering another platform for software development. The search for independence—platform independence in this case—drove widespread adoption of the technology inside the government. The government-owned Banco do Brasil (Latin America's largest bank), adopted Java in 1997, training more than 800 developers in their first Java training program. This is an early example of the search for freedom that happened before the OSI (and thus the open source definition) even existed, and long before any significant discussion of free software in the country.

Although many try to exclude Java developers from the FOSS movement, in Brazil it is sometimes hard to separate these two communities. Much of the development in the country is being done in Java today, including a large percentage of the FOSS development. Projects like JForum, JBanana, Prevayler, Bossa, eGen, Javali, Genesis, and Hotwork are just some examples of Java-based Brazilian FOSS projects that are in use throughout the country.

The early software livre movement was clearly driven by system administrators, security experts, web designers, social scientists, and politicians. They did one of the world's most effective pushes for the adoption of software livre and, if it had been only that, it would already be an impressive and fundamental contribution. However, they did more, and the software livre movement went far beyond the simple adoption push. They moved the software livre discussion into the country's hearth and involved politicians, lawyers, judges, social scientists, financing institutions, entrepreneurs, large- and small-business owners, and the government at large—in short, creating a viable ecosystem for software livre to flourish. Unfortunately, when it got time to discuss software development, they fell short. The push to use software livre was so strong that more concern was put into migration and digital inclusion strategies. Although there were very good FOSS developers and important projects, software development issues were given little attention initially. This is where the Brazilian Java community has been focusing its contribution, bringing software development back into the discussion.

Java is one of the most-used languages in the country and is the only one that is multiplatform and has strong support in both Windows and in GNU/Linux. So, it is understandable that most of the development discussion in FOSS happens around Java. There are many examples. Of the recent open source applications that have received financing from the Brazilian government, almost all are developed in Java, many running on top of free Java application servers like Tomcat and JBoss. At FOSS events in Brazil, it's easy to notice that most of the presentation proposals submitted that deal with development are Java related. Surprisingly, there's a lot less support and interest for important FOSS languages such as Perl, PHP, and Python.

Through the history of the software livre movement in Brazil, Java has played an important part, although at times it has been ignored and downplayed by many. Here are some other examples:

Direto

Developed as a replacement of Lotus Notes and Exchange, Direto is a web-based email and collaboration tool that handles calendar, address book, and other functionality. Developed by Procergs, the IT agency of Rio Grande do Sul (where FISL is held), Direto runs on GNU/Linux and Tomcat, with lots of other FOSS solutions. The development was mainly done in Java, and at the time it was released, it was the first initiative from a government agency to release a free software product. Because of the work done by one of the most respected free software developers in the country, Ricardo "Gandhy" de Mello (a Java developer), Direto was at some point able to run, with some limitations, on a free Java runtime. Today it probably would run on today's much better free runtimes if anyone cared to try. Direto is a strong example of how we can get one freedom, and then the next.

IRPFJava, the multiplatform income tax report application

For many years, to submit your yearly income tax report (IRPF) electronically, receive all the benefits of faster tax returns, and easily handle the complex forms, a Windows machine was needed. As a result, the government received many complaints from Mac and GNU/Linux users who had to resort to friends or accountants. This was an example of a government application that forced citizens to use a proprietary product. Receita Federal, the agency responsible for IRPF, rewrote the application to support multiple platforms, specifically GNU/Linux. IRPFJava became the first federal application targeted for large public consumption that was focused on supporting GNU/Linux. As expected, the developers were still using Windows and had no knowledge of GNU/Linux or the Mac: they relied on Java to support those and the many other OSes that were used to submit the reports. As another good example that shows the possibility of freedom in steps, efforts are now underway to use the free Java runtime to run the application.

Banco do Brasil, Caixa Economica Federal, Dataprev, Datasus, Procergs, and Serpro, to name but a few

These are public companies and agencies, and are among the largest developers of software inside the Brazilian government. All use Java heavily to guarantee vendor independence, and many were doing that long before the software livre movement took place in the country. Most have chosen to use Java-based FOSS tools and products because it made commercial and technological sense, and these projects are promoted today as success cases of software livre adoption. To show their support for software livre, many of these agencies refer to their use of

Java-based products such as Tomcat, JBoss, and Eclipse. In some of these places, Java-based tools are the only free software development tools being used. This is a clear sign of the importance of Java in software livre initiatives.

That is not to say that everything related to development that happens in the Brazilian software livre movement is necessarily related to Java. Far from it. The fact is that Java was there from the start, guaranteeing freedom inside the government, even before the software livre discussion turned mainstream. This should not be ignored or downplayed. And while most of the world is creating a chasm between Java and software livre, in Brazil Java is effectively being used as a lever to push software livre to higher grounds.

What to Do?

It is clear that Java and software livre work together. By combining these different approaches to freedom, we can have more freedom, not less. But many still feel that Java is not open source and try to leave Java out of this fundamental discussion, to the detriment of all.

There are many examples of initiatives to exclude Java as a development tool for FOSS projects, and even to exclude FOSS Java applications from FOSS events. I wonder how we can have a FOSS implementation of any standard if there's this kind of prejudice against standards that were not implemented as FOSS yet. Contrast this with the amount of effort put into providing drivers and even reverse engineering of proprietary patented software or protocols to establish links and integrate GNU/Linux and Windows, for instance. For the FOSS community, the effort to create a JVM provides a great integration strategy. The more code you write in Java today, the more you are able to run your applications in both Windows and GNU/Linux systems. Consider also the possibility of running your application on the recently announced open source Solaris OS, and you have more choice and more freedom. It's interesting, though, that for the most part, to run a GNU/Linux system you usually have a binary, but closed, standard underneath it: the Intel processor. Java is also a binary standard, although it normally sits on top of the OS. It is amazing that the FOSS community can't seem to see how allowing one and excluding the other is like throwing the baby out with the bath water.

The fact that many applications—lots of them under FOSS licenses—are written in Java means that more effort should be spent on having a 100% FOSS Java runtime available. Many people are working to enable this, but the general prejudice is still strong in the FOSS community, making it strong in the other direction from the Java community. These prejudices need to stop, and leaders from both camps need to come at terms and realize there's much to be gained from the joint pursue of freedom.

We Are Getting There

At a unique and historic meeting at the end of 2004, these two communities started to come together. Sponsored by Red Hat, the Free Runtime Summit joined companies and developers working toward a FOSS implementation of the J2SE standards, the last ones in the stack to have a complete open source implementation of Java. The meeting joined the leaders from the most important free Java projects—the GNU Classpath, the GNU Java Compiler, and the Kaffe VM—along with the JCP, the FSF, the OSI, the Apache Software Foundation, and the SouJava Java Users Group, and also representatives of several companies. The results of this important gathering are still to be seen, but it was discussed and agreed to move forward with the efforts of an implementation of the Java standards that have a FOSS license and that pass the compatibility tests. The meeting was a major step forward. For the first time, there's general agreement on the importance and viability of this implementation. This open collaboration and resolution of prejudices will prove very beneficial to both the Java and the FOSS communities.

The path Brazil has taken—going after freedom in its larger meaning, trying to bring together standards, multiplatform and software livre, to guarantee freedom of choice to developers, companies, and the government—is showing its viability day by day. The possibilities are promising, and we're working hard to create the freedom we need to innovate, to generate technology, and to strengthen our sovereignty. The Freedom to choose our technological future and to collaborate with the world will be our reward for the effort. That's the freedom we want, and we'll fight for that.

References

- *http://www.gnu.org.*
- *http://www.gnu.org/philosophy.*
- "Why 'Free Software' is better than 'Open Source'"; *http://www.gnu.org/philosophy/free-software-for-freedom.html.*
- *http://www.osi.org.*
- *http://www.jcp.org.*
- Richard Stallman, "Free But Shackled—The Java Trap"; *http://www.gnu.org/philosophy/java-trap.html.*

Beyond Open Source: Collaboration and Community

Section 2 moves beyond what we traditionally think of as open source and tackles the larger questions of the collaborative pattern, of which open source is but an instance.

We learn in essays by O'Reilly and Searls not just how open source is changing the surrounding technology landscape, but how the dynamics of that changing landscape are putting open source in a whole new context. Open source must adapt and evolve to continue to be relevant.

Of greater significance than the changes within the technology sector are the other endeavors that have learned from open source. Any creative enterprise that would benefit from increased collaboration can benefit from the lessons of open source. In essays by Jones, Hessel, and Sanger, we learn about some specific areas where open source principles are actively being applied.

Finally, we conclude with several essays that grapple with the big question of what is the form and practice of a collaborative community generally. Some of these essays approach the question using examples of specific communities (Kim; Bates and Stone). Others look not just at specific communities, but at emergent patterns of cooperation (Shah; Weber). These emerging patterns, more than any specific technological innovation that open source might yield, are fundamentally changing the world around us.

CHAPTER 15

Doc Searls

Making a New World

Open Sources was published in January 1999. That same month I became a full-time editor for *Linux Journal*, assigned to cover Linux in business. Over the next seven months, I also co-wrote (with Chris Locke, David Weinberger, and Rick Levine) *The Cluetrain Manifesto*, a rant that first took the form of a web site and later took the form of a book. Its subtitle was "The End of Business as Usual."

Cluetrain, like much of what I wrote for *Linux Journal* at the time, argued against bubble-headed marketing at a time when bubble-headed investing in "Linux companies" was growing to galactic dimensions. In August 1999, Red Hat had the largest IPO run-up in stock market history. In December 1999, VA Linux (now VA Software) set a new record, which remains unbroken. In January 2000, *Cluetrain* hit the bookstores and became a business bestseller. The bubble began to pop on January 17, and within a few months nearly everybody in a "Linux business" (including *Linux Journal*) suffered consequences ranging from dire to fatal. All businesses, including Linux ones, still feel the after-effects.

Yet the bubble was a red herring. It was off-topic in the extreme, because Linux was quietly being put to use in businesses everywhere, along with a growing suite of other open source infrastructural building materials. Linux was never about the stock market, or even about business. It was about something else—something that caused usage (especially in business) to grow regardless of whatever happened among commercial suppliers.

Today the spotlight is on a new set of Linux business leaders: IBM, HP, Novell, Oracle, Red Hat, and other large companies. We know they are leaders because they buy big booth space at LinuxWorld, pay employees to work on Linux code, make good products, sell to the growing market demand for Linux and open source goods, do good PR, and get lots of coverage in the press.

No offense, but they're red herrings too. They matter a lot, but they're not what Linux is about. They're not what open source is about, either, because Linux and open source are demand-side developments. They are all *what the demand side does to supply itself.*

Just like the environment where open source took root and grew: the Net.

If designing and building the Internet had been left up to the usual suspects, it never would have happened. Networking would still be a private affair, a grace of large vendors, each operating their own separate and barely interoperable networks. For an example of what that would be like, consider instant messaging. IM is a network service that never found its way into the Internet suite (unlike, say, hypertext, file transfer, email, and domain names). The situation for IM is not much different today than it was back in the 1980s, when "online services" like Compuserve, AOL, and Prodigy each had their own incompatible email systems. Today Yahoo!, Microsoft, AOL, and Apple all remain committed to closed proprietary IM systems that run *on* the Net, but are not *of* the Net, meaning they contribute nothing to the Internet's open, free, shared, and ubiquitous infrastructure. They are "platforms" supporting closed silos that trap and hold dependent inhabitants.

The platform and silo system is as old as computing. It's still with us, and won't go away quickly, if ever. But as a defining model for the software business, it is being replaced by a growing assortment of open standards and open source tools and building materials that together support far more business than they replace. Linux and its familiar LAMP suite (Linux, Apache, MySQL, PHP, Perl, Python, PostgreSQL, etc.) are the most obvious ones. SourceForge lists another 90,000, with dozens more added every day.

Nearly all of these tools and building materials were created by the demand side of the marketplace, to solve practical problems, and to provide useful infrastructural support for similar activities. The free and open way they contribute to the world is good for business. Whole businesses and business categories, old and new, are sited on bedrock composed of open standards and open source components. Growth horizons for these businesses and categories are unobstructed by dependencies on vendors and their platforms. Their environment is the wide world of the Net, not the inside of some vendor's silo. In fact, all platforms and silos in the computer business now find themselves in a subordinate position to the Net. To survive they have to operate in the wide new world of the Net.

That goes even for Microsoft, which built the largest and most widely used platforms and silos in computing history, with a monopoly in the most ubiquitous product category of all: personal computing.

Yet even Microsoft finds its vast monopoly, and all of its platforms and silos, forced to live in a new and larger world that:

- Nobody owns
- Everybody can use
- Anybody can improve

In "World of Ends" (*http://www.worldofends.com*), David Weinberger (*http://www.hyperorg.com/blogger*) and I initialized those three principles as NEA.

In the first two respects, this virtual world is like our physical one. Except for the patches of crust we call real estate, nobody owns the Earth or its atmosphere. And everybody can use the Earth's base infrastructural provisions: gravity, air, and filtered sunlight.

Exceptions can be found, of course. For example, MySQL is owned. Yet, as infrastructural building material, it doesn't matter whether MySQL is owned or not, because everybody can use it and anybody can improve it. MySQL's ownership matters only to those that choose to have a commercial relationship with the company. In that respect, MySQL doesn't behave as a traditional "platform" vendor, trying to lock customers and third parties into a dependent role in a private environment. Instead, it behaves more like a provider of building material for a construction project. MySQL, like other open source components, is modular stuff. It's made to work inside a larger context—a job, a design, an architecture, whatever—where each component does its job and not much more than that.

The list of open source building materials grows constantly. When I started writing for *Linux Journal*, there were a handful of familiar names: Linux, Apache, Sendmail, and Perl. That grew to become the LAMP suite (where the *M* represents MySQL and the *P* includes PHP and Python). Now the names dropped in a discussion of open source might also include Tomcat, Squid, Asterisk, JBoss, Eclipse, Jabber, ZeroConf, RSS, iPodder, or any of the 90,000+ projects on SourceForge alone.

That's because NEA's third principle—*anybody can improve it*—gives humans the power to *continue* making this new world. Think of this principle as do-it-yourself geology. We don't just play God here. We get to do His job.

The architecture of this world was first described in 1983 by J.H. Saltzer, D.P. Reed, and D.D. Clark in *End-to-End Arguments in System Design* (*http://www.hyperorg.com/blog*). Fourteen years later, *The Rise of the Stupid Network*, by David Isenberg, delivered a death sentence to the conceits of network centralizers. The Stupid Network was an end-to-end argument against AT&T's cherished belief in The Intelligent Network. David wrote, "A powerful leading indicator of the Stupid Network will arrive when entrepreneurs who have no vested interest in maintaining telephone company assumptions begin to offer profitable, affordable, widely available data services." A prophesy now fulfilled.

Craig Burton (*http://www.craigburton.com*) combines both ideas—*end-to-end* and *stupid*—by describing (*http://www.linuxjournal.com/article/4158*) the Internet as a hollow sphere, composed entirely of ends:

> I see the Net as a world we might see as a bubble. A sphere. It's growing larger and larger, and yet inside, every point in that sphere is visible to every other one. That's the architecture of a sphere. Nothing stands between any two points. That's its virtue: it's empty in the middle. The distance between any two points is functionally zero, and not just because they can see each other, but because nothing interferes with operation between any two points. There's a word I like for what's going on here: *terraform*. It's the verb for creating a world. That's what we're making here: a new world. Now the question is, what are we going to do to cause planetary existence? How can we terraform this new world in a way that works for the world and not just ourselves?

One current example is *podcasting*. The term first appeared in August 2004. When I wrote about it in IT Garage (an online sister to *Linux Journal*) in September 2004, a search for *podcasts* brought up 24 results on Google. Now the same search gets 778,000 results. By the time you read this, the number will probably be in the millions.

Podcasts are audio files distributed to subscribers' audio players (mostly iPods) via Really Simple Syndication (RSS), an XML dialect designed to serve as a syndication format for weblogs. The prime movers behind podcasting are Adam Curry and Dave Winer. Adam is a veteran broadcast personality (best known as an early MTV VJ) and a serial entrepreneur. Dave is a programmer, writer, and businessman whose fingerprints are on XML-RPC, SOAP, OPML, outlining, blogging, and other useful innovations, including RSS.

Podcasting grew rapidly after Adam created *iPodder* (*http://ipodder.sourceforge.net*), a script that automatically routes podcast feeds into iPods and other MP3 players, via computers. As a distribution system for audio (or any kind of media file), iPodder and its relatives are forming an infrastructural foundation for a whole new industry—one in which anybody can participate. And, by providing a limitless supply of talent, material, and low-cost any-to-any distribution, podcasting also offers boundless new opportunities for broadcasting, cable, satellite TV and radio, the record business, and lots of other industries—as well as to noncommercial institutions ranging from churches and civic organizations to public broadcasting and government.

Note that podcasting became a hot category without the help of a large company. Instead, it began with the demand side supplying itself.

Now watch for big companies to jump in, and for businesses of all sizes to start making money. And watch for most of that money to be made *because of* podcasting's open standards and open source components, instead of *with* them.

It will eventually become clear to everybody that far more money is being made *because of* open source than *with* open source. This is what we have to remember every time somebody asks, "How can you make money with (open source product)?" The answer is, "You don't make money *with* it. You make money *because of* it."

The *because of* principle is old hat in mature business categories, but it's new to the software business. Too many of us still want to see "business models" for all kinds of goods that don't belong on the income sides of balance sheets. Would you ask your telephone what its business model is? How about your front porch? Your driveway? Your clothes? Those things may *help us* make money; but they are not *how* we make money. Well, the same goes for open source products. They are a means to an end. You make money *because of* them, not *with* them.

It's also easy to forget that the most original sources in this new world are not technologies, but talented and productive human beings. We all know reputation is tremendously important in hacker culture, and that open source is required, literally, to substantiate reputations. It is less obvious that the same is true for every other talent that operates on the Net. Reputation grows fastest when the goods and services of creative minds are open to inspection, improvement, adoption, and reuse.

Take the case of English Cut (*http://www.englishcut.com*), the weblog of Thomas Mahon, bespoke Saville Row tailor, London. As I write this, Mahon's blog is a couple of weeks old, and he already has a *Technorati cosmos* (collection of current inbound links) north of 200 (up from 100 yesterday), which is remarkable (*http://www.technorati.com/cosmos*). Thomas Mahon is quickly becoming the most authoritative Saville Row tailor in the world—not because he's a terrific tailor (which he surely is), but because he also operates in the public marketplace we call the World Wide Web. There he converses with customers and fellow mavens about fabrics, drafting, cutting patterns, the trade, and (of course) his advantages over competitors. Thomas Mahon runs an open source business. Literally.

Not surprisingly, he has already attracted all the business he can stand. Again, *because of* his blog. Not *with* it.

Thomas was urged into blogging by one of his customers, Hugh MacLeod (*http://www.gapingvoid.com*), a cartoonist and professional marketing iconoclast whose blog is Gapingvoid.com. Hugh took the free ideas delivered by *The Cluetrain Manifesto*, and leveraged them into HughTrain (*http://www.gapingvoid.com/Moveable_Type/archives/000823.html*), a collection of wisdom that adds substantially to the open source–savvy marketing canon.

Thomas and Hugh are both referenced in this piece of wisdom (*doc.weblogs.com/discuss/msgReader$5489*) from Mike Warot (*bitgrid.blogspot.com*), which appeared in the comments section of my own blog (*doc.weblogs.com*) this morning:

> I think that division of labor and specialization are the ultimate human skill set. I see the invention of blogs, wikis, and other online communication as the most recent additions to the toolbox. We're going to increasingly bestow rewards

based on competence (as the Hughtrain and English Cut weblogs have done), but it won't be a direct reward.

It's going to be especially interesting to watch how the online expression and validation of trust and competence evolve. Cory Doctorow hinted strongly at this with the concept of Whuffie. I've seen, secondhand, how a friend of mine gets small but pervasive and persistent rewards, for having been the first to engineer a simple, open source, file-transfer protocol. There's not big money in it, but had he tried to make it proprietary, it won't have flown, and something else would have taken its place. My opinion is he's responsible for enabling a big chunk of the PC revolution through many tangents.

He invented something, gave it away (purely for the fun of it), and got some Whuffie in the long run.

Scary thought: we're going to see Whuffie become quantified (and even perhaps monetized) at some point. We have to plan for that eventuality.

This model will take the place of marketing for companies that are able to articulate and share their knowledge with others. I particularly find the English Cut weblog to be fascinating. I've learned quite a bit about the culture and technology of well-made suits. Should I happen to work my way into wealth, I might even become a customer. It certainly has increased my appreciation of the value of a Bespoke. He's adding a lot of value from my perspective.

Reputation matters. Authority matters. Google (perhaps the world's biggest example of how to make money *because of* open source) sorts search results by the PageRank (*http://www.google.com/technology*) system, which the company explains in this way:

> PageRank™ relies on the uniquely democratic nature of the Web by using its vast link structure as an indicator of an individual page's value. In essence, Google interprets a link from page A to page B as a vote, by page A, for page B. But, Google looks at more than the sheer volume of votes, or links a page receives; it also analyzes the page that casts the vote. Votes cast by pages that are themselves "important" weigh more heavily and help to make other pages "important."

That importance is what Cory Doctorow called *Whuffie* (*en.wikipedia.org/wiki/ Whuffie*), in his book *Down and out in The Magic Kingdom*. By whatever name—reputation, authority, brand value, Whuffie—we don't acquire it alone. Its value is bestowed by others. In fact, the same might be said for its substance.

Several years ago I was talking with Tim O'Reilly about the discomfort we both felt about treating *information* as a commodity. It seemed to us that information was something more than, and quite different from, the communicable form of knowledge. It was not a commodity, exactly, and was insulted by the generality we call "content."[1]

Information, we observed, is derived from the verb *inform*, which is related to the verb *form*. To inform is not to "deliver information," but rather, to *form* the other party. If you tell me something I didn't know before, I am changed by that. If I believe you and value what you say, I have granted you *authority*, meaning I have given you the right to *author* what I know. Therefore, *we are all authors of each other*. This is a profoundly human condition in any case, but it is an especially important aspect of the open source value system. By forming each other, as we also form useful software, we are making the world, not merely changing it.

Stewart Brand provides a helpful framework for understanding that world, with this "Layers of Time" model of civilization, from *The Long Now* (*http://www.longnow.org*), as shown in Figure 15-1.

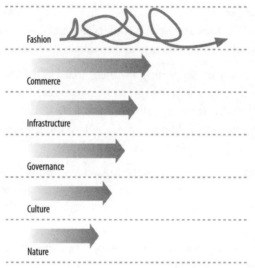

Figure 15-1. *Brand's Layers of Time*

Look at this as a layered section of surface on Craig Burton's hollow sphere—this "World of Ends" we call the Net.

At the bottom we find the end-to-end nature of the Net. It's also where we find Richard M. Stallman, the GNU project, the Free Software Foundation (FSF), and hackers whose

---

1   I had the same kind of trouble when I first started hearing everything one could communicate referred to as "content." I was a writer for most of my adult life, and suddenly I was a "content" provider. This seemed ludicrous to me. No writer was ever motivated by the thought that they were "producing content." Their products were articles, books, essays, columns, or (if we needed to be a bit more general), editorial. "I didn't start hearing about 'content' until the container business felt threatened," John Perry Barlow said.

interests are anchored in the nature of software, which they understand fundamentally to be free.

When Richard M. Stallman writes "everyone will be able to obtain good system software free, just like air," he's operating at the Nature level. He doesn't just believe software *ought* to be free; he believes its nature is to be free. The unbending constancy of his beliefs has anchored free software, and then open source development, since the 1980s. That's when the GNU tools and components, along with the Internet, began to grow and flourish.

The open source movement, which grew on top of the free software movement, is most at home one layer up, in Culture. Since Culture supports the Governance, the open source community devotes a lot of energy and thought to the subject of licensing. In fact, the Open Source Initiative (OSI) serves a kind of governance function, carefully approving open source licenses that fit its definition of open source. While Richard and the FSF, sitting down there at the Nature level, strongly advocate one license (the GPL or General Public License), the OSI has approved around 50 of them. Many of those licenses are authored by commercial entities with an interest in the governance that supports the infrastructure they put to use.

In fact, it was an interest in supporting business that caused the open source movement to break off of the free software movement. That break took place on February 8, 1998, when Eric Raymond wrote "Goodbye, 'free software'; hello, 'open source'" (*http://www.catb.org/%7eesr/open-source.html*). Here is where the Culture layer can clearly be seen moving faster, and breaking from, the Nature layer:

> After the Netscape announcement broke in January, I did a lot of thinking about the next phase—the serious push to get "free software" accepted in the mainstream corporate world. And I realized we have a serious problem with "free software" itself.
>
> Specifically, we have a problem with the term "free software" itself, not the concept. I've become convinced that the term has to go.
>
> The problem with it is twofold. First, it's confusing; the term "free" is very ambiguous (something the Free Software Foundation's propaganda has to wrestle with constantly). Does "free" mean "no money charged" or does it mean "free to be modified by anyone," or something else?
>
> Second, the term makes a lot of corporate types nervous. While this does not intrinsically bother me in the least, we now have a pragmatic interest in converting these people rather than thumbing our noses at them. There's now a chance that we can make serious gains in the mainstream business world without compromising our ideals and commitment to technical excellence—so it's time to reposition. We need a new and better label.

I brainstormed this with some Silicon Valley fans of Linux (including Larry Augustin of the Linux International board of directors) the day after my meeting with Netscape (Feb. 5th). We kicked around and discarded several alternatives, and we came up with a replacement label we all liked: "open source."

We suggest that everywhere we as a culture have previously talked about "free software," the label should be changed to "open source." Open source software. The open source model. The open source culture. The Debian Open Source Guidelines. (In pitching this to the corporate world I'm also going to be invoking the idea of "peer review" a lot.)

And, we should explain publicly the reason for the change. Linus Torvalds has been saying in "World Domination 101" that the open source culture needs to make a serious effort to take the desktop and engage the corporate mainstream. Of course he's right—and this re-labeling, as Linus agrees, is part of the process. It says we're willing to work with and co-opt the market for our own purposes, instead of remaining stuck in a marginal, adversarial position.

This re-labeling has since attracted a lot of support (and some opposition) in the hacker culture. Supporters include Linus himself, John "Maddog" Hall, Larry Augustin, Bruce Perens of Debian, and Phil Hughes of *Linux Journal*. Opposers include Richard Stallman, who initially flirted with the idea but now thinks the term "open source" isn't pure enough.

Bruce Perens has applied to register "open source" as a trademark and hold it through Software in the Public Interest. The trademark conditions will be known as the "Open Source Definition," essentially the same as the Debian Free Software Guidelines.

It's crunch time, people. The Netscape announcement changes everything. We've broken out of the little corner we've been in for twenty years. We're in a whole new game now, a bigger and more exciting one—and one I think we can win.

Seven years later, victory is all but complete. And not just for open source. While the free software movement has been, relatively speaking, a drag on the open source movement, the deep and abiding beliefs and commitments of free software advocates have anchored both movements and have helped software identified by both labels succeed.

Not coincidentally, the Culture on which this new world depends is hacker culture, about which Eric S. Raymond—a founder of the OSI—has written extensively (he edited both editions of *The Hacker's Dictionary*). Both he and Bruce Perens, another leading open source figure, have purposefully advocated open source to business for many years.

And although open source hackers tend to be more interested in business than free software hackers, both want Governance and Infrastructure that *support* business but are not *determined* by business—except when business works with the hacker community.

Hence OSI's license-approval process. While the number of open source licenses has been a source of some debate (almost everybody would rather see fewer licenses), it is important to note that the relationship between these layers is not the issue. The last thing anybody in the free software or open source movements wants is for anybody at the Commerce level to reach down into Governance to control or restrict Infrastructure that everybody relies upon. Even though that's exactly why large companies, and whole industries, hire lobbyists. More about that issue shortly.

Changing corporate culture to adapt to open source development methods is not easy. Dan Frye, who runs IBM's Linux development program, recently told me that IBM has worked hard to make its internal development efforts coordinate smoothly with Linux's. That way, when IBM "scratches its itches," the kernel patches that result have a high likelihood of acceptance. IBM has faith that its accepted patches are ones that are most likely to work for everybody and not just for IBM. This is a natural and positive way for infrastructure to grow.

And grow it has. The selection of commodity open source building materials is now so complete that most businesses have no choice but to use those components—or, in many cases, to recognize that IT personnel in their enterprises have been building their own open source "solutions" for some time.

That realization can come as a shock. Open source infrastructure inside companies often (perhaps usually—it's hard to tell) gets built without IT brass knowing about it. In many cases, internal open source development and use has had conditional approval by CIOs and CTOs. Whatever the course of open source growth, at a certain point a threshold is crossed, and companies suddenly know that open source is no longer the exception, but the rule. The result for IT is often something like "Oh God! Our sex has changed! What do we wear?"

Several years ago, when I showed the diagram in Figure 15-1 (along with others we'll visit later) to Rob Glaser, founder and CEO of RealNetworks, he made a remarkable observation: that the Internet revolution rocked the business world because, for the first time in history, Infrastructure changed faster than Commerce. "It was like the rug got pulled out from under everybody," he said.

On the one hand, this caused a great deal of excitement, some of which gassed up the dot-com investment bubble. On the other hand, it caused a great deal of fear, some of which got lobbied into the Digital Millenium Copyright Act (DMCA). In the first case, commercial interests were enthused by the new Internet infrastructure but failed to understand its deeper causes and principles. But at least they did no harm to the Nature, Culture, or Governance that produce and support Infrastructure. Not so with the entertainment industry and the lawmakers it successfully lobbied for passage of the DMCA. The DMCA is a prime example of how Commerce can lead Governance to screw up Infrastructure.

The DMCA was intended mostly to protect copyright holders from the ravages of Nature in a networked world where, as Richard Stallman put it in the GNU Manifesto (*http://www.gnu.org/gnu/manifesto.html*), "Copying all or parts of a program is as natural to a programmer as breathing, and as productive. It ought to be as free."

While its authors wanted the DMCA to protect their industries by limiting "piracy" (their word for illegal copying) of copyrighted works, in practice the DMCA has chilled free speech and scientific research, thwarted "fair use" (an established copyright permission), slowed the protected industries' adaptation to life in the networked marketplace, and crippled or prevented new business categories from emerging.

The case of Internet radio is instructive. The DMCA defined broadcasting on the Net as "performance," and digital copies as "perfect," regardless of their fidelity. It required "webcasters" to negotiate royalties with the recording industry, through a Copyright Arbitration Royalty Panel (CARP) administered by the U.S. Copyright Office. Led by the Recording Industry Association of America (RIAA), the CARP instituted royalty requirements so labyrinthine, difficult, and costly to webcasters that it effectively prevented the industry it purported to regulate. Had it been imposed on over-the-air broadcasting at the dawn of that industry, it would have strangled that baby in the cradle too.

One unintended consequence is podcasting. Today podcasters are growing in fertile markets that haven't been poisoned by the DMCA. "Podsafe music" (also known as "non-RIAA" music) is already a staple on podcasts. Once podcasting becomes sufficiently popular (and is perceived by business as something happening below the Fashion layer), look for the RIAA either to take advantage of it or to force leading new podcasting companies (or established broadcasting companies) to strike Faustian bargains with them, just so those companies can play RIAA-licensed music. Whatever happens, the DMCA's market poisons cripple or prevent far more business than it protects.

The Commerce level is inherently proprietary (you can't have business without property); but that doesn't need to be a problem for open source, or for anybody. That's because, in a mature and healthy industry, property claims are limited to each owner's own property. If you want to build platforms and silos, fine. Just don't expect to support whole categories, because there are lots of open and free infrastructural building materials already doing that.

The construction industry provides a good model for where we're going here. There are plenty of platforms and silos among construction materials and methods. For example, windows from Andersen, Eagle, and Marvin require their own branded replacement parts, or parts from "compatible" manufacturers. But since construction is a mature industry, with a countless variety of commodity materials made compatible by widely accepted standards, no one manufacturer can "own" the industry, or even a part of it, by popularizing a "platform" on which everybody else is required to build. It would be absurd to build a house on the "Weyerhaeuser platform" or the "Georgia-Pacific platform." Building materials are essentially modular. All your studs, joists, siding, wallboard, cinder block, roofing, flooring, lighting, and electrical materials are provided by manufacturers whose proprietary concerns generally don't extend beyond their own products and the purposes to which they are put.

Open source is a natural quality of most construction. Materials and methods are open to inspection, and copying. If a builder finds a better way to put up shingles, hang a door, nail a joist, or pour a foundation, they don't keep it a secret. They share it, and the knowledge gets passed along from crew to crew.

It's also true that many construction materials are full of patents and other intellectual property claims. Yet those claims are limited in scope by the modular nature of construction standards and practices. For example, you may choose to use proprietary door latches from one manufacturer throughout your house. Those latches may contain inventions patented or licensed by the manufacturer. But that manufacturer wouldn't think of extending those claims to the doors that contain those latches or to the whole house. To do so would be worse than absurd; it would be self-defeating.

It's not a coincidence that we already talk about software in terms of construction. We have "architectures," "builds," "designs," tools," "frameworks," "levels," "platforms," "components," and "structures." We have "sites" with "addresses and "locations" that are often "under construction."

The similarities between software and construction are so close in some ways that we can't help making sense of the former in terms of the latter. As cognitive science puts it, construction is a *conceptual metaphor* for software development and use. Lately George Lakoff (*http://www.rockridgeinstitute.org/people/lakoff*), the father of cognitive linguistics, has also done some borrowing from the same source. Instead of talking about "conceptual metaphors," he now *talks* about "frames" and "framing" (*http://www.rockridgeinstitute.org/projects/strategic/simple_framing*). In *Patterns of Software* (Oxford, Paperbacks, 1996), Richard Gabriel says, "Habitability is the characteristic of source code that enables programmers coming to the code later in its life to understand its construction and intentions and to change it comfortably and confidently." His ideal example is the New England farmhouse:

> The result is rambling, but each part is well suited to its needs, each part fits well with the others...The inhabitants are able to modify their environment because each part is built according to the familiar patterns of design, use and construction and because those patterns contain the seeds for piecemeal growth.

Stewart Brand says the same kind of farmhouse is an ideal example of "vernacular" construction:

> What gets passed from building to building via builders and users is informal and casual and astute. At least it is when the surrounding culture is coherent enough to embrace generations of experience.

Vernacular is a term borrowed since the 1850s by architectural historians from linguists, who used it to mean the native language of a region. It means "common" in all three senses of the word—widespread, "ordinary," and "beneath notice."

In terms of architecture, vernacular buildings are seen as the opposite of whatever is "academic," "high style," or "polite." Vernacular is everything not designed by professional architects—in other words, most of the world's buildings. Vernacular building traditions have the attention span to incorporate generational knowledge about long-term problems such as maintaining and growing a building over time. High-style architecture likes to solve old problems in new ways, which is a formula for disaster.

Vernacular buildings evolve. As generations of new buildings imitate the best of mature buildings, they increase in sophistication while retaining simplicity.

The opposite of vernacular is what Stewart calls "magazine" architecture: artistic, idealized, expensive, and made to impress rather than to operate. Here's another difference, described by Henry Glassie: "If a pleasure-giving function predominates, it is called art; if a practical function predominates, it is called craft."

It's hard to imagine anything more crafted and practical than a command-line interface. That insight comes to us from Neal Stephenson, the novelist and hacker, in his book *In the Beginning Was the Command Line* (HarperCollins, 1999). Here's how he describes the difference between the hacker-built Unix and operating systems architected in the magazine mode:

> The filesystems of Unix machines all have the same general structure. On your flimsy operating systems, you can create directories (folders) and give them names like Frodo or My Stuff and put them pretty much anywhere you like. But under Unix the highest level—the root—of the filesystem is always designated with the single character "/" and it always contains the same set of top-level directories:

/usr
/etc
/var
/bin
/proc
/boot
/home
/root
/sbin
/dev
/lib
/tmp

And each of these directories typically has its own distinct structure of subdirectories. Note the obsessive use of abbreviations and avoidance of capital letters; this is a system invented by people to whom repetitive stress disorder is what black lung is to miners. Long names get worn down to three-letter nubbins, like stones smoothed by a river.

...It is this sort of acculturation that gives Unix hackers their confidence in the system, and the attitude of calm, unshakable, annoying superiority...Windows 95 and Mac OS are products, contrived by engineers in the service of specific companies. Unix, by contrast, is not so much a product as it is a painstakingly compiled oral history of the hacker subculture.

Is it any surprise that, in the long run (which we've been in since long before and after Neal wrote *Command Line*), hacker culture has pushed up, like lava through cracks in the Earth, so much useful stuff to build on?

Vernacular construction, with its valuing of craft over art, tends to produce what Stewart Brand calls "Low Road" buildings. These tend to be "low-visibility, low-rent, no-style." He adds, "Most of the world's work is done in Low Road buildings...and even in rich societies the most inventive creativity, especially youthful creativity, will be found in Low Road buildings, taking full advantage of the license to try things."

Not coincidentally, Stewart's ideal Low Road building is MIT's late Building 20 (*http://www.eecs.mit.edu/building/20*). Known as "The Magical Incubator," Building 20 was home to countless scientific advances: radar, microwave, spectroscopy, quantum mechanics, atomic and molecular beams, masers and lasers, atomic clocks, radio astronomy, linear particle acceleration, magnetron phasing, fiber optics, digital data transmission, and much more. More significant, for the purposes of the open source narrative, is what Stewart Brand reports here: "The Tech Model Railroad Club on the third floor, E Wing, was the source in the early 1960s of most of the first generation of computer 'hackers' who set in motion a series of computer technology revolutions (still in progress)."

The term *hacker* is older than Moore's law, older than Unix, older than the whole software industry. Hacking has also persisted as a comparatively stable culture, while countless commercial "solutions" and fashions (moving to the top level there) have come and gone.

The construction industry makes a useful—even ideal—model for the software industry as it gradually matures because, while the software industry is a few decades old at best, construction is mature in the extreme. It might not be the oldest profession in the world, but it's probably the oldest industry. (Just ask the Masons.) It's also the largest industrial sector, with more than $3.5 trillion in revenues, worldwide. It understands and respects commodities (there's no fear and trembling about the "threat of commoditization"). It also finds plenty of ways to differentiate commodity products and make money with them. It understands the natural advantages and limits of patents and other intellectual property claims. It has lots of giant companies, but none that dominates like Microsoft does in personal computing, or like IBM used to do in mainframes. Nor does it have many (or any, as far as I know) large companies that make the 70–90% gross profit margins large companies in the software industry made in the days when there was no Net, and large vendors had to make their own infrastructure. (Or, in parlance that becomes more antique by the day, "be a platform provider".)

In fact, the ideals of modular and commoditized building materials have been around for a long time. Kim Polese, CEO of SpikeSource (disclosure: I'm an advisor to the company), is an industry veteran who labored toward the destinations of open source long before the means for getting there began to show up. "We tried a lot of good ideas," she says, "back when suppliers ruled the world: object-oriented programming. C++, CORBA, COM. Yet even our standards were isolated environments; no less monolithic than any vendor's silo. Our ideas—components, modularity, reusability—would have to wait for an open worldwide ecosystem to emerge."

So, we should stop and give credit where due to the vendors who worked in the meantime to build the proprietary infrastructures we call "platforms." Before the Net, and before a sufficient abundance of open source building materials appeared, there was often no choice. For some activities inside large enterprises, there still isn't much choice. If you're doing big-time Enterprise Resource Planning (ERP) or Business Process Management (BPM), there are no open source solutions out there. Still, the same used to be true of Customer Resource Management (CRM) and office private branch exchanges (PBXs)—to name two among many categories—but now that's changing with SugarCRM (*http://www.sugarcrm.com/home*) and Asterisk (*http://www.asterisk.org*).

So, the platform and silo system is still with us; it's just moving up the Civilization stack to the Commerce layer.

Smart software vendors who want to maintain their silos will still have to base them on free and open source infrastructure. That's what IBM is doing with Linux, which supports the company's proprietary DB2, Tivoli, and WebSphere products. It's what Apple did when it moved its whole silo from the decrepit Mac OS to Darwin, which is Appleized FreeBSD. And it's what Microsoft will do, eventually; but for now, we'll let that stand as prophesy.

The offerings are often mixed, so it gets hard (or annoying) to say what is open source and what is not. For example, while Apple contributes generously to the FreeBSD kernel (as well as to Apache, KDE, GNU, and other open source development communities), the OS X operating system Apple builds on BSD is highly proprietary. So is its popular iTunes software. In fact, iPods, for all their appeal, are essentially hardware extensions of iTunes software. They are a silo.

It would be a mistake, however, to dismiss Apple as a "proprietary" company. It is, but it also is not. Apple has an open source strategy. So do IBM, HP, Oracle, RealNetworks, Novell, Sun, SAP, and other large vendors that use open source strategies to support their proprietary offerings. All their strategies are different; but they are all based on an acceptance of open source as foundational infrastructure, on participation in open source development projects, and pm an appreciation for what open source provides to the world.

As long as we insist on treating open source and proprietary as polar opposites, we won't understand how complementary they can often be. Nor—if we are a company trying to succeed in a business world supported by open source—will we be able to come up with a useful understanding of how open source supports business, much less a strategy for putting that support to use.

For me, the wisest mind on this whole subject is Craig Burton. I first met Craig back in the 1980s, when he was one of the leading figures at Novell. As an old communications techie, I watched in amazement how thoroughly the networking vendors could screw up even a miraculously open standard like Ethernet. Even after IBM's PC became an industry standard, computer networking was a highly proprietary affair, in which customers faced baffling choices among local area networks (LANs) with names like DECnet, WangNet, OmniNet, Sytek, 3Com, Ungermann-Bass, Corvus, and IBM's Token Ring. There was thin and fat Ethernet, star and bus topologies, and a mess of noninteroperable standards and implementations, each based on what the trade called "pipes and protocols." Every vendor's set of pipes and protocols was a platform supporting a proprietary silo of products and services, sold exclusively to captive customers. Interoperability was a pie in the sky.

That whole paradigm was blown up when Novell, led by Craig Burton and Judith Clarke (she later married Craig and became Judith Burton), literally changed the networking conversation. When I first began to follow Novell, NetWare was a Motorola 68000-based file and print server. Craig brilliantly decided to move NetWare up a level, to become a Network Operating System (NOS) that worked independently both of hardware and of network pipes and protocols: essentially, as a networked filesystem for every PC, regardless of whose network connected those PCs. NetWare grew like wildfire. In the process, Craig and Novell reconceived networking as a set of services, and freed it from any notion of lower-level protocol and wiring dependency. When Craig, Judith, and Jamie Lewis later founded The Burton Group, their first achievement was formalizing the "network services model," which prevails to this day. When I said earlier that IM was not among the suite of standard Internet services, I was speaking in terms of that model.

Not long after I started with *Linux Journal*, I interviewed Craig a number of times, looking to get his point of view, both on open source and on the prospects for Linux (*http://www.searls.com/burton_interview.html*). He was greatly in favor of both, but he also had no patience with some of the rhetoric coming out of open source culture.

For example, he took issue with the notion that *open source* and *proprietary* are opposites. "There are collapsed distinctions here," he told me. "The opposite of open is not proprietary, but closed. The opposite of proprietary is not open, but public domain." If you "uncollapse" those distinctions, he said, and lay them out orthogonally, you get The Burton Matrix (see Figure 15-2).

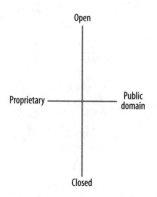

Figure 15-2. *The Burton Matrix*

Craig said that technical distinctions often collapse around moral sympathies. In the open source community, for example, "proprietary" and "closed" are considered *bad*, while "open" and "public domain" are considered *good* (see Figure 15-3).

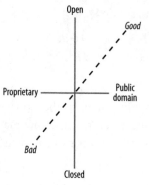

Figure 15-3. *Collapsed distinctions*

"However," he said, "if you remove morality as an issue, and spread out the two other distinctions that are collapsed, you have a good strategic framework for businesses to work with." To really take advantage of open source, he explained, you need to value ubiquity in your marketplace at least as much as you value scarcity in your product portfolio. In fact, your smartest move may be to take some of the products you're selling, and make them ubiquitous by moving them from proprietary/closed to open/public domain—literally, from scarcity to ubiquity (see Figure 15-4).

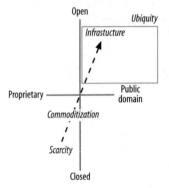

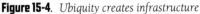

Figure 15-4. *Ubiquity creates infrastructure*

This is a form of commoditization (see Figure 15-5).

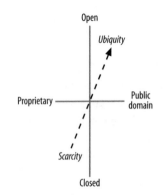

Figure 15-5. *The Open Source strategy: commoditize*

It's not the only one, of course. Most open source commodities are created from the start with the intention of putting them in the upper-right quadrant. But for businesses that want to create infrastructure and grow markets, this is one useful open source strategy. It's one of several a company can practice at the same time.

In IBM's case, the company adopted Linux (helping make it more ubiquitous), while also open sourcing Eclipse, moving it into the upper-right quadrant.

In Apple's case, the company open sourced nothing of its own, but used one of several other strategies: creation of a new standard (FireWire), adoption of an existing standard for the purpose of ubiquitizing it (USB, WiFi, ZeroConf/Rendezvous, MP3), and appropriation of an already developed codebase to save itself a lot of R&D work (FreeBSD, KHTML). Meanwhile, Apple has kept QuickTime and its growing portfolio of "i" applications on the proprietary side, even while opening them to free usage, essentially putting those in the upper-left quadrant.

When Tim O'Reilly had me show the Burton Matrix to Rob Glaser and Brian Behlendorf (Apache Foundation leader and founder/CTO of Collabnet), they helped me draw up the diagram shown in Figure 15-6 to explain how they were working together to open and ubiquitize many of the company's proprietary offerings (while also creating a development community).

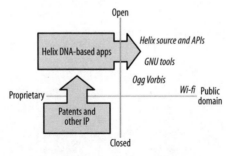

Figure 15-6. *Real's strategy*

Real's strategy, Rob said, was to move as much as possible from the lower to the upper left, and from the upper left to the upper right.

If we put the Burton Matrix beside the Long Now Foundation's "Layers of Time" model of Civilization, we see they have a common element: Infrastructure (see Figure 15-7).

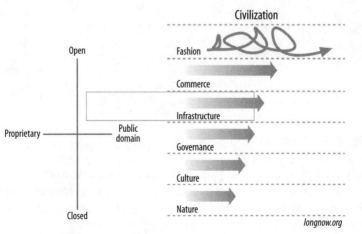

Figure 15-7. *Time layers and the Burton Matrix*

If we rock the Burton Matrix 45 degrees clockwise, superimpose it over the Civilization diagram, and cut out all but Infrastructure and Commerce, we have Figure 15-8.

Now we can start to see how the world ought to work (Figure 15-9).

Open source Infrastructure *supports* Commerce while Commerce *contributes* to open source Infrastructure.

We may or may not like what IBM, HP, Google, Apple, Sun, Oracle, Novell, and Red Hat build on the open source Infrastructure that supports them, but we have to

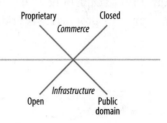

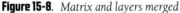

Figure 15-8. *Matrix and layers merged*

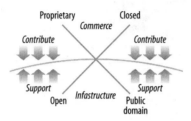

Figure 15-9. *How it oughta work*

respect the fact that all those companies contribute back to that Infrastructure. There is another word for this form of contribution (see Figure 15-10).

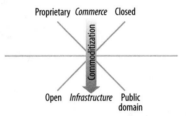

Figure 15-10. *Commoditization is contribution*

All of these examples, of course, pertain to technology companies. We don't see the same from Hollywood, which *uses* technology (including enormous quantities of Linux), but works against technology's interests, and with it the interests of the new world technology is busy creating (see Figure 15-11).

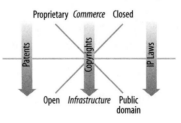

Figure 15-11. *How Hollywood sees it*

Unfortunately, there are a few technology companies that feel the same way. The most notable among these is SCO. Once a leading Unix vendor, SCO transformed itself into the sworn enemy of Linux and open source, first when it sued IBM, and later when it sued DaimlerChrysler and AutoZone. Regardless of the merits of SCO's cases (I think they are groundless, but that's not what matters here), the Fear, Uncertainty, and Doubt (FUD) effects have been enormous and undeniable.

It has never been easy to get IT personnel to talk about what they are up to in any case. It's not their job to speak for the company (that's up to PR and marketing), and they often don't want their bosses, competitors, or suppliers to know what they're doing with anything, including open source. But still, some would talk—enough so that editors like me could pull together enough facts to tell a fair story.

That changed with the SCO lawsuits against DaimlerChrysler and AutoZone. These suits had a highly adrenalizing effect on the legal departments of large companies. Formerly open channels of communication became tightly closed legal sphincters. For example, two out the four panelists for my Do-It-Yourself IT (DIY-IT) discussion at the Open Source Business Conference (OSBC) in April 2004 were no-shows. One was no mystery: the speaker showed up, but told me he wasn't allowed to speak. Although he wasn't allowed to give the reason, his expressions made the matter clear: the company didn't want anybody saying anything in public about Linux or open source. I didn't learn what happened to the other panelist, Phil Moore of Morgan Stanley, until later in the year, when he rose out of the audience during a panel discussion at the O'Reilly Open Source Convention and spoke bluntly about what was actually going on. He began:

> I work for the 38th largest company in the world, Morgan Stanley. We have a billion dollar IT budget. And we use a little of everything. Unfortunately. Excuse me, a LOT of everything. The trend I've seen in the last 10 years...is the exponential growth in the variety and the depth and breadth of installation of open source software in our infrastructure....What I'm seeing is that in the infrastructure, the core infrastructure, open source is going to take over, leaps and bounds....I'm predicting, right now, that by 2006 or 2007, we're going to be a 90% Linux shop.

At one point, Phil said, "We're still mostly a Solaris shop, but we are rapidly moving to Linux, though I'm not supposed to talk about that, for fear of being sued by SCO." Then he turned to Matt Asay (the Novell executive who ran OSBC) and added, "Which is the reason why I couldn't go to your conference, the OSBC. I wasn't allowed to go."

I ran those quotes, with Phil's permission, in my November 2004 column for *Linux Journal*. Phil no longer works at Morgan Stanley. He left voluntarily, but the fact that he's gone still speaks volumes. I want to thank him here for the honesty and courage it took for him to say what he did. Same goes for R0ml Lefkowitz (formerly) of AT&T Wireless, Roland Smith (formerly) of LSI Logic, Leon Chism of Orbitz, J.P. Rangaswami of Dresdner Kleinwort Wasserstein, and the rest of the handful of executives in large IT organizations who have talked to me factually and fearlessly about how open source—and Linux especially—are being put to good use in their companies.

Meanwhile, with a veil of silence over most IT departments, most of the publications covering IT are left with what comes easiest: writing about what vendors are doing. Or worse, treating open source projects as if they were vendors too. "*Microsoft v. Linux*: Who Will Win?" the headlines say. Again and again.

It's a false fight, and it always has been. We're making a world here. There's a limit to how well you can live in it and still ignore the fact that it exists. And how it got here.

CHAPTER 16

Tim O'Reilly

The Open Source Paradigm Shift

In 1962, Thomas Kuhn published a groundbreaking book titled *The Structure of Scientific Revolutions*. In it, he argued that the progress of science is not gradual, but rather (much as we now think of biological evolution), a kind of punctuated equilibrium, with moments of epochal change. When Copernicus explained the movements of the planets by postulating that they moved around the sun rather than the Earth, and when Darwin introduced his ideas about the origin of species, they were doing more than just building on past discoveries, or explaining new experimental data. A truly profound scientific breakthrough, Kuhn notes, "is seldom or never just an increment to what is already known. Its assimilation requires the reconstruction of prior theory and the re-evaluation of prior fact, an intrinsically revolutionary process that is seldom completed by a single man and never overnight."[1]

Kuhn referred to these revolutionary processes in science as "paradigm shifts," a term that has now entered the language to describe any profound change in our frame of reference.

Paradigm shifts occur from time to time in business as well as in science. And as with scientific revolutions, they are often hard fought, and the ideas underlying them not widely accepted until long after they were first introduced. What's more, they often have implications that go far beyond the insights of their creators.

---

1   Thomas Kuhn, *The Structure of Scientific Revolutions* (*http://www.press.uchicago.edu/cgi-bin/hfs.cgi/00/13220.ctl*), 7.

One such paradigm shift occurred with the introduction of the standardized architecture of the IBM personal computer in 1981. In a huge departure from previous industry practice, IBM chose to build its computer from off-the-shelf components, and to open up its design for cloning by other manufacturers. As a result, the IBM personal computer architecture became the standard, over time displacing not only other personal computer designs, but also over the next two decades, minicomputers and mainframes.

However, the executives at IBM failed to understand the full consequences of their decision. At the time, IBM's market share in computers far exceeded Microsoft's dominance of the desktop operating system market today. Software was a small part of the computer industry, a necessary part of an integrated computer, often bundled rather than sold separately. Those independent software companies did exist were clearly satellite to their chosen hardware platform. So, when it came time to provide an operating system for the new machine, IBM decided to license it from a small company called Microsoft, giving away the right to resell the software to the small part of the market that IBM did not control. As cloned personal computers were built by thousands of manufacturers large and small, IBM lost its leadership in the new market. Software became the new sun that the industry revolved around; Microsoft, not IBM, became the most important company in the computer industry.

But that's not the only lesson from this story. In the initial competition for leadership of the personal computer market, companies vied to "enhance" the personal computer standard, adding support for new peripherals, faster buses, and other proprietary technical innovations. Their executives, trained in the previous, hardware-dominated computer industry, acted on the lessons of the old paradigm.

The most intransigent, such as Digital's Ken Olsen, derided the PC as a toy, and refused to enter the market until too late. But even pioneers like Compaq, whose initial success was driven by the introduction of "luggable" computers, the ancestor of today's laptop, were ultimately misled by old lessons that no longer applied in the new paradigm. It took an outsider, Michael Dell, who began his company selling mail-order PCs from a college dorm room, to realize that a standardized PC was a commodity, and that marketplace advantage came not from building a better PC, but from building one that was good enough, lowering the cost of production by embracing standards, and seeking advantage in areas such as marketing, distribution, and logistics. In the end, it was Dell, not IBM or Compaq, that became the largest PC hardware vendor.

Meanwhile, Intel, another company that made a bold bet on the new commodity platform, abandoned its memory chip business as indefensible and made a commitment to be the more complex brains of the new design. The fact that most of the PCs built today bear an "Intel Inside" logo reminds us of the fact that even within a commodity architecture, there are opportunities for proprietary advantage.

What does all this have to do with open source software, you might ask?

My premise is that free and open source developers are in much the same position today that IBM was in 1981 when it changed the rules of the computer industry, but failed to understand the consequences of the change, allowing others to reap the benefits. Most existing proprietary software vendors are no better off, playing by the old rules while the new rules are reshaping the industry around them.

I have a simple test that I use in my talks to see if my audience of computer industry professionals is thinking with the old paradigm or the new. "How many of you use Linux?" I ask. Depending on the venue, 20% to 80% of the audience might raise their hands. "How many of you use Google?" Every hand in the room goes up. And the light begins to dawn. Every one of them uses Google's massive complex of 100,000 Linux servers, but they were blinded to the answer by a mindset in which "the software you use" is defined as the software running on the computer in front of you. Most of the "killer apps" of the Internet, applications used by hundreds of millions of people, run on Linux or FreeBSD. But the operating system, as formerly defined, is to these applications only a component of a larger system. Their true platform is the Internet.

It is in studying these next-generation applications that we can begin to understand the true long-term significance of the open source paradigm shift.

If open source pioneers are to benefit from the revolution we've unleashed, we must look *through* the foreground elements of the free and open source movements, and understand more deeply both the causes and the consequences of the revolution.

Artificial intelligence pioneer Ray Kurzweil once said, "I'm an inventor. I became interested in long-term trends because an invention has to make sense in the world in which it is finished, not the world in which it is started."[2]

I find it useful to see open source as an expression of three deep, long-term trends:

- The *commoditization* of software
- Network-enabled *collaboration*
- Software *customizability* (software as a service)

Long-term trends like these "three Cs," rather than the *Free Software Manifesto* or *The Open Source Definition*, should be the lens through which we understand the changes that are being unleashed.

---

2   Ray Kurzweil, Speech at the Foresight Senior Associates Gathering (*http://www.kurzweilai.net/articles/art0465.html?printable=1*), April 2002.

Software as Commodity

In his essay, "Some Implications of Software Commodification," Dave Stutz writes:

> The word commodity is used today to represent fodder for industrial processes: things or substances that are found to be valuable as basic building blocks for many different purposes. Because of their very general value, they are typically used in large quantities and in many different ways. Commodities are always sourced by more than one producer, and consumers may substitute one producer's product for another's with impunity. Because commodities are fungible in this way, they are defined by uniform quality standards to which they must conform. These quality standards help to avoid adulteration, and also facilitate quick and easy valuation, which in turn fosters productivity gains.

Software commoditization has been driven by standards, in particular by the rise of communications-oriented systems such as the Internet, which depend on shared protocols, and define the interfaces and datatypes shared between cooperating components rather than the internals of those components. Such systems necessarily consist of replaceable parts. A web server such as Apache or Microsoft's IIS, or browsers such as Internet Explorer, Netscape Navigator, or Mozilla, are all easily swappable, because to function, they must implement the HTTP protocol and the HTML data format. Sendmail can be replaced by Exim or Postfix or Microsoft Exchange because all must support email exchange protocols such as SMTP, POP, and IMAP. Microsoft Outlook can easily be replaced by Eudora, or Pine, or Mozilla mail, or a web mail client such as Yahoo! Mail for the same reason.

(In this regard, it's worth noting that Unix, the system on which Linux is based, also has a communications-centric architecture. In *The Unix Programming Environment*, Kernighan and Pike eloquently describe how Unix programs should be written as small pieces designed to cooperate in "pipelines," reading and writing ASCII files rather than proprietary data formats. Eric Raymond gives a contemporary expression of this theme in his book, *The Art of Unix Programming*.)

Note that in a communications-centric environment with standard protocols, both proprietary and open source software become commodities. Microsoft's Internet Explorer web browser is just as much a commodity as the open source Apache Web Server, because both are constrained by the open standards of the Web. (If Microsoft had managed to gain dominant market share at both ends of the protocol pipeline between web browser and server, it would be another matter! See "How the Web was almost won" [*http://salon.com/tech/feature/1999/11/16/microsoft_servers/print.html*] for my discussion of that subject. This example makes clear one of the important roles that open source does play in "keeping standards honest." This role is being recognized by organizations like the W3C, which are increasingly reluctant to endorse standards that have only proprietary or patent-encumbered implementations.)

What's more, even software that starts out proprietary eventually becomes standard-ized and ultimately commoditized. Dave Stutz eloquently describes this process in an essay titled "The Natural History of Software Platforms" (*http://www.synthesist.net/writing/software_platforms.html*):

> It occurs through a hardening of the external shell presented by the platform over time. As a platform succeeds in the marketplace, its APIs, UI, feature-set, file formats, and customization interfaces ossify and become more and more dif-ficult to change. (They may, in fact, ossify so far as to literally harden into hard-ware appliances!) The process of ossification makes successful platforms easy targets for cloners, and cloning is what spells the beginning of the end for plat-form profit margins.

Consistent with this view, the cloning of Microsoft's Windows and Office franchises has been a major objective of the free and open source communities. In the past, Microsoft has been successful at rebuffing cloning attempts by continually revising APIs and file formats, but the writing is on the wall. Ubiquity drives standardization, and gratuitous innovation in defense of monopoly is rejected by users.

What are some of the implications of software commoditization? One might be tempted to see only the devaluation of something that was once a locus of enormous value. Thus, Red Hat founder Bob Young once remarked, "My goal is to shrink the size of the operating system market." (Red Hat, however, aimed to own a large part of that smaller market!) Defenders of the status quo, such as Microsoft VP, Jim Allchin, have made statements such as "open source is an intellectual property destroyer," and paint a bleak picture in which a great industry is destroyed, with nothing to take its place.

On the surface, Allchin appears to be right. Linux now generates tens of billions of dol-lars in server hardware-related revenues, with the software revenues merely a rounding error. Despite Linux's emerging dominance in the server market, Red Hat, the largest Linux distribution company, has annual revenues of only $126 million, versus Microsoft's $32 billion. A huge amount of software value appears to have vaporized.

But is it value or overhead? Open source advocates like to say they're not destroying actual value, but rather, are squeezing inefficiencies out of the system. When compe-tition drives down prices, efficiency and average wealth levels go up. Firms unable to adapt to the new price levels undergo what the economist E.F. Schumpeter called "creative destruction," but what was "lost" returns manyfold as higher productivity and new opportunities.

Microsoft benefited, along with consumers, from the last round of "creative destruc-tion" as PC hardware was commoditized. This time around, Microsoft sees the com-moditization of operating systems, databases, web servers and browsers, and related software as destructive to its core business. But that destruction has created the opportunity for the killer applications of the Internet era: Yahoo!, Google, Amazon, and eBay—to mention only a few—are the beneficiaries.

And so I prefer to take the view of Clayton Christensen, the author of *The Innovator's Dilemma* and *The Innovator's Solution*. In a recent article in *Harvard Business Review*, he articulates "the law of conservation of attractive profits" as follows:

> When attractive profits disappear at one stage in the value chain because a product becomes modular and commoditized, the opportunity to earn attractive profits with proprietary products will usually emerge at an adjacent stage.[3]

We see Christensen's thesis clearly at work in the paradigm shifts I'm discussing here.[4] Just as IBM's commoditization of the basic design of the personal computer led to opportunities for attractive profits "up the stack" in software, new fortunes are being made up the stack from the commodity open source software that underlies the Internet, in a new class of proprietary applications that I have elsewhere referred to as "infoware" (*http://www.oreilly.com/catalog/opensources/book/tim.html*).

Sites such as Google, Amazon, and salesforce.com provide the most serious challenge to the traditional understanding of free and open source software. Here are applications built on top of Linux, but they are fiercely proprietary. What's more, even when using and modifying software distributed under the most restrictive of free software licenses, the GPL (*http://www.gnu.org/copyleft/gpl.html*), these sites are not constrained by any of its provisions, all of which are conditioned on the old paradigm. The GPL's protections are triggered by the act of software distribution, yet web-based application vendors never distribute any software: it is simply performed on the Internet's global stage, delivered as a service rather than as a packaged software application.

But more importantly, even if these sites gave out their source code, users would not easily be able to create a full copy of the running application! The application is a dynamically updated database whose utility comes from its completeness and concurrency and, in many cases, from the network effect of its participating users.

(To be sure, there would be many benefits to users were some of Google's algorithms public rather than secret, or Amazon's One-Click available to all, but the point remains: an instance of all of Google's source code would not give you Google, unless you were also able to build the capability to crawl and mirror the entire Web in the same way that Google does.)

---

3   Clayton Christensen, *Harvard Business Review*, Feb. 2004 (*http://www.tensilica.com/HBR_feb_04.pdf*).
4   I have been talking and writing about the paradigm shift for years, but until I heard Christensen speak at the Open Source Business Conference (*http://www.osbc2004.com*) in March 2004, I hadn't heard his eloquent generalization of the economic principles at work in what I'd been calling business paradigm shifts. I am indebted to Christensen and to Dave Stutz, whose recent writings on software commoditization have enriched my own views on the subject.

And the opportunities are not merely up the stack. There are huge proprietary opportunities hidden inside the system. Christensen notes:

> Attractive profits...move elsewhere in the value chain, often to subsystems from which the modular product is assembled. This is because it is improvements in the subsystems, rather than the modular product's architecture, that drives the assembler's ability to move upmarket towards more attractive profit margins. Hence, the subsystems become decommoditized and attractively profitable.

We saw this pattern in the PC market with most PCs now bearing the brand "Intel Inside"; the Internet could just as easily be branded "Cisco Inside."

But these "Intel Inside" business opportunities are not always obvious, nor are they necessarily in proprietary hardware or software. The open source Berkeley Internet Name Daemon (BIND) package used to run the Domain Name System (DNS) provides an important demonstration.

The business model for most of the Internet's commodity software turned out not to be selling that software (despite shrinkwrapped offerings from vendors such as Net-Manage and Spry, now long gone), but in services based on that software. Most of those businesses—the Internet Service Providers (ISPs), who essentially resell access to the TCP/IP protocol suite and to email and web servers—turned out to be low-margin businesses. There was one notable exception.

BIND is probably the single most mission-critical program on the Internet, yet its maintainer has scraped by for the past two decades on donations and consulting fees. Meanwhile, domain name registration—an information service based on the software—became a business generating hundreds of millions of dollars a year, a virtual monopoly for Network Solutions, which was handed the business on government contract before anyone realized just how valuable it would be. The "Intel Inside" opportunity of the DNS was not a software opportunity at all, but the service of managing the namespace used by the software. By a historical accident, the business model became separated from the software.

That services based on software would be a dominant business model for open source software was recognized in *The Cathedral & the Bazaar*, Eric Raymond's seminal work on the movement. But in practice, most early open source entrepreneurs focused on services associated with the maintenance and support of the software, rather than true software as a service. (That is to say, software as a service is not service in support of software, but software in support of user-facing services!)

Dell gives us a final lesson for today's software industry. Much as the commoditization of PC hardware drove down IBM's outsize margins but vastly increased the size of the market, creating enormous value for users and vast opportunities for a new ecosystem of computer manufacturers for whom the lower margins of the PC still made business sense, the commoditization of software will actually expand the software market. And as Christensen notes, in this type of market, the drivers of success "become speed to market and the ability responsively and conveniently to give customers exactly what they need, when they need it."[5]

Following this logic, I believe that the process of building custom distributions will emerge as one of the key competitive differentiators among Linux vendors. Much as a Dell must be an arbitrageur of the various contract manufacturers vying to produce fungible components at the lowest price, a Linux vendor will need to manage the ever-changing constellation of software suppliers whose asynchronous product releases provide the raw materials for Linux distributions. Companies like Debian founder Ian Murdock's Progeny Systems (http://progeny.com) already see this as the heart of their business, but even old-line Linux vendors such as SuSe and new entrants such as Sun tout their release engineering expertise as a competitive advantage.[6]

But even the most successful of these Linux distribution vendors will never achieve the revenues or profitability of today's software giants such as Microsoft or Oracle, unless they leverage some of the other lessons of history. As demonstrated by both the PC hardware market and the ISP industry (which, as noted earlier, is a service business built on the commodity protocols and applications of the Internet), commodity businesses are low margin for most of the players. Unless companies find value up the stack or through an "Intel Inside" opportunity, they must compete only through speed and responsiveness, and that's a challenging way to maintain a pricing advantage in a commodity market.

Early observers of the commodity nature of Linux, such as Red Hat's founder, Bob Young, believed that advantage was to be found in building a strong brand. That's certainly necessary, but it's not sufficient. It's even possible that contract manufacturers such as Flextronix, which work behind the scenes as industry suppliers rather than branded customer-facing entities, may provide a better analogy than Dell for some Linux vendors.

In conclusion, software itself is no longer the primary locus of value in the computer industry. The commoditization of software drives value to services enabled by that software. New business models are required.

Network-Enabled Collaboration

To understand the nature of competitive advantage in the new paradigm, we should look not to Linux, but to the Internet, which has already shown signs of how the open source story will play out.

---

5   Clayton Christensen, *Harvard Business Review*, Feb. 2004 (http://www.tensilica.com/HBR_feb_04.pdf.
6   From private communications with SuSe CTO, Juergen Geck, and Sun CTO, Greg Papadopoulos.

The most common version of the history of free software begins with Richard Stallman's ethically motivated 1984 revolt against proprietary software. It is an appealing story centered on a charismatic figure, and leads straight into a narrative in which the license he wrote—the GPL—is the centerpiece. But like most open source advocates, who tell a broader story about building better software through transparency and code sharing, I prefer to start the history with the style of software development that was normal in the early computer industry and academia. Because software was not seen as the primary source of value, source code was freely shared throughout the early computer industry.

The Unix software tradition provides a good example. Unix was developed at Bell Labs, and was shared freely with university software researchers, who contributed many of the utilities and features we take for granted today. The fact that Unix was provided under a license that later allowed AT&T to shut down the party when it decided it wanted to commercialize Unix, leading ultimately to the rise of BSD Unix and Linux as free alternatives, should not blind us to the fact that the early, collaborative development *preceded* the adoption of an open source licensing model. Open source licensing began as an attempt to preserve a culture of sharing, and only later led to an expanded awareness of the value of that sharing.

For the roots of open source in the Unix community, you can look to the research orientation of many of the original participants. As Bill Joy noted in his keynote at the O'Reilly Open Source Convention in 1999, in science, you share your data so that other people can reproduce your results. And at Berkeley, he said, we thought of ourselves as computer scientists.[7]

But perhaps even more important was the fragmented nature of the early Unix hardware market. With hundreds of competing computer architectures, the only way to distribute software was as source! No one had access to all the machines to produce the necessary binaries. (This demonstrates the aptness of another of Christensen's "laws," the law of conservation of modularity. Because PC hardware was standardized and modular, it was possible to concentrate value and uniqueness in software. But because Unix hardware was unique and proprietary, software had to be made more open and modular.)

This software source code exchange culture grew from its research beginnings, but it became the hallmark of a large segment of the software industry because of the rise of computer networking.

---

7   I like to say that software enables speech between humans and computers. It is also the best way to talk about certain aspects of computer science, just as equations are the best ways to talk about problems in physics. If you follow this line of reasoning, you realize that many of the arguments for free speech apply to open source as well. How else do you tell someone how to talk with their computer other than by sharing the code you used to do so? The benefits of open source are analogous to the benefits brought by the free flow of ideas through other forms of information dissemination.

Much of the role of open source in the development of the Internet is well known: the most widely used TCP/IP protocol implementation was developed as part of Berkeley networking; BIND runs the DNS, without which none of the web sites we depend on would be reachable; Sendmail is the heart of the Internet email backbone; Apache is the dominant web server; Perl the dominant language for creating dynamic sites; and so on.

Less often considered is the role of Usenet in mothering the Net we now know. Much of what drove public adoption of the Internet was in fact Usenet, that vast distributed bulletin board. You "signed up" for Usenet by finding a neighbor willing to give you a newsfeed. This was a true collaborative network, where mail and news were relayed from one cooperating site to another, often taking days to travel from one end of the Net to another. Hub sites formed an ad hoc backbone, but everything was voluntary.

Rick Adams, who created UUnet, which was the first major commercial ISP, was a free software author (though he never subscribed to any of the free software ideals—it was simply an expedient way to distribute software he wanted to use). He was the author of B News (at the time the dominant Usenet news server) as well as Serial Line IP (SLIP), the first implementation of TCP/IP for dial-up lines. But more importantly for the history of the Net, Rick was also the hostmaster of the world's largest Usenet hub. He realized that the voluntary Usenet was becoming unworkable and that people would pay for reliable, well-connected access. UUnet started out as a nonprofit, and for several years, much more of its business was based on the earlier Unix-Unix Copy Protocol (UUCP) dial-up network than on TCP/IP. As the Internet caught on, UUNet and others like it helped bring the Internet to the masses. But at the end of the day, the commercial Internet industry started out of a need to provide infrastructure for the completely collaborative UUCPnet and Usenet.

The UUCPnet and Usenet were used for email (the first killer app of the Internet), but also for software distribution and collaborative tech support. When Larry Wall (later famous as the author of Perl) introduced the patch program in 1984, the ponderous process of sending around nine-track tapes of source code was replaced by the transmission of "patches"—editing scripts that update existing source files. Add in Richard Stallman's GNU C compiler (gcc), and early source code control systems like RCS (eventually replaced by CVS and now Subversion), and you had a situation where anyone could share and update free software. The early Usenet was as much a "Napster" for shared software as it was a place for conversation.

The mechanisms that the early developers used to spread and support their work became the basis for a cultural phenomenon that reached far beyond the tech sector. The heart of that phenomenon was the use of wide area networking technology to connect people around interests, rather than through geographical location or company affiliation. This was the beginning of a massive cultural shift that we're still seeing today.

This cultural shift may have had its first flowering with open source software, but it is not intrinsically tied to the use of free and open source licenses and philosophies.

In 1999, together with Brian Behlendorf of the Apache project, O'Reilly founded a company called CollabNet to commercialize not the Apache product but the Apache *process*. Unlike many other OSS projects, Apache wasn't founded by a single visionary developer but by a group of users who'd been abandoned by their original "vendor" (NCSA) and who agreed to work together to maintain a tool they depended on. Apache gives us lessons about intentional wide-area collaborative software development that can be applied even by companies that haven't fully embraced open source licensing practices. For example, it is possible to apply open source collaborative principles inside a large company, even without the intention to release the resulting software to the outside world.

While CollabNet is best known for hosting high-profile, corporate-sponsored, open source projects like OpenOffice.org (*http://www.openoffice.org*), its largest customer is actually HP's printer division, where CollabNet's SourceCast platform is used to help more than 3,000 internal developers share their code within the corporate firewall. Other customers use open source-inspired development practices to share code with their customers or business partners or to manage distributed world-wide development teams.

But an even more compelling story comes from that archetype of proprietary software, Microsoft. Far too few people know the story of the origin of ASP.NET. As told to me by its creators, Mark Anders and Scott Guthrie, the two of them wanted to re-engineer Microsoft's ASP product to make it XML aware. They were told that doing so would break backward compatibility, and the decision was made to stick with the old architecture. But when Anders and Guthrie had a month between projects, they hacked up their vision anyway, just to see where it would go. Others within Microsoft heard about their work, found it useful, and adopted pieces of it. Some six or nine months later, they had a call from Bill Gates: "I'd like to see your project."

In short, one of Microsoft's flagship products was born as an internal "code fork," the result of two developers "scratching their own itch," and spread within Microsoft in much the same way as open source projects spread on the open Internet. It appears that open source is the "natural language" of a networked community. Given enough developers and a network to connect them, open source-style development behavior emerges.

If you take the position that open source licensing is a means of encouraging Internet-enabled collaboration, and focus on the end rather than the means, you'll open a much larger tent. You'll see the threads that tie together not just traditional open source projects, but also collaborative "computing grid" projects like SETI@home (*http://setiathome.ssl.berkeley.edu*), user reviews on Amazon.com, technologies like collaborative filtering, new ideas about marketing such as those expressed in The Cluetrain Manifesto (*http://www.cluetrain.com/book.html*), weblogs, and the way that Internet message boards can now move the stock market. What started out as a software development methodology is increasingly becoming a facet of every field, as network-enabled conversations become a principal carrier of new ideas.

I'm particularly struck by how collaboration is central to the success and differentiation of the leading Internet applications.

eBay is an obvious example—almost the definition of a "network effects" business—in which competitive advantage is gained from the critical mass of buyers and sellers. New entrants into the auction business have a hard time competing, because there is no reason for either buyers or sellers to go to a second-tier player.

Amazon is perhaps even more interesting. Unlike eBay, whose constellation of products is provided by its users and changes dynamically day to day, products identical to those Amazon sells are available from other vendors. Yet Amazon seems to enjoy an order-of-magnitude advantage over those other vendors. Why? Perhaps it is merely better execution, better pricing, better service, and better branding. But one clear differentiator is the superior way that Amazon has leveraged its user community.

In my talks, I give a simple demonstration. I do a search for products in one of my publishing areas, JavaScript. On Amazon.com, the search produces a complex page with four main areas. On the top is a block showing the three "most popular" products. Down below is a longer search listing that allows the customer to list products by criteria such as best-selling, highest-rated, by price, or simply alphabetically. On the right and the left are user-generated "ListMania" lists. These lists allow customers to share their recommendations for other titles related to the given subject.

The section labeled "most popular" might not jump out at first, but as a vendor who sells to Amazon.com, I know that it is the result of a complex, proprietary algorithm that combines not just sales but also the number and quality of user reviews, user recommendations for alternative products, links from ListMania lists, "also bought" associations, and all the other things that Amazon refers to as the "flow" around products.

The particular search that I like to demonstrate is usually topped by my own *JavaScript: The Definitive Guide*. The book has 192 reviews, averaging 4 1/2 stars. Those reviews are among the *more than 10 million* user reviews contributed by Amazon.com customers.

Now contrast that with the #2 player in online books, Barnesandnoble.com. The top result is a book published by Barnes & Noble itself, and there's no evidence of user-supplied content. *JavaScript: The Definitive Guide* has only 18 comments, and the order-of-magnitude difference in user participation mirrors the order-of-magnitude difference in sales.

Amazon doesn't have a natural network-effect advantage like eBay, but it has built one by architecting its site for user participation. Everything from user reviews, to alternate product recommendations, to ListMania, to the Associates program that allows users to earn commissions for recommending books, encourages users to collaborate in enhancing the site. Amazon Web Services, introduced in 2001, takes the story even further, allowing users to build alternate interfaces and specialized shopping experiences (as well as other unexpected applications) using Amazon's data and commerce engine as a back end.

Amazon's distance from competitors and the security it enjoys as a market leader is driven by the value added by its users. If, as Eric Raymond said in *The Cathedral & the Bazaar*, one of the secrets of open source is "treating your users as co-developers," Amazon has learned this secret. But note that it's completely independent of open source licensing practices! We start to see that what has been presented as a rigidly constrained model for open source may consist of a bundle of competencies, not all of which will always be found together.

Google makes a subtler case for the network-effect story. Google's initial innovation was the PageRank algorithm, which leverages the collective preferences of web users, expressed by their hyperlinks to sites, to produce better search results. In Google's case, the user participation is extrinsic to the company and its product, and so can be copied by competitors. If this analysis is correct, Google's long-term success will depend on finding additional ways to leverage user-created value as a key part of its offering. Services such as orkut (*http://www.orkut.com*) and Gmail (*https://gmail.google.com*) suggest that this lesson is not lost on them.

Now consider a counter-example. MapQuest is another pioneer that created an innovative type of web application that almost every Internet user relies on. Yet the market is shared fairly evenly among MapQuest (now owned by AOL), Maps.yahoo.com, and Maps.msn.com (powered by MapPoint). All three provide a commodity business powered by standardized software and databases. None of them has made a concerted effort to leverage user-supplied content, or engage its users in building out the application. (Note also that all three are enabling an Intel Inside–style opportunity for data suppliers such as NAVTEQ, now planning a multibillion-dollar IPO!)

Customizability and Software-as-Service

The last of my three Cs, customizability, is an essential concomitant of software as a service. It's especially important to highlight this aspect because it illustrates just why dynamically typed languages like Perl, Python, and PHP, so often denigrated by old-paradigm software developers as mere "scripting languages," are so important in today's software scene.

As I wrote in my 1997 essay "Hardware, Software and Infoware":

> If you look at a large web site like Yahoo!, you'll see that behind the scenes, an army of administrators and programmers are continually rebuilding the product. Dynamic content isn't just automatically generated, it is also often hand tailored, typically using an array of quick and dirty scripting tools.

The Architecture of Participation

I've come to use the phrase *the architecture of participation* to describe the nature of systems that are designed for user contribution. Larry Lessig's book, *Code and Other Laws of Cyberspace* (*http://www.code-is-law.org*), which he characterizes as an extended meditation on Mitch Kapor's maxim, "architecture is politics," made the case that we need to pay attention to the architecture of systems if we want to understand their effects.

I immediately thought of Kernighan and Pike's description of the Unix software tools philosophy (*http://tim.oreilly.com/articles/paradigmshift_0504.html*). I also recalled an unpublished portion of the interview we did with Linus Torvalds to create his essay for the 1998 book, *Open Sources* (*http://www.oreilly.com/catalog/opensources*). Linus too expressed a sense that architecture may be more important than source code. "I couldn't do what I did with Linux for Windows, even if I had the source code. The architecture just wouldn't support it." Too much of the Windows source code consists of interdependent, tightly coupled layers for a single developer to drop in a replacement module.

And of course, the Internet and the World Wide Web have this participatory architecture in spades. As outlined earlier in the section on software commoditization (*http://tim.oreilly.com/articles/paradigmshift_0504.html*), a system designed around communications protocols is intrinsically designed for participation. Anyone can create a participating, first-class component.

In addition, the IETF (*http://www.ietf.org*), the Internet standards process, has a great many similarities to an open source software project. The only substantial difference is that the IETF's output is a standards document rather than a code module. Especially in the early years, anyone could participate simply by joining a mailing list and having something to say, or by showing up at one of the three annual face-to-face meetings. Standards were decided by participating individuals, irrespective of their company affiliations. The very name for proposed Internet standards, Request for Comment (RFCs), reflects the participatory design of the Net. Though commercial participation was welcomed and encouraged, companies, like individuals, were expected to compete on the basis of their ideas and implementations, not their money or disproportional representation. The IETF approach is where open source and open standards meet.

And while there are successful open source projects like Sendmail, which are largely the creation of a single individual and have a monolithic architecture, those that have built large development communities have done so because they have a modular architecture that allows easy participation by independent or loosely coordinated developers. The use of Perl, for example, exploded along with CPAN (http://www.cpan.org), the Comprehensive Perl Archive Network, and Perl's module system, which allowed anyone to enhance the language with specialized functions, and make them available to other users.

The Web, however, took the idea of participation to a new level, because it opened that participation not just to software developers but to all users of the system.

It has always baffled and disappointed me that the open source community has not claimed the Web as one of its greatest success stories. If you asked most end users, they are most likely to associate the Web with proprietary clients such as Microsoft's Internet Explorer than with the revolutionary open source architecture that made the Web possible. That's a PR failure! Tim Berners-Lee's original web implementation was not just open source, it was public domain. NCSA's web server and Mosaic browser were not technically open source, but source was freely available. While the move of the NCSA team to Netscape sought to take key parts of the web infrastructure to the proprietary side, and the Microsoft-Netscape battles made it appear that the Web was primarily a proprietary software battleground, we should know better. Apache, the phoenix that grew from the NCSA server, kept the open vision alive, keeping the standards honest, and not succumbing to proprietary embrace-and-extend strategies.

But even more significantly, HTML, the language of web pages, opened participation to ordinary users, not just software developers. The "View Source" menu item migrated from Tim Berners-Lee's original browser, to Mosaic, and then on to Netscape Navigator and even Microsoft's Internet Explorer. Though no one thinks of HTML as an open source technology, its openness was absolutely key to the explosive spread of the Web. Barriers to entry for "amateurs" were low, because anyone could look "over the shoulder" of anyone else producing a web page. Dynamic content created with interpreted languages continued the trend toward transparency.

And more germane to my argument here, the fundamental architecture of hyperlinking ensures that the value of the Web is created by its users.

It's worth noting an observation made by Clay Shirky in a talk at O'Reilly's 2001 P2P and Web Services Conference (*http://conferences.oreillynet.com/p2p*) (now renamed the Emerging Technology Conference [*http://conferences.oreillynet.com/et2003*]), titled "Listening to Napster." There are three ways to build a large database, said Clay. The first, demonstrated by Yahoo!, is to pay people to do it. The second, inspired by lessons from the open source community, is to get volunteers to perform the same task. The Open Directory Project (*http://dmoz.org/about.html*), an open source Yahoo! competitor, is the result. (Wikipedia [*http://en.wikipedia.org/wiki/Main_Page*] provides another example.) But Napster (*http://www.napster.com*) demonstrates a third way. Because Napster set its defaults to automatically share any music that was downloaded, every user automatically helped to build the value of the shared database.

This architectural insight may actually be more central to the success of open source than the more frequently cited appeal to volunteerism. The architectures of Linux, the Internet, and the World Wide Web are such that users pursuing their own "selfish" interests build collective value as an automatic byproduct. In other words, these technologies demonstrate some of the same network effect as eBay and Napster, simply through the way that they have been designed.

These projects can be seen to have a natural architecture of participation. But as Amazon demonstrates, by consistent effort (as well as economic incentives such as the Associates program), it is possible to overlay such an architecture on a system that would not normally seem to possess it.

"We don't create content at Yahoo! We aggregate it," says Jeffrey Friedl, author of the book *Mastering Regular Expressions* and a full-time Perl programmer at Yahoo! "We have feeds from thousands of sources, each with its own format. We do massive amounts of 'feed processing' to clean this stuff up or to find out where to put it on Yahoo!" For example, to link appropriate news stories to tickers at Finance.yahoo. com, Friedl needed to write a "name recognition" program able to search for more than 15,000 company names. Perl's ability to analyze free-form text with powerful regular expressions was what made that possible.

Perl has been referred to as "the duct tape of the Internet," and like duct tape, dynamic languages like Perl are important to web sites like Yahoo! and Amazon for the same reason that duct tape is important not just to heating system repairmen but to anyone who wants to hold together a rapidly changing installation. Go to any lecture or stage play, and you'll see microphone cords and other wiring held down by duct tape.

We're used to thinking of software as an artifact rather than a process. And to be sure, even in the new paradigm, there are software artifacts, programs, and commodity components that must be engineered to exacting specifications because they will be used again and again. But it is in the area of software that is *not* commoditized, the "glue" that ties together components, the scripts for managing data and machines, and all the areas that need frequent change or rapid prototyping, that dynamic languages shine.

Sites like Google, Amazon, and eBay—especially those reflecting the dynamic of user participation—are not just products, they are *processes*.

I like to tell people the story of the Mechanical Turk, a 1770 hoax that pretended to be a mechanical chess-playing machine. The secret, of course, was that a man was hidden inside. The Turk actually played a small role in the history of computing. When Charles Babbage played against the Turk in 1820 (and lost), he saw through the hoax, but was moved to wonder whether a true computing machine would be possible.

Now, in an ironic circle, applications once more have people hidden inside them. Take a copy of Microsoft Word and a compatible computer, and it will still run 10 years from now. But without the constant crawls to keep the search engine fresh, the constant product updates at an Amazon or eBay, the administrators who keep it all running, the editors and designers who integrate vendor- and user-supplied content into the interface, and in the case of some sites, even the warehouse staff who deliver the products, the Internet-era application no longer performs its function.

This is truly not the software business as it was even a decade ago. Of course, there have always been enterprise software businesses with this characteristic. (American Airlines' Sabre reservations system is an obvious example.) But only now have they become the dominant paradigm for new computer-related businesses.

The first generation of any new technology is typically seen as an extension to the previous generations. And so, through the 1990s, most people experienced the Internet as an extension or add-on to the personal computer. Email and web browsing were powerful add-ons, to be sure, and they gave added impetus to a personal computer industry that was running out of steam.

(Open source advocates can take ironic note of the fact that many of the most important features of Microsoft's new operating system releases since Windows 95 have been designed to emulate Internet functionality originally created by open source developers.)

But now, we're starting to see the shape of a very different future. Napster brought us peer-to-peer file sharing, Seti@home introduced millions of people to the idea of distributed computation, and now web services are starting to make even huge database-backed sites like Amazon and Google appear to act like components of an even larger system. Vendors such as IBM and HP bandy about phrases like *computing on demand* and *pervasive computing*.

The boundaries between cell phones, wirelessly connected laptops, and even consumer devices like the iPod and TiVO are all blurring. Each now gets a large part of its value from software that resides elsewhere. Dave Stutz characterizes this as software above the level of a single device (*http://www.synthesist.net/writing/onleavingms.html*).[8]

Building the Internet Operating System

I like to say that we're entering the stage where we will treat the Internet as if it were a single virtual computer. To do that, we need to create an Internet operating system.

The large question before us is this: what kind of operating system is it going to be? The lesson of Microsoft is that if you leverage insight into a new paradigm, you will find the secret that will give you control over the industry, the "one ring to rule them all," so to speak. Contender after contender has set out to dethrone Microsoft and take that ring, only to fail. But the lesson of open source and the Internet is that we can build an operating system that is designed from the ground up as "small pieces loosely joined," with an architecture that makes it easy for anyone to participate in building the value of the system.

The values of the free and open source community are an important part of its paradigm. Just as the Copernican revolution was part of a broader social revolution that turned society away from hierarchy and received knowledge, and instead sparked a spirit of inquiry and knowledge sharing, open source is part of a communications revolution designed to maximize the free sharing of ideas expressed in code.

But free software advocates go too far when they eschew any limits on sharing, and define the movement by adherence to a restrictive set of software licensing practices. The open source movement has made a concerted effort to be more inclusive. Eric Raymond describes The Open Source Definition (*http://www.opensource.org/docs/ definition.php*) as a "provocation to thought," a "social contract...and an invitation to join the network of those who adhere to it."[9] But even though the open source movement is much more business friendly and supports the right of developers to choose nonfree licenses, it still uses the presence of software licenses that enforce sharing as its litmus test.

The lessons of previous paradigm shifts show us a subtler and more powerful story than one that merely pits a gift culture against a monetary culture, and a community of sharers versus those who choose not to participate. Instead, we see a dynamic migration of value, in which things that were once kept for private advantage are

---

8   Dave Stutz notes (in a private email response to an early draft of this piece), this software "includes not only what I call 'collective software' that is aware of groups and individuals, but also software that is customized to its location on the network, and also software that is customized to a device or a virtualized hosting environment. These additional types of customization lead away from shrinkwrap software that runs on a single PC or PDA/ smartphone and toward personalized software that runs 'on the network' and is delivered via many devices simultaneously."

9   From a private email response from Eric Raymond to an earlier draft of this paper.

now shared freely, and things that were once thought incidental become the locus of enormous value. It's easy for free and open source advocates to see this dynamic as a fall from grace, a hoarding of value that should be shared with all. But a historical view tells us that the commoditization of older technologies and the crystallization of value in new technologies is part of a process that advances the industry and creates more value for all. What is essential is to find a balance, in which we as an industry create more value than we capture as individual participants, enriching the commons that allows for further development by others.

I cannot say where things are going to end. But as Alan Kay once said, "The best way to predict the future is to invent it."[10] Where we go next is up to all of us.

Conclusion

The Open Source Definition and works such as *The Cathedral & the Bazaar* tried to codify the fundamental principles of open source.

But as Kuhn notes, speaking of scientific pioneers who opened new fields of study:

> Their achievement was sufficiently unprecedented to attract an enduring group of adherents away from competing modes of scientific activity. Simultaneously, it was sufficiently open ended to leave all sorts of problems for the redefined group of practitioners to resolve. Achievements that share these two characteristics, I shall refer to as "paradigms."[11]

In short, if it is sufficiently robust an innovation to qualify as a new paradigm, the open source story is far from over, and its lessons far from completely understood. Instead of thinking of open source only as a set of software licenses and associated software development practices, we do better to think of it as a field of scientific and economic inquiry, one with many historical precedents, and part of a broader social and economic story. We must understand the impact of such factors as standards and their effect on commoditization, system architecture and network effects, and the development practices associated with software as a service. We must study these factors when they appear in proprietary software as well as when they appear in traditional open source projects. We must understand the ways in which the means by which software is deployed change the way in which it is created and used. We must also see how the same principles that led to early source code sharing may impact other fields of collaborative activity. Only when we stop measuring open source by what activities are excluded from the definition, and begin to study its fellow travelers on the road to the future, will we understand its true impact and be fully prepared to embrace the new paradigm.

---

10   Alan Kay, spoken at a 1971 internal Xerox planning meeting, as quoted at *http://www.lisarein.com/ alankay/tour.html* (*http://www.lisarein.com/alankay/tour.html*).
11   Thomas Kuhn, *The Structure of Scientific Revolutions* (*http://www.press.uchicago.edu/cgi-bin/hfs.cgi/ 00/13220.ctl*), 10.

CHAPTER 17

Pamela Jones

Extending Open Source Principles Beyond Software Development

It starts with an idea.

Linus, for example, realized that if he put his kernel project online, people all around the world could work on it together, without having to be in the same building. They could quite literally write software in public that way, scattered around the world though they were.

Understanding such simple things changes the world sometimes.

But what about other areas? Is it possible to extend that same process to other kinds of work, or is it suitable only for software development? One thing can now be said for sure: legal research can be done that way. Groklaw is the proof of concept. But as I will explain, you need to tweak things just a bit.

I've done legal research for a living as a paralegal, and now I've done it with the world as a Groklaw volunteer, and I am therefore in a position to make comparisons. I think any company involved in any legal dispute that touches on technology could profit from using the open source method to tap into the community's group knowledge pool.

I'm a good researcher, and I do excellent work, but I know without a doubt that the input from thousands of readers made a huge difference in what Groklaw was able to accomplish in digging up helpful information in the SCO litigation.

How Did It Happen and How Does It Work?

When I began, it was just l'il ol' me. I had read Slashdot enough to know that while there was a high level of technical knowledge in some of the site's readers, the level of legal knowledge was low. I also saw there was a hunger to understand the law. Technical information that could influence the outcome of a lawsuit was available there, but it was not reaching the attorneys. And legal information that could help techies know what to dig up and helpfully provide was not readily available to the FOSS community.

At the beginning, I was trying to learn how to blog, because I had a job interview for a freelance assignment helping an attorney with his legal blog. You have to write something if you are blogging, so I decided to write about what I knew best, which is legal research. It felt private, like a diary, and I didn't think anyone would find what I was writing about or care much if they did.

I wrote to the air, thinking no one would read it anyway, and I horsed around, finding funny graphics for as many of the entries as I could, but it was just for fun, just to learn. I eventually chose to focus on the *SCO v. IBM* case because it appealed to my sense of humor and stirred my hatred of injustice, and because I knew quite a bit about the GPL, as it happened, and I knew SCO was going to fail on that part of its claims. I was also quite confident that Linus was not going to infringe on anyone's code on purpose. So, every day I'd add a little bit more to the story, as I saw news stories about the case and SCO's claims. I wrote as though I were talking with a good friend over dinner who asked me, "So, what's this SCO case all about? Is there any chance they could win?"

I didn't dumb anything down, because I wasn't thinking about an audience anyway, and I went into the research I was finding as deeply as possible. It did occur to me that I might find some things that would be helpful to Linus and to IBM—I figured IBM might have a service that scours the Net to find IBM-related stories—but that was the extent of my ambition. I knew most attorneys don't know a lot about computers or Linux or the GPL, and I knew a fair amount about them all, so I felt like I was throwing a message in a bottle out into the ocean, just hoping someone would find it and it would be useful.

After a couple of months, I got an email from a stranger, asking if I could please make the graphics smaller, because he was in Europe on dial-up and the blog took forever to resolve. He sent me instructions on how to do it.

Until that email, I had never bothered to read the stats on my site, even though Radio Userland, the software I started blogging with, provides them. I come from a family that has very little interest in computers, and I was used to people being emphatically uninterested in things I find fascinating, I didn't expect even my mother to read my blog. So, when I got the email, I was floored. I wrote back that I didn't know anyone was reading what I was writing, and he told me that lots of people read Groklaw, and that the community appreciated very much what I was doing. I was simply floored. I think I will remember that feeling until the day I die.

I looked at the stats and found out hundreds of people were reading what I was writing, apparently regularly. I turned on comments and, little by little, information began to be offered, particularly when I would ask for it, which I did more and more. Radio Userland also has stats on where your readers come from, what web site they visited before clicking on your site, and what I saw from that was that my readership was consistently growing, it was all word of mouth at that time, and the caliber of reader was very high. It included a high number of lawyers and programmers and professors at universities, from all over the world.

Finally came the idea. I had dug up some information about a Linux kernel author who made contributions to the kernel while an employee at Caldera, and when Slashdot put that article up on its site, the first time that had happened to Groklaw, the number of readers exploded. Even better, they had more information to offer. One would find an old press release, another an article from five years ago, another a speech by an executive, and so on.

When I saw that start to happen, I created a Legal Links page (*http://www.groklaw.net/staticpages/index.php?page=legal-links*), with links to legal resources, and started pointing to articles explaining things like copyright and patent law, things readers needed to understand so as to know which article or which press release detail mattered.

I realized then that Groklaw could be a bridge between the tech and the legal worlds, and that if I explained clearly how the court system works and what kind of evidence is valuable in that context, the community would find it, add what they knew, and it could work. The necessary pieces were in place. And I suddenly but totally understood the power inherent in this process of open, group effort. It felt like trying to ride a giant wave, as opposed to trying to turn on and then direct a stream of water in a particular direction. It had a life of its own, and my job was to try to follow the flow, not control it.

We found more and more. The readers built on each other's knowledge, and I learned that way too. My email level shot up also, as readers more and more began sending me tips and links and information. At the time, reporters were faithfully writing down every word the SCO folks spoke and reporting it as if it were all true, so I began reaching out to journalists. As we found information, I sent it to journalists, and some, to their credit, responded; some immediately, others over time.

Working as a Group

The first group project, in the sense of planned action, was to help everyone know how to write to journalists and editors so as to get good results. The FOSS community is not a group of phonies, and they tend to speak their minds. Also, a lot of us geeks are not socially skilled, so sometimes journalists would tell me how offensive the email they got had been. So, I put up on Groklaw examples of good letters, letters that did not offend, and the point was well received, so much so that I had two journalists remark that they never got any flames or nasty email from Groklaw readers.

At one point, we decided that someone should answer Darl McBride, CEO of SCO. He had written an open letter to the open source community. I asked if my readers felt like writing a response, and they did, so we worked on it online together, in the open. After all, his letter was addressed to us. Ideas would be left as comments, and then I'd incorporate them into the letter and post the next version; Then readers would suggest tweaks and more data, which I'd then incorporate and post the next version, until we were all satisfied. It took about two weeks. The Inquirer, which had been watching us create it, offered to post the letter and an accompanying collection of research supporting the points we had written on its web site.

This was very helpful, because by then, so many comments were being placed on Groklaw—hundreds more than any other site on Radio Userland—that the software was struggling, and we were afraid that if we got any more traffic, we'd simply melt off the Internet. That letter led to another growth spurt of Groklaw members, and at around that point, we simply had to move to larger quarters, and ibiblio graciously invited us on board after a Groklaw member wrote them to petition on our behalf.

It was still the early days, back in the fall of 2003, and we hadn't yet attracted many trolls or astroturfers, which is why it all worked so well. We were a group of like-minded people, all striving toward a common goal. No one cared a bit about credit, only results, and it was refreshing, even if it was one of the hardest things I've ever done.

What did I learn? That there truly is wisdom in crowds, and that you can rely on someone in such a group thinking of everything you truly need. Also, that some-body has to be willing to work harder than everyone else and be the final arbiter, or nothing ends up getting done. Later in Groklaw's development, there were other lessons to be learned.

Dealing with the Disrupters

As Groklaw became more popular and began winning recognition, along came the deliberate disrupters. I got my Ph.D. in trolls and astroturfer, you might say, so I'll share with you some things I learned in the University of Trolldom and Astroturfing, because it has a bearing on whether an open legal research project will succeed or fail.

Here's what I know. Trolls are mean. I can't stress that enough. If they see you trying to go to the right, they push to the left. Then they place comments whining that you won't go to the left or insist you ought to be going left but are going right when you shouldn't be. It doesn't matter at all that you are correct in wanting to go to the right. It doesn't matter that it's your decision to make. It doesn't matter that they are inter-fering with the work you've set out to accomplish. They are spoilers, and the bigger the blotch they leave on your page, the better they like it. There is nothing to do with a troll but delete his comments when you are sure trolling is the purpose. If you are weak in the knees and can't bring yourself to do that, trolls will destroy your open group project. It's that simple and clear. They enjoy destroying what you want to do.

When the open source project is legal research, you also must expect that the side you are working against will show up. They won't be wearing an ID tag. They are essentially spies. Here's how you will know: they work harder than anyone at first, and when they think you are lulled, they try to destroy your reputation and maybe your life, if they have the resources to do so. The interval when they are helping, however, has one purpose only: to gather information to use against you later and to form relationships with your volunteers, so they can undermine from within. It's absolutely essential to identify and either eject or corral such individuals. I can't explain how to do this in great detail until after the SCO wars are over, but it's not impossible to do. Of course, you need a strong stomach and a bit of a tinfoil hat.

I will give just one example. The very first such individual showed up when Groklaw was very new. He began by attacking Linux, then pretended to have an epiphany thanks to Groklaw, and then tried to stir my readers into unhelpful actions. For example, he suggested that everyone send Darl McBride certified letters protesting his actions. Certified letters. Right: SCO would have everyone's address.

Another time, he suggested everyone go to court in Groklaw T-shirts and take PDAs and phones to record the session, which they could stream to Groklaw live. That, of course, would have been a problem. First, the T-shirts would have made partici-pants look undignified, but it would also have made them easily identifiable. This was not helpful. And recording a court session is a violation of court rules. I could just see the headline, so I had to put the kibosh on that fast. I did so by deleting his comments.

I know someone will put up a web site all about this now, but I don't care. You need to know that such things will happen, and you must be ruthless in making sure such individuals don't take over. They will try. Some will be fooled and will criticize you for stomping on the spy's ideas, which he will offer with so much mock sincerity, it isn't hard to comprehend how others accept it at face value. You can't explain pub-licly that you researched the individual and are reasonably sure they are a spy, and you must just take your lumps. Let them put up web sites. In the end, what matters most is that they are isolated.

Astroturfers are sometimes of that same mind, but usually they just want to steer the conversation their way. They don't want to be kicked off, so they are subtler. We have had a number of astroturfers. I call them the "I used to love Groklaw, but" crowd. Some of them look, at first, like spies. They work harder than anyone, take in all the info they can about you and how you work, and then—diverging from the spy path—they try to steer your project their way. If they fail, at some manufactured moment they publicly find fault with you and your work and loudly make their grievances known to the world, using your own web site and others to try to destroy your reputation. Sometimes they'll put up whole web sites about it. They're mean too, but it's just a job. Nothing personal. They just play-act the emotion.

All three groups will, sad to say, appeal to some of your readers. They deliberately manufacture issues they know will draw followers. If you leave their comments on your site, they will take over the conversation and readers will leave in disgust. If you moderate them away, they will loudly proclaim their love for freedom of speech, and some will join them, not realizing they are being played like violins. The purpose is to destroy your project and make sure it never succeeds. This is something that rarely happens in Linux kernel development, and in my experience it requires tweaking the open source process just enough to keep getting your work done.

Deciding what goes into the Linux kernel is a breeze in comparison to deciding whose ideas can be trusted in doing legal research. You must trust your instincts, and it is one of the most important reasons the majority can *never* rule when doing research.

The Difference Between Doing Legal Research in Public and Writing Software in Public

I mention all this because if you are doing legal research in public, sometimes you can't say in the open all you might say privately or if doing legal research for a firm. Parties are in litigation really don't want to show the other side their cards until trial, as you may have observed in the protracted discovery wars between IBM and SCO.

You may have an idea for an avenue to research, for example, but you don't know what the result of your research might be. If it is negative, it might not be wise to present the news that you are researching this area until you know what you have found. Sometimes information that seems negative, upon deeper digging, turns out to be helpful, and you very much might want to wait to tell the world all about it. It isn't a matter of hiding information; it's more a question of timing and presentation.

Someone in a legal research project has to know what to keep private and what to make public. There is a great deal at stake, and the outcome can be affected by the decisions you make. That simply never happens in developing the kernel. So as time went on, I built up a feel for whom I could trust in an inner circle of advisers, for both legal and technical research. No one person can do everything, so spreading out the responsibility is vital.

I view the most beneficial structure for such work as a kind of pyramid, where anyone is free to contribute at the bottom of the structure, but as the information moves up the chain, it finally has to go through one or a few at the top of the pyramid. In my experience, that person or persons must be able to say no and mean it, come what may. They must know enough about legal work to intelligently decide what should and should not be published and in which direction to take the work next. And they have to have thick skin, because criticism is sure to come from those who wish to turn the process upside down and have all decisions made by some kind of democratic vote. Linus doesn't even do things that way, but some will be *sure* he does and will try to make you follow such a setup.

Perhaps that works in other fields, but in my experience it doesn't in legal research. It probably *would* work beautifully if all the volunteers were lawyers, paralegals, and professors. Or it might work if your geek contingent didn't vote on legal issues, only tech issues. Otherwise, you are doomed, because it is hard for those who aren't legally trained to realize just how complex the law really is, and when they learn a little, they sometimes think they know enough to begin running the process. A little knowledge can actually be worse than none at all, especially when accompanied by a lack of humility. I could write three chapters on this subject, but I'll spare you.

The reverse is true for me with tech decisions. I know I am not the expert there, so I never make those decisions. I trust reliable lieutenants to decide such things, and I listen to my readers very carefully.

It's the same with deciding which stories to mention and which to ignore. Part of Groklaw's purpose is antiFUD, but there is so much of it, what do you cover? I've learned to trust my readers' opinions on this, and if I get a lot of email about a story, I know it matters, even if I didn't think so originally. So, there is a kind of group decision-making.

In many ways, it's not unlike the kernel process, but there are elements of necessary secrecy in legal research that you don't have in programming. No one is likely to sue you for what you post about the kernel, but someone very well may over open legal research.

For example, we tried a second public group project—a summary page—and the troll-astroturf contribution was so high, I was afraid of being sued, because they left outrageous comments that I frantically scurried to get rid of, and they presented ideas that—while sounding superficially plausible—were actually designed to take the work in a direction that would undermine the effectiveness.

Eventually, we had to take the work private, which was not a huge problem, because by then I knew who was skilled at this kind of thing. Groklaw is a meritocracy. I leave the structure loose, so anyone can volunteer to do anything they feel like doing, but over time, I notice whose work is most useful, and others usually agree.

Still, it's an unfortunate thing that we had to do that project behind the scenes, because we had to limit ourselves to only those we already knew, which is not desirable in an open project. The workaround I've found is to do the fundamental work with known and trusted volunteers, and then post the results for comment and tweaking by the public at large. That keeps the door open to some brainiac newcomer, which you want, but it doesn't let spies and disrupters ruin things, which you don't want.

Why and When It Works

I don't think the process would work as well for a less, shall we say, inspiring case. Volunteers responded because they seriously cared about the outcome, not because they found learning to do legal research fascinating. I have gotten a lot of email about enjoying the learning, actually, but I also know that SCO was an inspiration. For some, watching an attack on Linux is like watching someone kick Dorothy's dog, Toto: people get mad and want to do something about it. You don't get the same response in all cases or by paying people. There isn't enough money in the world to pay me for the amount of work I donated to Groklaw, the nights without sleep, the anxiety, or the jerks I had to deal with sometimes, if I may speak plainly.

But it isn't by any means the only case I or my readers care about. Patents and standards also interest the FOSS community and should there eventually be a patent infringement attack on Linux or GNU/Linux, as I believe there will be, I know for sure that the community will react and be available to help. I hope and expect that Groklaw will be ready to be useful again, perhaps in doing prior-art searches, for example, which could definitely be done completely in the open, in contrast to the legal research in the SCO case.

I also know that if a company had a tech issue and needed to tap into the Groklaw group mind, they could simply place the issue as a comment, and readers would tell them whatever they knew. The encouragement on Groklaw is that you provide either a URL or personal experience to back up the thoughts you've expressed, so that anyone can follow the thread and prove or disprove it. That is vital. It lets everyone know that what they comment about is important, that they must stand behind what they write and be responsible to be careful.

The power of applying open source principals to legal research is real. I've lived it, and I feel it. It worked because no one knows as much as all of us together. There is no law firm in the world that can afford to hire the numbers of researchers Groklaw made available. And a small group of trained paralegals would not have been able to find all the technical information that we at Groklaw found together.

So the bottom line is this: as long as there is the heart and the will to do it, the open source process is effective in doing legal research. If you would like to experience it in action, come and join in the fun.

CHAPTER 18

Andrew Hessel

Open Source Biology

Open source software (OSS) has played a central role in the growth of the Internet and increases in economic importance each year. It has rapidly changed the face of computing, with server side companies like Sun Microsystems, to end-user companies such as Adobe, to full platform/service companies like IBM incorporating open source into their offerings. With this success, open source is poised to diversify its influence. One experiment is open source biology (OSB), the idea that biological products such as drugs, vaccines, or pest-resistant crops, can be developed using open intellectual property (IP) models.

Academic science, like open source, supports the belief that knowledge evolves best when ideas, data, and methods are freely shared, and each contributor can build on the works of others. Universities have housed and promoted scientific thought for more than 1,000 years, creating a public commons. In contrast, alchemy is the forerunner of modern business. Today, with academic research a valuable economic good, weighing the societal benefits of freeing or protecting IP is a pressing challenge. In no scientific discipline is this more important than biology, central to all living things.

Commercial biotechnology was founded on the premise that strong IP protection was necessary. However, after nearly three decades, a sustainable industry has not yet been achieved. Public mistrust of the genetic technology persists. Now, with biology facing a paradigm shift, one where synthetic DNA will replace conventional manipulations, genetic engineering is converging with software engineering. OSB, guided by lessons from open software development, could result in a new, economically supportable route to biological products.

The Rise of Modern Biotechnology

The success of the Manhattan Project brought university research to national attention at the end of World War II. Recognizing the economic and defensive value of this work, the project's director, Vannevar Bush, produced a report for President Roosevelt titled "Science—The Endless Frontier," encouraging greater federal support for public research. This document led to the creation of both the National Institutes of Health (NIH) and the National Science Foundation (NSF), now the main agencies that support life science studies.

Through the late 1960s, biological science was conducted almost exclusively within academia and had few ties to business. The commercial value of biology was unrecognized. Drugs were chemicals—and pharmaceutical innovation had stalled in the absence of new targets. Chemistry and engineering, not biology, represented the majority of industry partnerships. Where university-industry relationships did exist, activity was generally low. Technology transfer occurred primarily via corporations hiring university graduates or academic consultants.

In the early 1970s, a new generation of life science companies began to appear, with some focused on developing DNA technologies. Although DNA was discovered in 1953, it had remained a curiosity of chemists. The amino acids (the fundamental building blocks of proteins) corresponding to genetic code were not determined until 1965, and few techniques existed to "read" or edit the chemical instructions. A breakthrough came in 1973 when biologists Stanley Cohen (Stanford) and Herbert Boyer (UCSF) developed a practical way to manipulate DNA constructs. Their method for *recombinant DNA technology*, published in the *Proceedings of the National Academy of Sciences* later that year, described how fragments of DNA could be directly cloned and expressed in other cells.

With recombinant techniques, snippets of the molecule could be "cut" from one genome and "pasted" into the DNA of another with enzymatic tools. Common microbes like the gut bacterium *Escherichia coli* could be transformed into miniature factories, able to make biochemical products difficult or impossible to synthesize using standard chemistries. DNA created an efficient way to develop biologicals, including vaccines, viral components, or even complex proteins like hormones or antibodies. Heavily touted in the scientific and popular media, genetic engineering created great expectations for the future.

News of this technology reached Neils Reimer, the director of Stanford University's patenting and licensing efforts. Earlier, Reimer had developed a novel IP capture scheme intended to grow licensing revenues. Under his plan, IP was solicited proactively, with any resulting royalties split equally (1/3 each) between the submitting researcher, the researcher's department, and the university. With scientists benefiting directly, IP submissions increased significantly. Pleased with the results, Stanford went on to create a formal IP development service, the University Office of Technology Licensing, one of the first dedicated technology transfer offices in the country.

Reimer recognized how attractive the new DNA technology would be to industry. Cohen gave his permission to proceed with a patent, but true to academic principles, disavowed any personal share of proceeds. Reimer's application to the U.S. Patent and Trademark Office (USPTO) became the center of scientific and public controversy. Apart from the safety of genetic engineering, concerns included opposition to the patenting of a general research method, questions over the patentability of life forms and genes (eventually affirmed by the Supreme Court in *Diamond v. Chakrabarty*), and how university commercial activities might threaten free inquiry. Complicating matters, government policies that addressed the ownership of inventions made using federal funds were vague.

Eventually, commercial interest outweighed safety and regulatory concerns, and U.S. patent 4,237,224 was granted to the two universities in December of 1980. Two other applications related to the technology, collectively known as the *Cohen-Boyer recombinant DNA cloning patent,* were also issued and describe some of the fundamental tools for the sciences of molecular biology and genetics.

Commercial biotechnology began with a handshake deal. In 1975, venture capitalist Robert Swanson met Boyer at a bar near the UCSF campus. Over drinks, they formed a plan to create a company to sell gene-based medicines. They incorporated Genentech (*Genetic Engineering Tech*nology) the following year, with each making an initial investment of $500 in the firm. Two years later, the company had successfully cloned and expressed the gene for human insulin, a remarkable achievement for the day. When Genentech shares soared at IPO in 1980, they initiated a wave of speculative activity that carried another dozen biotechnology firms to the market over the next 24 months.

The biotech boom quickly transformed the congenial, open world of biological research into a genetic gold rush. Overnight, academic scientists were thrust into the role of executives and businessmen. Naïve, brash, and fueled by VC cash, they competed against each other to identify and express medically important genes. Still skeptical of the technology, pharmaceutical companies watched from the sidelines as Genentech cemented its early lead, licensed insulin to Eli Lilly, and brought recombinant Human Growth Hormone to market independently in 1985. In less than a decade, a credible threat to established chemistry-based pharmaceuticals had emerged from nowhere.

In virgin commercial space, biotechnology grew rapidly, along with a host of supportive companies. Firms scrambled to identify and characterize new genes—potential drug targets and a scarce, nonrenewable resource—funneling millions into parallel research streams. With a new tool for dissecting cellular biochemistry that allowed the molecular basis of human disease and health to be explored with precision, academic research also flourished. Other genetic technologies soon followed in the wake of recombinant DNA, including polymerase chain reaction (PCR)—a method of amplifying minute amounts of DNA. The rate of innovation in life science moved closer to that of the semiconductor industry.

New legislation was created to streamline the transfer of IP between the public and private domains. Significantly, the Bayh-Dole act of 1980, drafted to encourage private investment for the commercial development of academic discoveries, allowed institutions to file patents on inventions resulting from federally funded research. Research became a new source of revenue for universities. Schools established or expanded technology transfer offices, which grew from 25 or 30 throughout the country in 1980 to more than 250 today, and began to actively seek commercially attractive ideas. Biology figured prominently in this search. According to recent statistics, currently 10 of the top 25 holders of U.S. DNA-based patents are universities, research institutions, or the U.S. government itself.

How to best manage biotech IP was an open question. With gene sequences potentially worth billions, and the validity of biotech patents untested, aggressive IP capture was encouraged, if only as a defensive measure. This position has been reinforced over time, and is now widely reflected in industry practices and statistics. According to a survey of biotech patenting trends published in the October 2004 issue of *Nature Biotechnology*, only 42 DNA-based patents were approved by the USPTO in 1981; by 2001, this figure had swollen to 4,463—although numbers have fallen back to roughly 3,500 since this peak. The Biotechnology Industry Organization (BIO) maintains that strong IP is essential not only to the success of biotechnology companies, but also to their survival.

Biotechnology delivered on its promise to bring new innovation and wealth. Today more than 200,000 people are employed directly by the industry, and companies have appeared to fill every technological and market niche. Biotechnology has led to many new medicines, diagnostics tools, and consumer products, including food, textiles, and enzymes. It has also stimulated new innovation in the traditional pharmaceutical and agricultural companies, all of which today incorporate biotechnologies into their research and development programs.

Universities also participated in the prestige and economic rewards of biotech. Biological research has blossomed throughout the academic world. By helping the first biotech startups get their footing, universities have formed close relationships with these now-established firms. These alliances were seeded in part by the nonexclusive licensing of Cohen-Boyer methods, eventually leading to more than 400 companies' purchasing rights. Licensing proved a rich source of university discretionary funds, with Cohen-Boyer alone returning about $250 million to UCSF and Stanford over the 17 years the patents remained in force.

Intellectual Property and Growing Challenges

Biotechnology now touches on virtually every facet of human culture and technological achievement and is a rising economic force throughout the world. All Western countries (and many developing ones) have specifically targeted life science as an engine of long-term economic growth. The public and scientific expectation for what biotechnology can or will accomplish has yet to peak, in part because the technology is still relatively complex and confined to professional circles.

Life science research has enjoyed healthy expansion within academia, supported in part by the NIH, whose annual budget increased from $3 billion in 1980 to over $27 billion in 2003. Reflecting this growth, the primary result of research—scientific publications—continues to mushroom. Pubmed, a journal database service maintained by the National Library of Medicine, adds 7,000 new science and medicine citations each week and now indexes more than 15 million listings from 18,000 journals. The volume of scientific data has driven scientists into increasingly fine specializations in an effort to remain current with recent literature.

The biotechnology industry has also enjoyed rapid expansion, fueled by pharmaceutical research and development (R&D) spending that climbed from $1.5 billion in 1980 to more than $20 billion in 2002. More than 1,450 companies now operate in the U.S.; those publicly traded have a market capitalization of about $300 billion. Industry revenues have increased from $11 billion in 1994 to almost $39 billion in 2003, mainly from the sale of drugs—a consumer market that continues to expand. The Congressional Budget Office (CBO) reported in 2003 that prescription drug expenditures rose at an inflation-adjusted rate of 14.5% between 1997 and 2002 to surpass $160 billion annually, outpacing all other health spending categories.

Yet, despite these strong results, biotechnology is not the picture of perfect health. In defiance of R&D spending trends, drug output has slowed over the last decade. Applications to the Food and Drug Administration (FDA), the agency responsible for evaluating new pharmaceuticals, fell from a high of 131 in 1996 to only 72 in 2003, while approvals of new molecular entities (NMEs) fell 60% over the same period, dropping from 53 to 21. Even with the growth observed in drug sales, the biotechnology industry as a whole remains in the red, recording $50 billion in losses since 1994. The top 50 companies account for the bulk of the industry's market capitalization and revenues.

While research seems to be thriving, development—in life science, a process exclusive to the commercial domain—is struggling. Although often grouped together as "R&D," research and development are actually very different processes. Research tends to be relatively unstructured and produces new observations, often summarized in scientific publications. It attracts free thinkers, explorers, and risk takers. In contrast, development attempts to transform a research discovery into a finished product, ready for sale. Drug development can take years to advance a molecule through the series of phased clinical trials (ranging from I to III) meant to determine basic safety, dosages, and efficacy necessary before seeking the FDA's approval for sale. Most drugs never exit this "pipeline." If they do, they enter an ongoing postmarket analysis (Phase IV) that in part monitors for rare or long-term effects. Development thus attracts careful, detail-oriented, process-driven individuals intent on minimizing risks.

Pushing a drug through development requires massive investment and commitment. Wyeth R&D president, Robert Ruffolo Jr., estimated in 2003 that R&D charges now range between $1.2 to 1.4 billion, while others place this figure anywhere between $400 million and $800 million. There is no way to be sure of the true cost, as companies closely guard these figures, which are used to justify new drug pricing. Whatever the exact numbers, the cost of drug development continues to rise at about 12% to 14% each year, well in excess of inflation. Given finite financial resources, the increasing cost of developing a new drug is the main bottleneck between promising research and new therapeutics. Only a small fraction of research, public or private, will ever enter the development pipeline.

IP practices contribute to this constriction. Only heavily protected molecules are likely to be backed by investors and developed. Competitive pressures have also fostered a secretive mindset and produced a mass of patent claims renowned for its complexity. This "thicket" impedes collaboration and materials transfer with other companies and universities, slowing R&D. IP sculpts the overall form of biotech companies. In an effort to reduce intellectual friction while retaining proprietary control, companies are driven to bring outside groups "in-house" through purchase or hire. In part, this has resulted in successive waves of merger and acquisition (M&A) activity, consolidating the biotech and pharma industries to produce giant, global organizations.

While sales and marketing efficiencies may result from consolidation, little proof that size only can yield R&D efficiencies remains scarce. Research cannot be mandated, at any price. Meanwhile, candidate drugs in development—although selected with the utmost care—may fail at any point in the pipeline. Given these risks, the ability to remain flexible and make unbiased decisions would seem crucial, but large R&D organizations can display considerable inertia and be hard to steer. Research may be slow to transfer to development, while failing projects in development may linger in the pipeline, burning cash. Industrial scientists also face intellectual isolation, with little exposure to ideas or peers outside company walls. Finding the right balance for successful research and development has been difficult for companies—one factor in why life commercial R&D has yet to demonstrate any clear economies of scale.

Meanwhile, expanding corporate bulk narrows the range of development choices that make economic sense. Large companies often set their sights on *blockbusters*—molecules with the potential to bring in $1 billion or more in annual sales. This makes the choice of what candidates to advance into development critical—a multibillion-dollar, multiyear commitment with risks and rewards different for each molecule. Even FDA approval for sale does not eliminate risk exposure, since drugs can be withdrawn if serious complications are discovered—an outcome certain to produce a flurry of class action lawsuits. Accordingly, the industry lobbies that strong IP is necessary for companies to recoup R&D costs, have cash to expand R&D activities, and also accumulate defensive legal reserves.

At the other end of the corporate spectrum, small biotech companies also struggle with IP. Nimble and highly motivated, most struggle to manage cash "burn" just to survive among the big industry players. They face not only high R&D costs, but also substantial legal fees, as they work to create new products or technologies. With limited cash and only a small number of patents in their IP portfolios, they produce little competitive pressure in the industry. Most remain speculative investments with almost no opportunity to independently market drugs. To persist, many companies form a symbiosis with big pharma, while others offer themselves as prey—innovative fodder for those with the resources to consume them.

Meanwhile, the richest and most plentiful source of low-cost innovation for companies—academic research—is fast drying up. Universities, while still friendly to commercial interests, better understand the value of their IP and have become shrewd negotiators. Technology transfer negotiations require more time—and end up costing much more money. In response to these complications, deals and collaborations with individual researchers have fallen out of favor, in preference to comprehensive "blanket" alliances. These sweeping arrangements, however, are much less attractive to the universities, particularly in the face of mounting reports of conflict of interest. The ideal economic balance for IP transfer from the public to private domain remains to be found.

The present drug development paradigm thus appears to suffer from economic challenges that have yet to be solved. Perhaps the most worrisome of these problems is the industry's failure, despite great internal effort, to decrease drug development costs. Without a turnaround in this metric, no reversal of drug output or consumer pricing trends can result—a mounting concern as Western society ages and demands more healthcare. Increasing tensions is the recall of several heavily marketed drugs that may have shown dangerous side effects even in early testing. Not only has this damaged consumer trust that companies will make safety the top priority, but it has also called in question the FDA's practices and relationship with industry. It has even forced a reevaluation of the financial risks and liabilities associated with large-market blockbuster drugs.

Open Source Biology

With present pharmacoeconomic trends unlikely to change in the immediate future, the path toward a sustainable drug industry remains as elusive. There is also a widening understanding that today's pharmaceutical companies—focused on disease management—may not have consumers' best interests in mind. Today there are few incentives for companies to improve any drug (at least until patent protection nears expiration) or to develop biological technologies that might lead to either prevention or cure.

In this light, some have begun to openly question whether there are other viable paths to drug development. At the heart of alternative routes is IP management. Since the passage of Bayh-Dole, scientists have been presented with two options for sharing research: publication; or patent and license with optional publication. The latter choice, heavily favored for discovery with commercial potential, has resulted in the current biotechnology industry. The alternative—open, unrestricted publication—has never been seriously considered a path to commercial development in the life sciences.

Now, open source, mainly used to develop computer software, has yielded strong evidence that open development may in fact be economically viable. OSS projects, including the Linux operating system and Apache Web Server, have become prominent examples that open strategies can result in robust, commercial-grade offerings. Open source has also emerged as an economic force, resulting in the formation of new companies like Red Hat, or adding revenues to the top line of others, like Sun and IBM. This success has encouraged speculation that similar results could be produced in biological development, if OSB could be made to work.

Recently, lawyers Stephen Maurer and Arti Rai and computational biologist Andrej Sali published a paper titled "Finding cures for tropical diseases: Is open source an answer?" to discuss how OSB might work. They suggest that OSB could organize many small research and development efforts toward the manufacture and testing of tropical disease drugs, reducing the final point-of-sale cost. Whether such a scheme would work in practice is unknown. However, there is no *a priori* reason why non-software products like drugs cannot be made using open source methods: it is not unreasonable that a community of open drug developers could produce open drugs. The unanswered question is whether, without IP, this development could be made economically sustainable enough to attract investors.

Any R&D effort, drugs or software, will consume resources that have real dollar costs. OSS developments are economically sustainable in part because these costs are kept very low. Geographic location, time zones, and physical facilities are not factors. Similarly, legal fees, product distribution costs, communication charges, and travel costs are also essentially zero. Few salaries are paid. OSS works because overhead is minimized while the aggregated value of donated developer time keeps growing over time. OSB faces a different economic reality. Any life sciences project, even an open source one, would come attached with physical constraints and very large costs. Laboratories are required. Millions of dollars of reagents, equipment, and testing are necessary. Realistically, before any OSB effort could yield a commercial product, the real dollar cost of biological R&D would need to be greatly reduced.

The Internet is helping to do this. The Web has already dropped the direct cost of doing scientific research, while also encouraging IP freedom. It has become an important repository for scientific information, much of it accessible openly and for free. Open access journal sites like the Public Library of Science (PLoS) and BioMedCentral now deliver peer-reviewed articles online at no charge and without copyright restrictions. Databases of DNA sequence data, human variation data, and, more recently, clinical trial results are available online. Sophisticated tools that link research datasets and support complex queries are beginning to appear. Science Commons, recently launched by the nonprofit Creative Commons, hopes to further interaction by making it easier for scientists, universities, and industries to share data and other IP. Overall, the Internet now allows most individuals, professionals or not, and even those in developing nations, free access to a wealth of high-quality scientific information. The main challenge for OSB to work, then, is to translate this research data into sustainable real-world open development projects.

New development strategies are beginning to emerge. Although not strictly open source, the company OneWorld Health appears to have found one successful path to reducing both IP and development charges. Based in San Francisco and billing itself as the first U.S. nonprofit pharmaceutical company, OneWorld assembles donated IP, expertise, and funds to further therapeutic development for diseases common in the Third World. It is working on drugs for malaria, leishmaniasis, and Chagas disease (a parasitic disease that can lead to heart failure), among others, and expects to launch its first product in 2005. However, while OneWorld's efforts are to be praised, its model is limited to drug molecules and markets not considered interesting to its proprietary partners.

The Biological Innovation for Open Society initiative (BIOS, *http://www.bios.net*), the brainchild of plant biologist Richard Jefferson, is also working to reduce the cost of biological development, and is willing to challenge proprietary groups to do so. Launched in 2004 and supported by a Rockefeller grant and technology from IBM, BIOS provides researchers with tools to share, manage, and navigate biotech IP, with an eye to facilitating open agricultural biotechnology. Keen on open source, Jefferson intends to create a patent commons and seed it with a broad method that allows plant researchers, public or private, to sidestep proprietary gene transfer technologies that restrict genetically modified (GM) crop development. Crops created with this community IP would be more affordable by growers throughout the world and be easier to manage than proprietary offerings. Meanwhile, the open patent commons would provide a defensive shield against proprietary challenges to their use.

Yet neither of these development models resembles the archetype OSS project, with an online platform, simple IP structure, and low overhead. For this reason, support is being found for a simpler model, a direct way for open source biology to follow in the footsteps of open software. The idea is to treat DNA, the foundation of virtually everything biological, for what it already is widely recognized in biology to be—a programming language. Virtually anything related to biology on this planet, living or not, can be reduced to this common denominator: a sequence of DNA bases that specifies its form and metabolism. DNA in a cell is no different from the 0s and 1s in a computer program. DNA is biological source code.

If OSS works, and DNA is software, don't reinvent: adapt. Allow genetic engineering to be done in the same way that software is engineered today, on computers with specialized software tools. In this way, OSB could closely parallel the strategies and, perhaps, realize the same advantages of open source software. Furthermore, since the DNA molecule can be a commercial product unto itself, and can direct biological synthesis of many bioproducts in vivo, circumventing the need for large production facilities, genetics can shorten the distance between research and development considerably. Because of these features, DNA code holds great potential to make OSB a reality, and also the possibility of developing a wide range of open biological products economically. Today a new science called *synthetic biology* is allowing researchers to move beyond mere speculation of this potential to practically test these ideas in reality.

Synthetic Biology and Genomic Programming

Since 1972, genetic engineering has been performed using the Cohen-Boyer recombinant techniques. These methods require DNA molecules to be extracted from cells and physically rearranged into new genetic designs—a process not unlike writing a letter "ransom note" style. Done in the lab, proficiency in this work typically requires an advanced degree with practical experience, a lot of equipment and reagents, and considerable time. Even relatively mundane procedures can take experienced technicians many months of tedious work, visible only by indirect methods. Compared to other modern engineering efforts—for example, microprocessor, aircraft, or building design, today performed in computer environments—genetic engineering remains a crude, manual process.

The emergence of synthetic biology (SB) changes everything. Founded on automated chemistries that permit long-chain DNA to be synthesized *de novo*, SB is a platform of software tools used to design and test artificial DNA molecules. It is an output device for bioinformatic software, and provides scientists with a way to write DNA sequences, not just read and comprehend them. The technology greatly lowers the barriers to genomic work: anyone with access to a computer can effectively create or edit DNA with exquisite precision. Overall, by transforming DNA into a biological programming language, SB represents the biggest improvement in genetic technology since Cohen-Boyer. It advances biological design into the digital age.

More than just bringing new speed and convenience to genetics, SB brings genetic scientists an alternative to unrestricted publication or patent. It is a creative tool, one that both proprietary companies and academic researchers will use to design DNA code. However, the technology brings an opportunity to reevaluate how the resultant IP should be protected. Today patent is used almost exclusively for biotech IP, including gene sequences—but synthetic biology makes new DNA designs into authored products like software. This type of IP is most often protected by copyright. Copyright would be inexpensive and easy to use, and would dovetail well with the application open source licenses, offering attractive IP benefits. However, without historical or legal precedent, there is no way to know how genetic copyrights would change R&D, or whether they would even be recognized as valid.

With the close similarity to software programming, synthetic biology gives OSB modeled after OSS a good chance of success. OSB could adapt the open source concepts, tools, licenses, and business models that already exist. Already, dozens of bioinformatic tools have been released under open source licenses, and software development platforms like SourceForge and Tigris could be easily modified to support DNA codewriting. Overall, for OSB based on SB to produce biological products, it would need to overcome only two main obstacles. First, it would need to attract an open developer community. Second, the genetic programs developed in the digital domain would need to be made affordably testable in real-world laboratories.

Open source synthetic biology will presumably find some support among genomic and bioinformatic scientists, many of whom currently release both data and tools openly on the Internet. However, Drew Endy, an assistant professor at MIT, is not taking any chances. He is actively seeding a new generation of biological programmers by teaching students how to build custom bacteria. Using presynthesized DNA dubbed "biobricks," or *de novo* code, Endy has created the biological equivalent of many electronic parts, including transistors, LEDs, and photosensors. Biobricks can be assembled in various ways to new create biological circuits, with bacterial cells the test breadboard. Biobricks form the foundation for MIT's multisite graduate student challenge, meant to encourage new synthetic designs and raise interest in synthetic techniques. The strategy is working: Endy's efforts have received wide attention in the technology press, and MIT's first conference on synthetics brought together more than 300 participants.

Endy is also a strong supporter of OSB, placing the biobricks standard registry in the public domain, a move he hopes will encourage others to use the technology and to share their own components. There is concern that unless an open ideology can be fostered, researchers might choose to patent each individual component, making biological programming a legal quagmire. Already, synthetic switches to turn genes on and off have been patented. Engineers Rob Carlson and Roger Brent, also early adopters of synthetic technologies, have warned of choosing a proprietary path and slowing innovation. In a white paper sent to DARPA, the advanced research agency of the U.S. military and an early backer of synthetic development, the pair argued that the development of a public domain "kernel" in synthetic biology could avert the negative consequences of having knowledge useful to the design of living organisms held proprietary. They maintain that biology conducted in an open manner would be, like open software, "robust and adaptive, providing for a more secure economy and country."

Great advantages could result if OSB can seed developer communities with keen interest in writing biological software. The sharing of genetic program designs openly should quickly lead to novel designs. The ability to engineer life on a computer desktop, not in a laboratory, should dissolve interdisciplinary boundaries and bring many new ideas into the biological sciences. Importantly, it allows genomic projects to aggregate and organize large numbers of developers. Online genomic development communities could blossom into virtual R&D organizations that dwarf those even of big pharma, yet be far more sustainable, open, and empowered. With inclusive membership and open data, "hobbyist" researchers—increasingly valuable contributors to astronomy, physics, and other sciences—would also enjoy the opportunity to participate meaningfully in collaborative genetic projects.

Meanwhile, the second obstacle to OSB producing a biological product, discovering inexpensive ways for genomic designs to be testable in the real world, is self-resolving. The per-base cost of long-chain DNA synthesis is dropping rapidly as commercial DNA providers compete for research customers. Today constructs that are viral size can be produced affordably. If current trends continue, human genome-size constructs will be realistic, both technologically and financially, by 2010. The economics of making commercial products with synthetics should become more attractive over time, if we can just learn how to write good code.

The Risk of Biological Hacking

Open source synthetic biology could result in a broad base of genomic skills in society and lead to low-cost gene-based commercial products. However, some people worry that convenient biological programming raises the chance that amateurs, hackers, or even terrorists will use the same tools to develop malicious genetic designs, either on purpose or by accident. While *in silico* genetic experimentation arguably poses little risk (the information remains within the digital domain), gaining access to synthetic DNA, or the equipment to make synthetic DNA, is not difficult, even for private individuals. The equipment for a functional DNA lab can be bought on eBay and would fit in a basement, kitchen, or garage. With the complete genomes for dozens of viruses publicly available—including Ebola, Marsburg, and SARS—a biological incident involving a synthetic virus may be a matter of *when*, not *if*. The proof of concept has already been demonstrated: in 2002, researchers at SUNY Stony Brook assembled an infectious synthetic poliovirus using mail-order DNA fragments.

The threat of a carefully engineered bioweapon unlike anything found in nature is thus real and significant. A CIA document titled "The Darker Bioweapons Future" published in 2003 cites a panel of experts that note "the effects of some of these engineered biological agents could be worse than any disease known to man." This panel also noted that genomics is entering "an explosive growth phase" and that "the resulting wave of knowledge will evolve rapidly and be so broad, complex, and widely available to the public that traditional intelligence means for monitoring WMD development could prove inadequate." These warnings make clear that the consequences of hacking DNA can be greater than those of hacking computer code. DNA programs, if they are chemically synthesized, will share physical reality with us. If released into the environment, the genetic information cannot be easily deleted or traced. Unwanted genetic distribution is already a problem in agricultural biotechnology, underscoring the fact that this is not a purely theoretical problem. Additionally, unlike in the digital world, nature cannot be "rebooted" if we make a serious mistake or encounter unexpected problems.

Yet, despite these concerns, synthetic DNA itself does not pose a *new* risk to society or the environment. Conventional laboratory methods of mutating and selecting organisms for enhanced pathogenicity have existed for decades, suggesting that those intent on using organisms for malicious purposes are probably already well equipped to do so. Genetic engineering is too powerful a technology to banish or outlaw, and it is already too late to suppress synthetic technologies: the underlying chemistries have been available for more than 20 years. Synthetic DNA will act mainly as an innovative accelerant, affecting all biotechnological applications, positive or negative. With synthetic technologies, the appearance of designer pathogens tuned to defeat our immune systems will not appear overnight, but we can no longer risk being complacent. For maximum safety, we need to broadly foster genetic awareness and skills in society, if only to better deal with rapidly evolving natural threats like SARS, Asian bird flu, or West Nile virus.

This makes the decision of whether to support OSB a critical one that extends beyond IP or economics. OSB may prove necessary as a means to assimilate the body of genetic information as a whole, an inherent advantage unlikely to be matched by more focused proprietary groups. No one company has the resources to understand the full complexity of DNA, the most difficult programming language for humans to comprehend. Because we didn't create the language or the computing environment, we must reverse engineer our understanding—taking systems apart one organism at a time, one cell type at a time, and finally one gene at a time. Putting all this data together again to get the big picture is like making a giant jigsaw puzzle. It requires cooperation, not fragmentation, to get perspective. By this rationale, the use of open genomics may bring far greater safety and security to synthetics. It should minimize fundamental design errors while also maximizing the responsive capability to unexpected challenges, including natural and engineered threats. Open computer software is also widely regarded as being more secure and more adaptive to threats than proprietary offerings.

Future Trends in Open Source Biology

Synthetic DNA appears poised to stimulate a new wave of genomic innovation, one closely aligned to software and using similar programming concepts. The decision of whether to support OSB will have long-term ramifications on the environment, the economy, and human health. There are thousands of biological products that could be designed collaboratively and produced inexpensively using open source synthetic DNA—

including vaccines, proteins, and gene therapies. If sufficient resources can be identified to bring these products to market without transferring them to proprietary interests—for example, through open IP partnerships with generic drug manufacturers or HMOs, government loans, or the collection of personal donations—the commercial sale of first-run therapeutics at generic prices could result, and not just for tropical medicines.

While real-world data and experiences will have to be collected before any conclusions can be drawn, the economics of open development may prove very attractive for synthetic genetics. Nature lends some support to this idea. DNA is openly shared in the physical world. The molecule is able to cross easily between species and is surprisingly plastic even *within* species or individuals. Not only this, but the grammar and syntax of the genetic language have remained conserved across all species throughout evolution. Nature doesn't waste energy with needless complications. If has retained a common genetic language and supported free exchange because this was less expensive than any other alternative.

If OSS proves any guide, synthetic DNA code will evolve fastest in the digital domain if allowed to be free. Efforts are underway in the scientific community so that a fair test of this idea can be made. Allies are being sought in government, law, finance, industry, and the general public to marshal support for open biological development projects. The idea is alluring to many people, in part because of the increasing awareness, use, and approval of open source software. Just as open software now provides individuals with greater software choice, OSB may one day offer individuals another source of safe and affordable gene-based technologies, therapeutics, and other bioproducts.

While OSB may not be immediately attractive, and perhaps may even be threatening, to some biotechnology companies, the benefits may be very appealing to others, especially those with limited resources. Given how few biotechnology companies can successfully produce and market any product, the value of proprietary IP may be considerably overvalued in life science—especially when sufficient R&D opportunities exist to prevent competitive overlaps. Companies may find that switching to open source allows them access to public information, institutions, and a scientific community usable or unapproachable by their closed peers. Open companies are able to exchange ideas or materials on core technologies easily, without threatening development within their particular specializations.

In addition to encouraging and amassing shared biological innovation, the decision to support OSB could produce immediate cost savings or benefits. Researchers, whether in academia or open companies, would conserve valuable research time not writing patent applications, while retaining the ability to freely publish and profit (albeit nonexclusively) from their innovations. Open companies would reduce legal costs, freeing cash to further develop programs. Universities would also save money on IP maintenance while retaining the ability to commercialize life science innovations. Meanwhile, government agencies involved in IP—including the patent office, the FDA, and the courts—could find their workloads easier to manage if open source use grows. This will be particularly important with biotechnology IP if, as expected, synthetic leads to an acceleration of research discovery. In any case, the use of patent and strong IP would still remain available as an option if deemed more appropriate.

OSB could also appreciably shrink public distrust related to genetic development. The public is growing more aware of the issues that surround GM foods, genetic testing, stem cell technologies, gene doping, and gene therapies. While OSB will not solve the differences of opinion that make these topics controversial, open source ensures that each individual has equal access to information and also a voice in what products or technologies are developed. Open developments succeed because there is a demand for the product and sufficient community support, be it skill or money, to bring them to market. This shifts the agenda away from proprietary interests toward the needs and demands of consumers—a shift likely to be strongly supported by those requiring new drugs or therapeutics.

OSB also fits nicely with emerging trends in health and medicine. Pharmaceutical development is expected to grow increasingly personalized in the future, drawing on recognition that individual variation—genetic, environmental, and behavioral—plays a large role in human health and disease. Pharmacogenomic efforts are underway to allow better disease appearance prediction (facilitating preventative steps or treatments) and to determine how any particular patient will respond to a given drug. In time, therapies tailored to small patient subgroups, even individuals, will become favored—although present economic trends do not support this direction. Unable to make large-market drugs sustainable, personalized medicine cannot be accomplished by today's drug industry. However, gene-based medicine offers a tantalizing solution: change the *informational content* of each drug, not the drug itself. With DNA-based treatments, or gene therapies, the chemical entity remains the same, while the biological effects it can produce in cells changes. DNA may prove an efficient drug, inexpensive to make and modify, with a range of delivery options.

Synthetic design software, connected to an open, integrated dataset, could quickly evolve to facilitate the fabrication of custom health solutions for small populations or individuals. This capability will likely bring about large changes to the way drugs are currently tested in clinical trials: customized drugs could be tested only on the individuals for whom they were designed, who would presumably be agreeable to their use. New medicines could be quickly delivered to those in need. Even without broad clinical trials, gene-based drugs should prove very safe and reliable as test data is accumulated. Flaws with a genetic design would be excluded by software from happening again, and any complications would naturally be isolated to very small patient populations. Open source personalized medicine could also bring some legal relief to drug companies. With an open, software-based drug development system, class action lawsuits against synthetic drug manufacturers would be highly unlikely except in cases of outright negligence or fraud.

In summary, the concept of OSB is highly compelling. Focusing collaborative energies on the rapid and inexpensive development, synthesis, and testing of innovative biological products is a commanding vision. The potential of open genomics, with countless applications in medicine, agriculture, and environmental protection, is enormous—but so are the challenges to society. Familiar ideas and structures may need to be discarded before forward steps can be taken. Large changes in how scientific information is shared, new drugs are designed and tested, and knowledge is protected are sure to come. As we advance toward this future, tapped into a vast global web of information, we may find ourselves worrying much less about the ownership of old ideas and more about how to generate the next new one.

Eugene Kim

CHAPTER 19

Everything Is Known

In Gabriel Garcia Marquez's epic novel, *One Hundred Years of Solitude*, Aureliano Buendia spends his days and nights in an old alchemical laboratory, isolated from the rest of the world, his nose buried in books. While Aureliano mostly keeps to himself and says little, he occasionally surprises members of his family or his small group of friends with a fact or insight they would have expected from someone more worldly. When they ask him how he could possibly have known about something, he responds simply, "Everything is known."

It's a fascinating thesis, one that not only highlights our individual ignorance, but also suggests a different approach to pursuing knowledge. When we are faced with something that is new to us, we often forge ahead blindly, learning and creating as we go along. We usually call this process "innovation." Our use of this word suggests that we place high value on uniqueness, on things we think have not happened before. However, how often are we qualified to determine whether something is new? How much do most of us know about our past or about the past of others? More importantly, if something is profound and relevant and significant, who cares if it's new? It's far more interesting if it's been discovered repeatedly in many different contexts. We should take note of important new ideas, but we should cherish ideas that prove their importance over and over again.

Many people find the emergent, collaborative aspects of free and open source software fascinating and compelling. However, before we start labeling anything as "innovative," we ought to seek what is already known by examining emergent collaboration in other contexts and by identifying recurring patterns there. If we can find patterns that occur over and over again in all these contexts, we can apply this knowledge to other areas. Simply put, these patterns can improve our ability to work together regardless of the domain, a skill on which the future of our world depends.

This chapter examines two stories: the PACT compiler project in the mid-1950s— the first collaborative software effort to transcend organizational boundaries—and the recovery effort at the site of the World Trade Center following the tragedy on September 11, 2001. On the surface, these stories seem very different, but closer examination reveals several common patterns, patterns that are also found in successful free and open source software projects. These patterns suggest principles for facilitating emergent collaboration in many different contexts, from software development to grass-roots politics.

The PACT Project

If you examine open source projects closely, you will discover a deeply embedded culture of collaboration, a culture that is embodied in the community's tools and practices. Programmers gather in a variety of contexts, both physically (conferences, user group meetings, code sprints) and virtually (mailing lists, IRC channels). Their toolkits are stocked with software for sharing, documenting, and collectively authoring code. The very notion of reusable source code is an implicit invitation to collaborate. Most programmers take this culture for granted, but these tools and norms did not always exist. They can be traced as far back as 1954, to an early collaborative compiler project known as PACT.

The onset of the Cold War in the late 1940s created a burgeoning aerospace industry in Southern California, and the companies that emerged suffered from a common affliction: a lack of good programmers. Computers were barely a decade old at this point, and programming was difficult and expensive. The notion of sharing source code was laughable, partially because the notion of source code barely existed. There were no high-level programming languages. Most people wrote software in machine language, which meant that software written for one type of machine could not be transferred to another. Although some people wrote homegrown assemblers and interpreters to ease their programming burden, there were no standards, so code sharing was still impossible. Without high-level languages, programming was expensive. Without the ability to share code, collaborating on software projects outside of one's organization was technically infeasible.

To make matters worse, there were only about 1,000 programmers in the United States at the time (Campbell-Kelly, 193). As a result, companies regularly raided each other's talent, which led to greater secrecy within companies about their computer-related work and a general tightfistedness with their prized programmers. While the technical barriers to collaboration were large, the culture of secrecy was the biggest impediment. R. Blair Smith, an IBM salesman, observed this culture firsthand in the early 1950s. He watched each of his customers struggle with the same expense and difficulty of developing software, and he decided that the overlap in effort was silly. If these companies would just set aside their differences and work together to simplify programming for everyone, everyone would win. As obvious as this seemed to Smith, he was realistic. Collaboration made sense, but it wasn't going to happen unless the culture of secrecy shifted radically.

Smith decided to throw a party. He invited all of his customers to dinner on November 15, 1952, at the Santa Inez Inn in Santa Monica, California (Mapstone, 367). The evening did not begin well. His guests were not used to interacting with their competitors, and the tension was thick. Smith decided something drastic was in order, so he decided to do something that neither he nor any of his IBM colleagues had ever done before. He bought drinks for his guests, compliments of IBM. Several rounds later, the tension had been replaced by good spirits, and the party began in earnest.

Little was gained other than a new sense of camaraderie, consensus regarding a second gathering, and the largest expense sheet Smith ever submitted to IBM. Smith stopped picking up the dinner and drinks tabs, but the meetings continued. The group was expanded, and a name was chosen—the Digital Computers Association (DCA). Speakers were invited to discuss various technical topics, but the real value of the meetings stemmed from the camaraderie. The atmosphere was jovial and irreverent, and the predinner cocktails quickly became the most important aspect of the meeting. As with the first gathering, little of tangible value was accomplished. However, two important things did happen. First, the members made a tacit gentleman's agreement not to raid each other's programmers (Carlson, 66). Second, bringing people with similar interests from different companies together made them recognize the commonality of their computing problems and the possible benefits of intercompany collaboration.

The first to test the possibility of these benefits were two DCA stalwarts, Jack Strong and Frank Wagner, both from North American Aviation. Concerned as always with the high cost of programming and the performance overhead of interpretive systems, Strong and Wagner wanted to explore automatic programming as a way of addressing these issues. On November 16, 1954, Strong and Wagner organized a meeting held at Douglas Aircraft Company's El Segundo plant (Melahn, 267). With representatives from five different companies—North American Aviation, Douglas Aircraft Company, IBM, Ramo-Woolridge, and the RAND Corporation—in attendance, Strong and Wagner proposed a cooperative effort to develop an automatic coding system for the IBM 701. The group agreed to establish two committees to undertake the project, the Policy Committee and the Working Committee, with each participating company committing one full-time representative to the project. Calling themselves Project for the Advancement of Coding Techniques, or PACT, the group decided to meet and work at the RAND Corporation, which was considered neutral territory among these competing interests.

The working committee immediately ran into trouble with language and culture. Wesley Melahn, a participant from RAND, wrote, "People didn't really know their neighbors' systems well enough to be able to talk about them intelligently, much less to understand the subtle pressures that made small points seem important enough to argue about" (Melahn, 268). Paul Armer, another important contributor, added:

> The members of the working committee of PACT spent several weeks in mutual education, because at first they had to overcome the "our way is best" attitude and also a serious language problem. That this mutual education led to mutual admiration and respect for other people's abilities is testified to by the final report of the PACT-I working committee. I quote from their *Primary* recommendation: "The spirit of cooperation between member organizations and their representatives during the formulating of PACT-I has been one of the most valuable resources to come from the project. It is essential that this spirit of cooperation continue with future project plans" (Armer, 125).

The collaboration ultimately resulted in the first software developed cooperatively by employees of several different companies: a working compiler that went through two versions, PACT-I and PACT-II. The group also delivered a series of papers at the 1955 Meeting of the Association of Computing Machinery (ACM) in Philadelphia. More importantly, PACT was a watershed. As a standard compiler that many companies used, it removed the technical barrier to collaborating on software projects. PACT also demonstrated the feasibility of a cooperative coding project and affirmed the value of cooperation, transforming the culture of business computing and creating a climate conducive to further collaboration.

Some of the similarities between PACT and today's free and open source projects are obvious. For example, neutral space is critical for encouraging emergent collaboration. R. Blair Smith's party would not have had the same effect had he worked for one of those aerospace firms rather than for IBM. Similarly, it was no accident that PACT participants chose to meet at RAND. Neutral space manifests itself in a number of ways in free and open source projects. When Linus Torvalds, the creator of the wildly successful Linux operating system, graduated from the University of Helsinki, he chose not to accept a job at one of the many emerging Linux distribution companies to maintain his neutrality. Although companies are not required to give up ownership of their code to make their software open source, many choose to transfer their copyright to neutral bodies, such as the Free Software Foundation and the Apache Software Foundation.

For collaboration to emerge, there must be a culture of collaboration. Prior to PACT, that culture did not exist. Fostering it required many gatherings and a large drink bill, courtesy of IBM. Those gatherings did not result in concrete deliverables, but they helped establish shared understanding and shared language. The culture of collaboration within successful free and open source projects is so deeply engrained among their participants, it's often taken for granted. Projects that simply release source code under an open source license in the absence of this culture usually fail.

The World Trade Center Recovery Effort

R. Blair Smith's dinner party in 1952 helped catalyze a shift in culture among the business computing community, but it still took two years before it manifested in a concrete project. The terrorist attacks on the morning of September 11, 2001 caused a far more immediate and visceral transformation. The collapse of the World Trade Center not only killed 3,000 people, it also brought fear and uncertainty to the entire United States. Nobody knew who was responsible, how it had happened, or worst of all, what to expect next. The only thing we knew was that there had been an attack and that our lives would never be the same.

That was all the residents of New York needed to know. As soon as the towers collapsed, those who were at the site began a concerted rescue and recovery effort. The impulse to help after such a horrific tragedy was not surprising, but the collaborative process that emerged from this initial impulse was remarkable. William Langewiesche, a correspondent for *The Atlantic*, was on hand from the beginning, and in his book, *American Ground*, he described the intense motivation shared among those who participated in the recovery:

> Throughout the winter and into the spring the workers rarely forgot the original act of aggression, or the fact that nearly 3,000 people had died there, including the friends and relatives of some who were toiling in the debris. They were reminded of this constantly, not only by the frequent discovery of human remains, and the somber visits from grieving families, but also by the emotional response of America as a whole, and the powerful new iconography that was associated with the disaster—these New York firemen as tragic heroes, these skeletal walls, these smoking ruins as America's hallowed ground. Whether correctly or not, the workers believed that an important piece of history was playing out, and they wanted to participate in it—often fervently, and past the point of fatigue. From the start that was the norm (Langewiesche, 9).

The recovery effort consisted of two parts: recovering the remains of the dead, and restoring the site. This latter task was literally and figuratively enormous. Each tower had stood 110 stories high, almost a quarter of a mile each. The collapse had created a 1.5-million-ton labyrinth of steel and concrete over 17 acres, with hazards hidden everywhere. The six-story-deep foundation and the entire New York subway system was in danger of being flooded by the Hudson River, thanks to a damaged subway tube and a fragile slurry wall protecting the foundation. Most dangerous of all was the underground chiller plant, which had cooled all of the buildings and which the collapse had rendered inaccessible. The plant contained 168,000 gallons of freon, and nobody knew the status of that gas. If the freon containers had somehow survived the collapse and if the workers inadvertently damaged it in the recovery process, the gas would immediately escape, displacing the air and suffocating those working underground. Additionally, if that gas caught fire, it would become a mustard gas–like chemical that would pose a threat to the entire city. As if navigating through the wreckage carefully for their own safety and the safety of the city weren't enough, the workers also had to sift through it carefully for remains of the victims and for potential evidence that could shed light on the tragedy.

Thousands of volunteers had converged onto the scene immediately following the attacks, looking to help in whatever way they could. However, those volunteers could do little, and after the first few days, the vast majority of them were turned away. The only volunteers who were allowed to stay were members of the Red Cross and Salvation Army who fed the workers. The recovery effort required expertise and plenty of heavy machinery. There were national and city plans for responding to disasters of all sorts, including terrorism, that prescribed the procedures and hierarchies for requisitioning the necessary resources. However, those organizational charts were scrapped, and the plans were never executed. Langewiesche explained:

> The problems that had to be solved were largely unprecedented. Action and invention were required on every level, often with no need or possibility of asking permission. As a result, within the vital new culture that grew up at the Trade Center site even the lowliest laborers and firemen were given power. Many of them rose to it, and some of them sank. Among those who gained the greatest influence were people without previous rank who discovered balance and ability within themselves, and who in turn were discovered by others (Langewiesche, 11).

Kenneth Holden and Michael Burton, top officials of New York's Department of Design and Construction (DDC), quickly emerged as the leaders of the recovery effort. The afternoon following the collapse, the two met and started enlisting resources within the DDC to help with the rescue effort. Langewiesche wrote:

> Holden and Burton responded tactically, with no grand strategy in mind. At the police headquarters they discovered a telephone in a room off the temporary command center—a chaotic hall filled with officials struggling to get organized—and they began making calls. No one asked them to do this, or told them to stop. One of the deputy mayors there had formally been given the task of coordinating the construction response, but with little idea of how to proceed, he had so far done nothing at all. Holden and Burton themselves were operating blind, groping forward through the afternoon with only the vaguest concept of the realities on the ground. The DDC's previous experience with emergencies had been limited to a sinking EMS station in Brooklyn, caused by a water-main break, and a structural failure at Yankee Stadium, one week before baseball's opening day (Langewiesche, 88).

The fire and police departments and other government agencies were also frantically coordinating their own efforts, with little communication between each group. It was clear that these groups could not continue working in isolation, and that a collaborative process would be in order. However, the city decided to scrap its emergency response plan in favor of the process that was already emerging. At that point, the DDC had already bypassed the standard emergency response procedure, and Holden and Burton had already in many ways taken charge of the effort. The city made it official, giving the DDC, the fire department, and the police department joint control over the cleanup effort.

As with the PACT project, neutral territory played an important role in defusing traditional politics and allowing new roles to emerge. Burton commandeered a kindergarten classroom at a nearby school and held twice-daily meetings there. Because of the urgency of their task, they did not keep minutes or write memos. Those meetings were the primary nexus of communication, and everyone involved—about 20 government agencies and several contractors—had representatives there. Recording devices were banned to encourage participants to voice their opinions freely. According to Langewiesche, "Some of the participants were accomplished people with impressive resumes, but within the inner world of the Trade Center site it hardly mattered what they had done before. However temporarily, there was a new social contract here, which everyone seemed to understand. All that counted about anyone was what that person could provide now" (Langewiesche, 113).

The recovery effort was largely complete after six months, although it did not officially end until a few months later. The entire 1.5 million tons of wreckage had been properly examined and transported to a landfill on Staten Island. The site was no longer hazardous, and the city was ready to build something new. Incredibly, despite all of the dangers, not a single person died during the recovery process.

Comparing the World Trade Center recovery to free and open source projects may seem far-fetched at first, but there are important similarities. For instance, most of the recovery effort workers were financially compensated, although the plan for doing so was not predetermined. Once the recovery process became apparent, city officials made sure the appropriate people were paid. Similarly, writing software requires tremendous expertise, and for a project to be sustainable, it must be able to retain its experts. With free and open source projects, the methods of compensation are not always explicit, nor are they always monetary, but they are there.

The most evident pattern from the World Trade Center recovery was an orientation among participants toward action. In the free and open source software community, there is a common saying: "scratch your own itch." That attitude also pervaded the recovery effort. No one waited for instructions from above. People simply did what had to be done. At the same time, the recovery effort was not devoid of process. Not only were the participants action oriented, but they also were well coordinated, thanks to the twice-daily meetings and culture of openness. Similarly, successful free and open source projects employ effective and open lines of communication and coordination.

Facilitating Emergent Collaboration

The similarities among PACT, the World Trade Center recovery, and free and open source projects reveal several important patterns and lessons for facilitating emergent collaboration. I say *facilitate*, not *create*, because emergence implies an element of surprise and a lack of intention and control. Process and organization exist, but are not imposed. Patterns such as neutral territory create an environment that is permissive, not prescriptive, and participants who are action oriented thrive in these environments.

The defining characteristic of the World Trade Center recovery was that action trumped everything else. The problem was constantly changing, and thus, the process had to evolve as well. Asking for permission didn't make sense in this environment, because those on the ground understood the problem better than anyone holding traditional authority. City leaders recognized this and deliberately chose to rubber-stamp the process that emerged, instead of imposing a written plan that was clearly not suited to the problem. Similarly, PACT did not happen because the leaders of the different companies came together and decided to collaborate on a project. PACT happened because those who would most benefit from collaboration—the software developers—began to trust each other and identified a common need. These same developers agreed on a representative governance model with a consensus-based decision-making approach. While not as freewheeling as the Trade Center recovery process, it still emphasized exploration by all of its members, and its leaders were especially proactive. The PACT process had to be slower and more deliberative by design, because the motivation to collaborate was nascent. Participants had to see collaboration work before they became comfortable with it, which meant that the process evolved slowly.

Another prerequisite for emergent collaboration is an emphasis on open, effective communication. Effective communication begins with shared language. When the PACT project began, different participants had different understandings of the same words, and communication was rendered ineffective. Before the PACT participants could even start thinking about software, they had to understand each other's world views and language. Shared language was not an issue with the World Trade Center recovery. Everyone had a vivid picture of what had happened, and what needed to be done. However, while the recovery effort was highly individualistic, coordination was imperative, and hence, those twice-daily meetings among all participants were critical. Banning recording devices encouraged participants to express their feelings openly. Good communication is the hallmark of the best free and open source projects. Anyone may participant in a forum, and core participants respond to questions quickly. Active projects often summarize online discussion on a regular basis so that others can follow high-volume discussions without being overwhelmed. Most importantly, there is a shared language among those who participate, some of which is embodied in the various free and open source licenses.

All of these elements are necessary for collaboration to emerge, but they are not sufficient. Culture is critical. The strong desire to collaborate among those who participated in the World Trade Center recovery was a direct result of its unique circumstances and scale. With PACT, the incentives and environment were not powerful enough to establish a culture of collaboration, and rightfully so. You cannot reasonably expect a software project to inspire the same emotion as an act of terror and the tragic loss of life. Instead, the instigators of PACT had to nurture that culture every step of the way, beginning by bringing the groups together to break bread and to discover commonalities for themselves. The seedlings of community—camaraderie among those with shared interests and goals—catalyzed a culture of collaboration, and every instance of successful collaboration further reinforced that culture.

We have limited control over emergent collaboration, and we must adjust our expectations accordingly. However, the patterns we discover by examining stories such as PACT and the World Trade Center recovery help us understand how we can facilitate emergent collaboration. These stories were very different, yet their commonalities are striking, especially in light of what we already know about free and open source software development. Perhaps everything is not known, but we can learn much from knowing what is. The most important lesson is that the best way to understand collaboration is to collaborate.

Acknowledgments

Thanks to Chris Dent, H. Jessica Kim, and Mark Stone for their insightful comments.

References

Paul Armer, "SHARE—A Eulogy to Cooperative Effort." *Annals of the History of Computing* 2, no. 2, April 1980: 122–129.

Martin Campbell-Kelly and William Aspray, *Computer: A History of the Information Machine* (New York, NY: BasicBooks, 1996).

Walter M. Carlson, "The Life and Times of the Digital Computers Association (DCA)." *Annals of the History of Computing* 18, no. 2, Summer 1996: 63–66.

William Langewiesche, *American Ground: Unbuilding the World Trade Center* (New York, NY: North Point Press, 2002).

Bobbi Mapstone and Morton I. Bernstein, "The Founding of SHARE: NCC '80 Pioneer Day." *Annals of the History of Computing* 2, no. 4, Fall 1980: 363–372.

Wesley S. Melahn "A Description of a Cooperative Venture in the Production of an Automatic Coding System." *Journal of the ACM* 3, Oct. 1956: 266–271.

CHAPTER 20

Larry Sanger

The Early History of Nupedia and Wikipedia: A Memoir

An impassioned debate has been raging, particularly since about the summer of 2004, concerning the merits of Wikipedia and the future of free online encyclopedias. This discussion has not benefited by much detailed, accurate consideration of the origins of Wikipedia and of its parent project, Nupedia. Yet those origins are crucial to forming a proper judgment of the current state and best future direction of free encyclopedias.

Wikipedia as it stands is a fantastic project; it has produced enormous amounts of content and thousands of excellent articles, and now, after just four years, it is getting high-profile, international recognition as a new way of obtaining at least a rough and ready idea about many topics. Its surprising success may be attributed, briefly, to its free, open, and collaborative nature.

This has been my attitude toward Wikipedia practically since its founding. But in late 2004, I wrote an article critical of certain aspects of the Wikipedia project, "Why Wikipedia Must Jettison Its Anti-Elitism,"[1] which occasioned much debate. I have also been quoted, as co-founder of Wikipedia, in many recent news articles about the project, making various other critical remarks. I am afraid I am getting an undeserved reputation as someone who is opposed to everything Wikipedia stands for. This is completely incorrect. In fact, I am one of Wikipedia's strongest supporters. I

---

1   Posted December 31, 2004, at *http://www.kuro5hin.org/story/2004/12/30/142458/25.* All URLs in this chapter were accessed April 5, 2005.

am partly responsible for bringing it into the world (as I will explain), and I still love it and want only the best for it. But if a better job can be done, a better job *should* be done. Wikipedia has shown fantastic potential, and it is open content—and so if the project has aspects which will keep it from being the maximally authoritative, broad, and deep reference that I believe it could be, I firmly believe that the world has the right to, and should, improve upon it.

Wikipedia's predecessor, which I was also employed to organize, was Nupedia. Nupedia aimed to be a highly reliable, peer-reviewed resource that fully appreciated and employed the efforts of subject-area experts, as well as the general public. When the more free wheeling Wikipedia took off, Nupedia was left to wither. It might appear to have died of its own weight and complexity. But, as I will explain, it could have been redesigned and adapted—it could have, as it were, "learned from its mistakes" and from Wikipedia's successes. Thousands of people who had signed up and who wanted to contribute to the Nupedia system were left disappointed. I believe this was unfortunate and unnecessary; I always wanted Nupedia and Wikipedia *working together* to be not only the world's largest but also the world's *most reliable* encyclopedia. I hope that this memoir will help to justify this stance. Hopefully, too, I will manage to persuade some people that collaboration between an expert project and a public project is the correct approach to the overall project of creating open content encyclopedias.

I am not writing to request that Nupedia be resuscitated now, as nice as that would be. But I would like to tell the story of Nupedia and the first couple of years of Wikipedia as I remember it. I present this as a memoir—a personal view—not as an authoritative history. The "overall project of creating open content encyclopedias" is something about which I have been writing since at least 2001. For example, in July 2001, while still working on both Wikipedia and Nupedia, I wrote, "If some other open source project proves to be more competitive, then it should and will take the lead in creating a body of free encyclopedic knowledge."[2] Since Wikipedia is open content and hence may be reproduced and improved upon by anyone, I have always been cognizant that it might not end up being the only or best version. My personal devotion has always been to the ideal project as I have envisioned it, not necessarily to particular incarnations of Nupedia or Wikipedia; and I think this attitude is fully consistent with the (very positive) spirit of open source collaboration generally.

This being said, let me also emphasize strongly that, throughout this discussion, I am *not* suggesting that Wikipedia needs to be *replaced* with something better. I do, however, think that it needs to be *supplemented* by a broader, more ambitious, and more inclusive vision of the overall project.

---

2   "Britannica or Nupedia? The Future of Free Encyclopedias," posted July 25, 2001, at *http://www.kuro5hin.org/story/2001/7/25/103136/121.*

Some Recent Press Reports

This memoir seems all the more important to publish now because the early history of Nupedia and Wikipedia has been mischaracterized in the press recently. If there were only a few inaccuracies, which made no difference, I would be happy to leave well enough alone. But some of the mischaracterizations I've seen do make a difference. They give the public the impression that Nupedia failed because it was run by snobbish experts whose standards were too high. As I will make clear, that is not correct. One might also gather from some reports that the idea for Wikipedia sprang fully grown from Jimmy Wales' head. Jimmy, of course, deserves enormous credit for investing in and guiding Wikipedia. But a more refined idea of how Wikipedia originated and evolved is crucial to have, if one wants to appreciate fully why it works now, and why it has the policies that it does have.

For example, in the November 1, 2004 issue of *Newsweek*, in "It's Like a Blog, But It's a Wiki,"[3] reporter Brad Stone writes:

> [Jimmy] Wales first tried to rewrite the rules of the reference-book business five years ago with a free online encyclopedia called Nupedia. Anyone could submit articles, but they were vetted in a seven-step review process. After investing thousands of his own dollars and publishing only 24 articles, Wales reconsidered. He scrapped the review process and began using a popular kind of online Web site called a "wiki," which allows its readers to change the content.

This capsule history is, of course, very brief and so should be expected not to have every relevant detail. But some of the claims made here are not just vague, they are actually misleading, and so several clarifications are in order:

- The article makes it sound as if Jimmy were the only person making the relevant decisions. That is incorrect; the Nupedia system (indeed, seven steps) was established via negotiation with Nupedia's volunteer Advisory Board, mostly Ph.D. volunteers, who served as editors and peer reviewers. I articulated our decisions in Nupedia's "Editorial Policy Guidelines."[4] Jimmy started and broadly authorized it all, but as to the details, he really had little to do with them.

- Nupedia's Advisory Board might be surprised to learn that Jimmy "scrapped the review process." Jimmy was certainly disappointed with the process (as were many people), and he did not actively support it after 2001 or so. But in fairness to the people actually working on Nupedia, the fact is that work on Nupedia gradually petered out in 2001–2002. I in particular was stretched thin—in 2001, I was both chief organizer of Wikipedia and editor in chief of Nupedia—and my

---

3   *http://www.msnbc.msn.com/id/6298340/site/newsweek.*
4   A version of "Nupedia.com Editorial Policy Guidelines" from 2001 can be found at *http://web.archive.org/web/20010607080354/www.nupedia.com/policy.shtml.*

own slowing work on Nupedia was obvious to all active Nupedia contributors. It might be better to say that Nupedia withered due to neglect—which was largely due to a lack of sufficient funds for paid organizers—which was as much due to the bursting of the Internet bubble as anything else.

- Also, to the best of my knowledge, the "thousands of his own dollars" invested in these projects were, if I am not very mistaken, the dollars of Bomis.com, which is jointly owned by three partners: Jimmy, Tim Shell, and Michael Davis. (The money for Wikipedia now comes from donations.) But again, Jimmy was the prime motivating force within Bomis.

- Moreover, Nupedia had fewer than 24 articles when Wikipedia launched, being not quite a year old at that time. The idea of adapting wiki technology to the task of building an encyclopedia was mine, and my main job in 2001 was managing and developing the community and the rules according to which Wikipedia was run. Jimmy's role, at first, was one of broad vision and oversight; this was the management style he preferred, at least as long as I was involved. But, again, credit goes to Jimmy alone for getting Bomis to invest in the project and for providing broad oversight of the fantastic and world-changing project of an open content, collaboratively built encyclopedia. Credit also of course goes to him for overseeing its development after I left, and guiding it to the success that it is today.

A March 2005 *Wired Magazine* article by Daniel Pink also got a number of things wrong, despite being, in other respects, an excellent article:[5]

> With Sanger as editor in chief, Nupedia essentially replicated the One Best Way model. He assembled a roster of academics to write articles. (Participants even had to fax in their degrees as proof of their expertise.) And he established a seven-stage process of editing, fact-checking, and peer review. "After 18 months and more than $250,000," Wales said, "we had 12 articles."
>
> Then an employee told Wales about Wiki software. On January 15, 2001, they launched a Wiki-fied version and within a month, they had 200 articles. In a year, they had 18,000....Sanger left the project in 2002. "In the Nupedia mode, there was room for an editor in chief," Wales says. "The Wiki model is too distributed for that."

This too needs clarifications:

- The "roster of academics" (the aforementioned Nupedia Advisory Board) was not limited to academics; they were experts in their fields, in any case. Moreover, they were editors and peer reviewers; the general public was able to propose and write articles on subjects about which they had some knowledge.[6]

---

5 "The Book Stops Here," *http://www.wired.com/wired/archive/13.03/wiki.html*.
6 Consult the 2001 assignment policy if you are interested: *http://web.archive.org/web/20010607080354/http://www.nupedia.com/policy.shtml#assignment*.

- It is incorrect to say that participants had to fax their degrees as proof of their expertise; we did verify bona fides by matching the names and email addresses of editors and reviewers with a web page—often, but not always, an academic web page. Indeed there was one (but only one) case that I recall in which I asked someone, who had no web page or any other easy way to prove who he was, to fax a degree. Verifying bona fides seemed like a good idea especially when initially building what was to be an academically respectable project.

- Again, I did not establish the editorial process alone; I had considerable assistance (for which I am still grateful) from Nupedia's excellent Advisory Board.

- And as I wrote on July 25, 2001 for Kuro5hin,[7] Nupedia had "just over 20" articles—not 12—after 18 months. We always suspected that we would wind up scrapping our first attempts to design an editorial system, and that we would learn a great deal from those first attempts; and that's essentially what happened. But Nupedia could have evolved, and would have, had we continued working on it.

- The second paragraph begins, "Then an employee told Wales about Wiki software." I don't know how Jimmy first learned about wikis, but as I will explain, I proposed to him and to the Nupedia community at large that we start a wiki-based encyclopedia.

- The context of the line "Sanger left the project in 2002"—particularly with Jimmy quoted as saying, "In the Nupedia mode there was room for an editor in chief"— makes it sound as if I were let go specifically because I was working only on Nupedia and was no longer needed for that. In fact, I was working on Wikipedia far more at the time than Nupedia, and the reason for my departure from both projects was that Bomis was, like virtually all dot-coms, losing money. They could not afford to pay me; I was told that I was the last of several newer Bomis employees to be laid off on account of the tech recession. But Wikipedia indeed was able to continue on without me, and I agreed even at the time that Wikipedia could survive without me, and that it had become essentially "unmanageable."

In view of such problematic reporting, considering the rather good chance that Wikipedia will become historically important, and considering that the planners of related projects might find some value in this, I want to tell my story as I remember it. This memoir covers only the first few years of the project. I have followed the project fairly closely and with interest after my departure, but silently and from the sidelines.

Nupedia

I'm going to begin with Nupedia. The origin of Wikipedia cannot be explained except in that context. Moreover, the Nupedia project itself was very worthwhile,

---

7 "Britannica or Nupedia?" op. cit.

and I think it *might* have been able to survive, as I will explain. Finally, some errors regarding Nupedia have been passed around although they are little more than unfounded rumors. It is unfortunate that the thousands of hours of excellent volunteer work done on Nupedia should be thus disrespected or grossly misunderstood. I personally will always be grateful to those initial contributors who believed in the project and our management, worked hard for a completely unproven idea, and laid the groundwork for the growing institution of open content projects.

In 1999, Jimmy Wales wanted to start a free, collaborative encyclopedia. I knew him from several mailing lists back in the mid-'90s, and in fact we had already met in person a couple of times. In January 2000, I emailed Jimmy and several other Internet acquaintances to get feedback on an idea for what was to be, essentially, a blog. (It was to be a successor to "Sanger and Shannon's Review of Y2K News Reports," a Y2K news summary that I first wrote and then edited.) To my great surprise, Jimmy replied to my email describing his idea of a free encyclopedia, and asking if I might be interested in leading the project. He was specifically interested in finding a philosopher to lead the project, he said. He made it a condition of my employment that I would finish my Ph.D. quickly (whereupon I would get a raise)—which I did, in June 2000. I am still grateful for the extra incentive. I thought he would be a great boss, and indeed he was.

To be clear, the idea of an open source, collaborative encyclopedia, open to contribution by ordinary people, was *entirely* Jimmy's, not mine, and the funding was entirely by Bomis. I was merely a grateful employee; I thought I was very lucky to have a job like that land in my lap. Of course, other people had had the idea; but it was Jimmy's fantastic foresight actually to invest in it. For this the world owes him a considerable debt. The actual development of this encyclopedia was the task he gave me to work on.

I arrived in San Diego in early February 2000 to get to work. One of the first things I asked Jimmy was how free a rein I had in designing the project. What were my constraints, and in what areas was I free to exercise my own creativity? He replied, as I clearly recall, that most of the decisions should be mine; and in most respects, as a manager, Jimmy was indeed very hands-off. I spent the first month or so thinking very broadly about different possibilities. I wrote quite a bit (that writing is now all lost—that will teach me not to back up my hard drives) and discussed quite a bit with both Jimmy and one of the other Bomis partners, Tim Shell.

I maintained from the start that something really could not be a credible *encyclopedia* without oversight by experts. I reasoned that, if the project is open to all, it would require *both* management by experts *and* an unusually rigorous process. I now think I was right about the former requirement, but wrong about the latter, which was redundant; I think that the subsequent development of Wikipedia has borne out of this assessment.

One of the first policies that Jimmy and I agreed upon was a "nonbias" or neutrality policy. I know I was extremely insistent upon it from the beginning, because neutrality has been a hobbyhorse of mine for a very long time, and one of my guiding principles in writing "Sanger's Review." Neutrality, we agreed, required that articles should not represent any one point of view on controversial subjects, but instead fairly represent all sides. We also agreed in rejecting an alternative that (for a time) Tim and some early Nupedians plugged for: the development, for each encyclopedia topic, of a series of different articles, each written from a different point of view.

I believed, moreover, that a strongly collaborative and open project could not survive if its contributors were not "personally invested" in the project, and that this required some input and management by its users. It was very early on that I decided that Nupedia should have an Advisory Board—editors, and peer reviewers, who would together agree to project policy—and that the public should have a say in the formulation of policy.

An early incarnation of Nupedia's Advisory Board was in place by summer of 2000 or so. It was made up of the project's highly qualified editors and reviewers, mostly Ph.D. professors but also a good many other highly experienced professionals. Eventually the Advisory Board agreed to an extremely rigorous seven-step system. A lot of the details of the Nupedia policy and processes were *proposed* by me, but then tweaked and elaborated by others, and the policy was not published as project policy until we had a quorum of editors and peer reviewers who could fully discuss and approve of a policy statement. Even so, our policy overlooked a fundamental problem. We should not have assumed that such a complex system could be navigated patiently by many volunteers.

I spent significant time recruiting people for Nupedia, emailing new arrivals, posting to mailing lists, giving interviews, and so on. I had had some experience publicizing Internet projects when I worked on several philosophy discussion groups as a graduate student in the 1990s and I knew that getting many willing and active participants was difficult but important. I even had an administrative assistant for six months in 2000 and 2001, Liz Campeau, whose sole job was to recruit people to work on Nupedia and then Wikipedia. I think a large part of the reason Wikipedia got off the ground so quickly and so well is that it was started by Nupedians, who were then a very large base of people who wanted to work on an encyclopedia, and who had many definite ideas about how it should be done. Roughly 2,000 Nupedia members were subscribed to the general announcement list in January 2001 when Wikipedia launched. We operated the system initially using email and mailing lists, while planning and finalizing process details. That lasted from spring through fall 2000. I think our first article ("atonality" by Christoph Hust), that made it entirely through the system, was published in June or July 2000. To move the system to a completely web-based one, there was, of course, a great deal of design and programming to do. So in fall of 2000, I worked a lot with a programmer (Toan Vo) and the

Bomis sysadmin (Jason Richey) to transfer the system from a clunky mailing list system to the Web. But by the time the web-based system was ready it had become obvious to Jimmy and me that the seven-step editorial process would move too slowly, even when managed on the Web. But Magnus Manske later, in 2001, made some very nice additions to the Nupedia system.

Some institutional traditions begin easily but die hard. Nupedia's Advisory Board was reluctant to seriously consider a simpler system, despite months of coexistence and uncomfortable comparison between Nupedia and Wikipedia. Nupedia editors and peer reviewers had a very strong commitment to rigor and reliability, as did I. Moreover, as Wikipedia became increasingly successful in 2001, Jimmy asked me to spend more and more time on it, which I did; Nupedia suffered from neglect. It wasn't until summer of 2001 that I was able to propose, get accepted, and install something we called the Nupedia Chalkboard. This was a wiki which was to be closely managed by Nupedia's staff. It offered both a simpler way to develop encyclopedia articles for Nupedia, and a way to import articles from Wikipedia. Established practices are hard to break, and the Chalkboard went largely unused. The general public simply used Wikipedia if they wanted to write articles in a wiki format, while most Nupedia editors and peer reviewers were not persuaded that the Chalkboard was necessary or useful.

By early winter 2001, Nupedia had published approved versions of only about 25 articles, although there were dozens of draft articles at various stages in process. I was finally able to persuade the Advisory Board to move the system to a much simpler two-step process, virtually identical to that used to run many academic journals: articles would be submitted to an editor; the editor would, if the article seemed good enough, forward it to a reviewer for acceptance or rejection; if accepted, the article would be posted. We also contemplated various ways of allowing public comment, moderation, and editing of posted articles. I believe this new, simpler system would have produced thousands of articles for Nupedia very quickly. The Nupedia community was certainly interested and motivated. The Advisory Board was gradually accepting that the system's complexity was the main obstacle to getting more articles into and through the system.

Unfortunately, Nupedia's new system arrived too late. This system should have been adopted in the winter of 2001–2002. At the same time, Wikipedia was demanding as much attention as I could give it, and I had little time to implement the new Nupedia system. I am quite sure we could have started Nupedia in early 2002 had we made the time. But Bomis lost the ability to pay me and, newly unemployed, I did not have the time to lead Nupedia as a volunteer. I did not entirely lose hope on Nupedia, however.

The Origins of Wikipedia

In the fall of 2000, Jimmy and I were in agreement that Nupedia's slow productivity was probably going to be an ongoing problem and that there needed to be a way, moreover, in which ordinary, uncredentialed people could participate *more easily*. Uncredentialed people *could* (and did) participate in Nupedia, particularly as writers and copy editors, but it was challenging for most of them to get articles through the elaborate system. We had a huge pool of talent, motivated to work on an encyclopedia but not motivated enough to work on Nupedia, going to waste.

It was my job to solve these problems. I wrote multiple detailed proposals for a simpler, more open editing system and I ran them by Jimmy. His reply to all of them was that it would require too much programming, and he couldn't afford to pay more high-priced programmers. In retrospect, of course, I realize that we could have found a way to enlist volunteers to develop the system. Jimmy and I both probably knew that at the time; unfortunately, we didn't pursue it.

While I was thinking hard about how to create a more open system with minimal setup requirements, I had dinner with an old Internet friend of mine, Ben Kovitz. Ben had moved to town for a new job and we were out at a Pacific Beach Mexican restaurant, talking about jobs, tech stuff, and philosophy (Ben, Jimmy, and I all knew each other from those philosophy mailing lists on which we were active). Ben explained the idea of Ward Cunningham's WikiWikiWeb[8] to me. Instantly I was considering whether wiki would work as a more open and simple editorial system for a free, collaborative encyclopedia, and it seemed exactly right. The more I thought about it, without even having seen a wiki, the more it seemed obviously right. Immediately I wrote a proposal—unfortunately, lost now—in which I said that this might solve the problem and that we ought to try it. Given that setting up a wiki would be very simple and would not require hiring a programmer, Jimmy could scarcely refuse. He liked the idea but was initially skeptical—properly so, as I was, despite my excitement.

Wiki advocates often point out[9] that Wikipedia is nonstandard as a wiki. This is partly because we began just with the very basic wiki concept and not so much of the culture. Wiki culture is very distinctive. Wiki pages can be started and edited by anyone, but in Thread Mode[10] (as in "the thread of this discussion"), the dialogue becomes complex. In that case, or when consensus is reached, or when positions have hardened, it is considered a good idea to "refactor"[11] pages (a term borrowed

---

8 For an introduction, see the "Welcome Visitors" page of WikiWikiWeb: *http://www.c2.com/cgi/wiki?WelcomeVisitors*.

9 Usemod.com, a wiki about wikis, has many articles that introduce the old-fashioned idea about wikis. See "WikiPedia Is Not Typical," *http://www.usemod.com/cgi-bin/mb.pl?WikiPediaIsNotTypical*.

10 "Thread Mode," *http://www.c2.com/cgi/wiki?ThreadMode*.

11 "What Is Refactoring," *http://www.c2.com/cgi/wiki?WhatIsRefactoring*.

from programming)—i.e., to rewrite them, taking into account the highlights of the dialog. Then the dialog might be represented in "Document Mode."[12] Opinions are very welcome on a typical wiki. There are *many* other collective habits that make up typical wiki culture; these are only a few.

However, I denied the necessity of organizing Wikipedia according to these precise principles. To be sure, a few other participants wanted Wikipedia to adopt wiki culture wholesale so that it would be "just another wiki," and they had some small influence over the direction of the project. Still, I viewed wiki software as simply a tool, a way to organize people who want to collaborate. I saw no necessity whatsoever to partake in all aspects of the idiosyncratic culture that happened to be associated with the advent of this very generally applicable tool, since we were engaged in a very specific sort of project with very specific requirements. This caused some consternation among some wiki advocates, who appeared to think that Wikipedia should, or inevitably would, become just another wiki, somehow necessarily partaking of typical wiki culture. Ward Cunningham's prediction,[13] when Jimmy asked him whether wiki software "could successfully generate a useful encyclopedia," was: "Yes, but in the end it wouldn't be an encyclopedia. It would be a wiki." As I said in reply: "Wikipedia has a totally different culture from this wiki, because it's pretty single-mindedly aimed at creating an encyclopedia. It's already rather useful as an encyclopedia, and we expect it will only get better."

Typical wiki culture aside, wiki software does encourage, but does not strictly require, extreme openness and decentralization: openness, since page changes are logged and publicly viewable, and pages may be further changed by anyone; and decentralization, because for work to be done, there is no need for a person or body to assign work, but rather, work can proceed as and when people want to do it. Wiki software also *discourages* the exercise of authority, since work proceeds at will on any page, and on any large, active wiki it would be too much work for any single overseer or limited group of overseers to keep up. These all became features of Wikipedia.

My initial idea was that the wiki would be set up as part of Nupedia; it was to be a way for the public to develop a stream of content that could be fed into the Nupedia process. I think I got some of the basic pages written—how wikis work, what our general plan was, and so forth—over the next few days. I wrote a general proposal for the Nupedia community, and the Nupedia wiki went live January 10. The first encyclopedia articles for what was to become Wikipedia were written then. It turned out, however, that a clear majority of the Nupedia Advisory Board wanted to have nothing to do with a wiki. Again, their commitment was to rigor and reliability, a concern I shared with them and continue to have. They evidently thought a wiki

---

12 "Document Mode," *http://www.c2.com/cgi/wiki?DocumentMode.*
13 "Wiki Pedia," *http://www.c2.com/cgi/wiki?WikiPedia.*

could not resemble an encyclopedia at all, that it would be too informal and unstructured, as the original WikiWikiWeb was, to be associated with Nupedia. They of course were perfectly reasonable to doubt that it would turn into the fantastic source of content that it did. Who could reasonably guess that it would work? But it did work, and now the world knows better.

Wikipedia's First Few Months

We decided to relaunch the wiki under its own domain name. I came up with the name "Wikipedia," a silly name for what was at first a very silly project, and the newly independent project was launched at Wikipedia.com on January 15, 2001. It was a ".com" at first because, at the time, we were contemplating selling ads to pay for me, programmers, and servers. It was easy to deprecate ".com" in favor of ".org" in 2002, after Jimmy was able to assure users that Wikipedia would never run ads to support the project.

I took it to be one of my main jobs to promote Wikipedia, and this resulted in a steady influx of new participants. I wrote on the Wikipedia announcement page January 24, "Wikipedia has definitely taken [on] a life of its own; new people are arriving every day and the project seems to be getting only more popular. Long live Wikipedia!" By the end of January, we reportedly[14] had 600 articles; there were 1,300 in March, 2,300 in April, and 3,900 in May. Not only was the project growing steadily, but the rate of growth was also increasing.

Wikipedia started with a handful of people, many from Nupedia. The influence of Nupedians was crucial early on. I think, especially, of the tireless Magnus Manske (who worked on the software for both projects), our resident stickler Ruth Ifcher, and the very smart poker-playing programmer Lee Daniel Crocker—to name a few. All of these people, and several other Nupedia borrowings, had a good understanding of the requirements of good encyclopedia articles, and they were intelligent, skilled writers. The direction that Wikipedia ought to go in seemed obvious to us all, in terms of what sort of content we wanted. But what we did not have worked out in advance was how the community should be organized, and (not surprisingly) that turned out to be the thorniest problem. Still, because the project started with these good people, and we were able to adopt, explain, and promote good habits and policies to newer people, the Nupedian roots of the project helped to develop a robust, functional, and successful community. As to project leadership or management, we began with me, Jimmy, and Tim Shell; Tim mostly stopped participating after the first few months.

---

14 "Wikipedia: Size of Wikipedia," *http://en.wikipedia.org/wiki/Wikipedia:Size_of_Wikipedia.*

The many rank-and-file users did the heavy lifting, and if there had not been a reasonable consensus among them about what the project should look like, it just wouldn't have happened. In any collaborative project, it is the contributors who are responsible for the outcome. Those early adopters should feel proud of themselves, because they were essential in shaping a thing of beauty and usefulness.

I recall saying casually, but repeatedly, in the project's first nine months or so, that experts and specialists should be given some particular respect when writing in their areas of expertise. They should be deferred to, I thought, unless there was some clear evidence of bias. In those first months, deference to expertise was a policy that at least *I* usually insisted upon, but not strongly or clearly enough. It was nearly a year after the project began that I finally articulated this view as a policy to consider.[15] Perhaps this was because, indeed, most users *did* make a practice of deferring to experts up to that time. "This is just common sense," as I wrote, "but sometimes common sense needs to be spelled out!" What I now think is that that point of common sense needed to be spelled out quite a bit sooner and more forcefully, because in the long run, it was *not* adopted as official project policy, as it could have been.

Some questions have been raised about the origin of Wikipedia policies. The tale is interesting and instructive. We began with no (or few) policies in particular and said that the community would determine—through a sort of vague consensus based on its experience working together—what the policies would be. The very first entry on a "rules to consider" page[16] was the "Ignore All Rules" rule (to wit: "If rules make you nervous and depressed, and not desirous of participating in the wiki, then ignore them entirely and go about your business"). This is a "rule" that I personally proposed. I thought we first needed experience with wikis before we could have rules about wikis. Even more importantly at that point, we needed participants more than we needed rules. As the project grew and the requirements of its success became increasingly obvious, I became ambivalent about this particular "rule" and then rejected it altogether. As one participant later commented, "this rule is the essence of Wikipedia."[17] That was certainly *never* my view; I always thought of the rule as being a temporary and humorous injunction to participants to add content instead of being distracted by (then) relatively inconsequential issues about how exactly articles should be formatted, etc. In a similar spirit, I proposed that contributors be bold in updating pages.[18]

---

15 "Deferring to the experts," *http://meta.wikimedia.org/wiki/Deferring_to_the_experts.*
16 "RulesToConsider," *http://web.archive.org/web/20010307211833/www.wikipedia.com/wiki/ RulesToConsider.*
17 "Wikipedia talk:Ignore all rules," *http://en.wikipedia.org/wiki/Wikipedia_talk:Ignore_all_rules.*
18 "Be bold in updating pages," *http://web.archive.org/web/20011111150732/www.wikipedia.com/ wiki/Be_bold_in_updating_pages.*

I also, for similar reasons, specifically disavowed any title; I was organizing the project but I did not want to present myself as editor in chief. I wanted people to feel comfortable adding information without having to consult anything like an editor. Participation was more important, I felt.

As we set it up, Wikipedia *did* have some minimal wiki cultural features: it *was* wide open and extremely decentralized, and (provisionally, anyway) featured very little attempt to exercise authority. Insofar as I was able to organize it at all, I guided the project through force of personality and what "moral authority" I had as co-founder of the project. Jimmy and I agreed early on that, at least in the beginning, we should not eject anyone from the project except perhaps in the most extreme cases. Our first forcible expulsion (which Jimmy performed) did not occur for many months, despite the presence of difficult characters from nearly the beginning of the project. Again, we were learning: we wished to tolerate all sorts of contributors to be well situated to adopt the wisest policies. However, this provisional "hands-off" management policy had the effect of creating a difficult-to-change tradition, the tradition of making the project *extremely* tolerant of disruptive (uncooperative, "trolling") behavior. And as it turned out, particularly with the large waves of new contributors from the summer and fall of 2001, the project became very resistant to any changes in this policy. I suspect that the cultures of online communities generally are established pretty quickly and then are very resistant to change, because they are self-selecting; that was certainly the case with Wikipedia, anyway.

So, I could only attempt to shame any troublemakers into compliance; without recourse to any genuine punitive action, that was the most I could do. In the first eight months of the project, this was *usually* sufficient for me to do my job. Wikipedia began as a good-natured anarchy, a sort of Rousseauian state of digital nature. I always took Wikipedia's anarchy to be provisional and purely for purposes of determining what the best rules, and the nature of its authority, should be. What I, and other Wikipedians, failed to realize is that our initial anarchy would be taken by the next wave of contributors as the very essence of the project—how Wikipedia was "meant" to be—even though Wikipedia could have become anything we the contributors chose to make it.

This point bears some emphasis: Wikipedia became what it is today because, having been seeded with great people with a fairly clear idea of what they wanted to achieve, we proceeded to make a series of free decisions that determined the policy of the project and culture of its supporting community. Wikipedia's system is neither the only way to run a wiki, nor the only way to run an open content encyclopedia. Its particular conjunction of policies is in no way natural, "organic," or necessary. It is instead artificial, a result of a series of free choices, and we could have chosen differently in many cases; and choosing differently on some issues might have led to a project *better* than the one that exists today.

Though it began as anarchy, there were quite a few policies that were settled within the first six months. This required some struggle, especially on my part. Since the project was a wiki, some participants thought that there should be no rules at all. But it was made clear from the beginning that we intended Wikipedia to be an *encyclopedia,* and so we pushed for at least those rules that would help define and sustain the project as an encyclopedia.

For instance, throughout the early months, people added various content that seemed less than encyclopedic. Many people seemed to confuse encyclopedia articles with dictionary entries, and eventually I wrote a page called "Wikipedia is not a dictionary."[19] As people found new ways not to write encyclopedia articles, I started "What Wikipedia is not":[20] I and others would note on an article's discussion page that some content did not belong in an encyclopedia, and then underscored the point by adding an entry to the "What Wikipedia is not" page. To take another example, Wikipedia was not to be a place for publishing original research. In fact, this is a policy that had been settled upon and even enforced in Nupedia days; enforcing it actually led to the departure of Nupedia's erstwhile Classics editor sometime in 2001.

Many of our first controversies were over these restrictions. At the time, I had enough influence within the community to get these policies generally accepted. And if we had not decided on these restrictions, Wikipedia might well have ended up, like many wikis, as *nothing in particular.* But since we *insisted* that it was an encyclopedia, even though it was just a blank wiki and a group of people to begin with, it became an encyclopedia. There is something simple, yet profound about that. I also like to think that we helped to show the world the potential that wikis have.

Another policy that was instituted early on was the nonbias or neutrality policy. This was borrowed from the Nupedia project[21] and was made a Rule to Consider—in a very early version, the policy was put this way:

> Avoid bias: Since this is an encyclopedia, after a fashion, it would be best if you represented your controversial views either (1) not at all, (2) on *Debate, *Talk, or *Discussion pages linked from the bottom of the page that you're tempted to grace, or (3) represented in a fact-stating fashion, i.e., which attributes a particular opinion to a particular person or group, rather than asserting the opinion as fact. (3) is strongly preferred.

---

19 "Wikipedia:Wikipedia is not a dictionary," *http://en.wikipedia.org/wiki/Wikipedia:Wikipedia_is_not_a_dictionary.*

20 "Wikipedia:What Wikipedia is not," *http://en.wikipedia.org/wiki/Wikipedia:What_Wikipedia_is_not.*

21 "Nupedia.com Editorial Policy Guidelines," Version 3.31 (November 16, 2000), Part III, "General Nupedia Policies," *http://web.archive.org/web/20001205000200/http://www.nupedia.com/policy.shtml#III.*

Jimmy then started a specialized policy page he called "Neutral Point of View."[22] I confess I don't much like this name as a name for the policy, because it implies that to write neutrally, or without bias, is actually to express a *point of view,* and, as the definite article is used, a *single* point of view at that. "Neutrality," "neutral," and "neutrally" are better to use for the noun, adjective, and adverb. But the acronym "NPOV" came to be used for all three, by Wikipedians wanting to seem hip, and then the unfortunate "POV" came to be used when the perfectly good English word "biased" would do.

In addition to these, I suggested a number of other rules. I believe I am responsible for the original formulations of a lot of the article-naming conventions, as well as the conventions of bolding the title of the article, starting articles with full sentences, making article titles uncapitalized, and much else. I think these policies were just a matter of common sense for anyone who understood what a good encyclopedia should be like. And of course I was not the only person proposing conventions. Moreover, actual project policy, or community habits, succeeded in being established only by being followed and supported by a majority of participants. It was then, we said, that there was a "rough consensus" in favor of the policy. And consensus, we said, is required for a policy actually to be considered project policy. For our purposes, a "consensus" appeared to consist of (1) widespread common practice, (2) many vocal defenders, and (3) virtually no detractors.

But that way of settling upon policy proposals—viz., by alleged consensus—did not scale, in my opinion. After about nine months or so, there were so many contributors, and especially brand-new contributors, that nothing like a consensus could be reached, for the simple reason that condition (3) in the previous paragraph was never achievable: there would after that always be somebody who insisted on expressing disagreement. There was, then, a nonscaling policy adoption procedure, and a crying need to continue to adopt sensible policies. This led to some serious problems in the community. But first, something more positive.

Why Wikipedia started working

This is a good place to explain why Wikipedia actually got started and why it worked. The explanation involves several factors, some borrowed from the open source movement, some borrowed from wiki software and culture, and some more idiosyncratic:

Open content license
> We promised contributors that their work would always remain free for others to read. This, as is well known, motivates people to work for the good of the world—and for the many people who would like to teach the whole world, that's a pretty strong motivation.

---

22 "NeutralPointOfView," *http://web.archive.org/web/20010416035757/www.wikipedia.com/wiki/ NeutralPointOfView.* For the current version, see *http://en.wikipedia.org/wiki/Wikipedia:Neutral_ point_of_view.*

Focus on the encyclopedia

We said that we were creating an encyclopedia, not a dictionary, etc., and we encouraged people to stick to creating the encyclopedia and not use the project as a debate forum.

Openness

Anyone could contribute. Everyone was specifically made to feel welcome (e.g., we encouraged the habit of writing on new contributors' user pages, "Welcome to Wikipedia!" etc.). There was no sense that someone would be turned away for not being bright enough, or not being a good enough writer, or whatever.

Ease of editing

Wikis are easy for most people to figure out. In other collaborative systems (like Nupedia), you have to learn all about the system first. Wikipedia had an almost flat learning curve.

Collaborate radically; don't sign articles

Radical collaboration, in which (in principle) anyone can edit any part of anyone else's work, is one of the great innovations of the open source software movement. On Wikipedia, radical collaboration made it possible for work to move forward on all fronts at the same time, to avoid the big bottleneck that is the individual author, and to burnish articles on popular topics to a fine luster.

Offer unedited, unapproved content for further development

This is required if one wishes to collaborate radically. We encouraged putting up their unfinished drafts—as long as they were at least roughly correct—with the idea that they can only improve if there are others collaborating. This is a classic principle of open source software. It helped get Wikipedia started and helped keep it moving. This is why so many original drafts of Wikipedia articles were initially of poor quality, and also why it is *surprising* to the uninitiated that many articles have turned out very well indeed.

Neutrality

A firm neutrality policy made it possible for people of widely divergent opinions to work together, without constantly fighting. It's a way to keep the peace.

Start with a core of good people

I think it was essential that we began the project with a core group of intelligent, good writers who understood what an encyclopedia should look like, and who were basically decent human beings.

Enjoy the Google effect

We had little to do with this, but had Google not sent us an increasing amount of traffic each time they spidered the growing web site, we would not have grown nearly as fast as we did.

That's pretty much it. The focus on the encyclopedia provided the task, and the open content license provided a natural motivation: people work hard if they believe they are teaching stuff to the world. Openness and ease of editing made it easy for new people to join in and get to work. Collaboration helped move work forward quickly and efficiently, and posting unedited drafts made collaboration possible. The fact that we started with a core of good people from Nupedia meant that the project could develop a functional, cooperative community. Neutrality made it easy for people to work together with *relatively* little conflict. And the Google effect provided a steady supply of "fresh blood"—who in turn supplied increasing amounts of content.

Nearly all other project rules were either optional, or straightforward applications of these principles. The project probably would still have succeeded nicely even if it had moderated or tweaked some of these principles. For instance, radical openness—that is, being open even to those who brazenly flouted and disrespected the project's mission, was surely not necessary; after all, without them, the project would have been more welcoming to the *many* people who felt they could not work with such difficult people. And if we had required people to sign in, that would not have made very much difference (although it probably would have made some in the beginning; the project wouldn't have grown as fast). Of course, we didn't have to use the GNU FDL[23] for the license. Certainly, we did not need to set the community up initially as an anarchy governed by some vague consensus: instead, we could have adopted a charter from the very start. The project could have been managed quite differently; there could have been specially designated and well-qualified editors. The project could have officially encouraged and deferred to experts. An article approval process could have been adopted without threatening the principle of posting unedited content for collaboration. Certainly, many of the later bells and whistles—the arbitration committee, a three-revert rule, having administrators with the particular configuration of rights they have, etc.—were not absolutely necessary to adopt in the precise forms they took. These differences would not have threatened the basic principles that made the project work.

The basic principles that explain why Wikipedia could start working—and still does work—are relatively simple, few in number, and above all, general. The more specific principles that Wikipedia adopted were a matter of historical accident. There was a great deal of "wiggle room." Those intent on studying or replicating the Wikipedia model would do well to bear that in mind.

A Series of Controversies

So much for the very early history of Wikipedia; the next phase involved rapid growth and some serious internal controversies over policy and authority. If Wikipedia's basic

---

23 That is, the GNU Free Documentation License. Can be read at *http://www.gnu.org/copyleft/fdl.html*. By 2000–2001, this license was the biggest thing going, as far as open content licenses were concerned; Creative Commons, at *http://creativecommons.org*, did not get started until 2001.

policy was settled upon in the first nine months, its culture was solidified into something closer to its present form in the nine months after that.

The project continued to grow. We had 6,000 articles by July 8; 8,000 by August 7; 11,200 by September 9; and 13,000 by October 4. Consulting the web site logs, we noted a Google effect: each time Google spidered the web site, more pages would be indexed; the greater the number of pages indexed, the more people arrived at the project; the more people involved in the project, the more pages there were to index. In addition to this source of new contributors, Wikipedia was Slashdotted several times and had large influxes of new users, particularly after two articles I wrote for Kuro5hin were posted on Slashdot: "Britannica or Nupedia? The Future of Free Encyclopedias" (July 25, 2001)[24] and "Wikipedia is wide open. Why is it growing so fast? Why isn't it full of nonsense?" (September 24, 2001).[25]

This growth brought difficult challenges. Some of our earliest contributors were academics and other highly qualified people, and it seems to me that they were slowly worn down and driven away by having to deal with difficult people on the project. I hope they will not mind that I mention their names, but the two that stick in my mind are J. Hoffman Kemp[26] and Michael Tinkler,[27] a couple of Ph.D. historians. They helped to set what I think was a good precedent for the project in that they wrote about their own areas of expertise, and they contributed under their own, real names. The latter has the salutary effect of making the contributor more serious and more apt to take responsibility for his contributions. They are also very nice people, but they did not "suffer fools gladly." Consequently, they wound up in some silly disputes that would have driven less patient people away instantly. So, there was a growing problem: persistent and difficult contributors tend to drive away many better, more valuable contributors; Kemp and Tinkler were only two examples. There were many more who quietly came and quietly left. Short of removing the problem contributors altogether—which we did only in the very worst cases—there was no easy solution under the system as we had set it up. And I am sorry to have to admit that those aspects of the system that led to this problem were as much my responsibility as anyone else's. Obviously, I would not design the system the same way if given the chance again.

As a result, I grew both more protective of the project and increasingly sensitive to abuse of the system. As I tried to exercise what little authority I claimed, as a corrective to such abuse, many newer arrivals on the scene made great sport of challenging my authority. One of the earliest challenges happened in late summer 2001. The front page of Wikipedia—then open to anyone to edit, like any other page on the project—was

---

24 Op. cit.
25 http://www.kuro5hin.org/story/2001/9/24/43858/2479.
26 "User:JHK," http://en.wikipedia.org/wiki/User:JHK.
27 "User:MichaelTinkler," http://en.wikipedia.org/wiki/wiki.phtml?title=User:MichaelTinkler.

occasionally vandalized with infantile graffiti. Someone then tried to make an archive of the vandalism that had been done to the front page of Wikipedia. I maintained that to make such an archive would be to encourage such vandalism, so I deleted the archive. This occasioned much debate. Then a user made the archive a subpage of his own user page—and user pages were generally held to be the bailiwick of the user. Consequently I deleted that subpage, which occasioned a further hue and cry that, perhaps, I was abusing my authority. The vandalism-enshrining user in question proceeded to create a "deleted pages" page, on which the deleted vandalism archives were listed, as if to accuse me of trying to act without public scrutiny—but this was, of course, perfectly acceptable to me. At the time, I thought this controversy was just as silly as it will sound to most people reading this. I thought that I needed only to "put my foot down" a little harder and, as had happened for the first six months of the project, participants would fall into line. What I did not realize was that this was to be only the first in a long series of controversies. The ultimate upshot of these was to undermine my own moral authority over the project and to make the project as safe as possible for the most abusive and contentious contributors.

Throughout this and other early controversies, much of the debate about project policy was conducted on the wiki itself. Other debates were conducted on mailing lists, Wikipedia-L[28] and then later for the English language project, WikiEN-L.[29] In addition, people had taken to putting their own essays on Wikipedia, as subpages of their user pages. These too were occasioning debate. It seemed to me, and many other contributors, that this debate was distracting the community from our main goal: to create an encyclopedia. Consequently I proposed[30] that we move the debate to another wiki that was to be created specifically for that purpose—what became known as the "meta-wiki."[31] This proposal was very widely supported, so we set it up.

As it happened, the meta-wiki became even more uncontrolled than Wikipedia itself, and for many months was continually infested with contributions by people that can only be called "trolls."[32] That epithet came to be discouraged, however, for reasons soon to be explained. The existence of trolls was a problem we felt we should tolerate—and deal with only *verbally*, not with harsh penalties—for the sake of encouraging the broadest amount of participation. In the first years, only the worst trolls were expelled from the project. I do not know whether this policy has been changed as a result of the operation of the much-later installed Arbitration Committee.[33]

---

28 http://mail.wikipedia.org/pipermail/wikipedia-l.
29 http://mail.wikimedia.org/pipermail/wikien-l.
30 "Moving commentary out of Wikipedia," posted November 3, 2001, http://meta.wikimedia.org/wiki/Moving_commentary_out_of_Wikipedia.
31 Wikipedia Meta-Wiki, http://meta.wikimedia.org/wiki/Main_Page.
32 "Internet troll," http://meta.wikimedia.org/wiki/Main_Page.
33 "Wikipedia:Arbitration Committee," http://en.wikipedia.org/wiki/Wikipedia:Arbitration_Committee.

There are obvious reasons that the meta-wiki proved harder to control. First, it had no specific purpose, other than to host project debate and essays that do not belong on the main wiki—which was not enough to make anyone care very much about it. Second, because many people did not care what happened on the meta-wiki, they did not do the very necessary weeding[34] that takes place on Wikipedia. Besides, as the meta-wiki was a repository of *opinion,* people felt less comfortable editing or deleting what was, after all, only opinion.

What happened was that project policy discussions moved almost exclusively to the project mailing lists.[35] There is a reason why this was a superior solution to having much debate on an uncontrolled, "unmoderated"[36] wiki. On a wiki, contributions exist in perpetuity, as it were, or until they are deleted or radically changed. Consequently, anyone new to a discussion sees the first contribution first. So, whoever starts a new page for discussion also, to a great extent, sets the tone and agenda of the discussion. Moreover, nasty, heated exchanges live on forever on a wiki, festering like an open wound, unless deliberately toned down afterward; if the same exchange takes place on a mailing list, it slips mercifully and quietly into the archives.

At about the same time that we decided to start the meta-wiki, and soon after the vandalism archive affair, I was thinking a great deal about Wikipedia's apparent anarchy, and I wrote an essay titled "Is Wikipedia an experiment in anarchy?"[37] This and the discussion that ensued tended to ossify positions with regard to the authority issue: I and a few others agreed that Jimmy and I should have special authority within the system, to settle policy issues that needed settling. Jimmy was relatively quiet about this issue. This was probably because his authority, unlike mine, was generally accepted. By November or December of 2001, Wikipedia was growing fast, and became the subject of regular news reporting, even by the likes of *The New York Times* and MIT's *Technology Review.* After the two major Slashdottings[38] earlier in the year, we knew that large influxes of members could change the nature of the project, and not necessarily for the better. If there were some major news coverage—an evening news story in the U.S., for example—there might be *many* new people who would need to be taught about Wikipedia's standards and positive cultural aspects. So, I proposed what I thought was a humorously named "Wikipedia Militia"[39] which would manage new (and very welcome) "invasions" by new contributors. By this time, however, there was a small core

---

34 "The Art of Wikipedia Weeding," posted September 26, 2001, *http://meta.wikimedia.org/wiki/ The_art_of_Wikipedia_weeding.*

35 "Wikipedia:Mailing lists," *http://en.wikipedia.org/wiki/Wikipedia:Mailing_lists.*

36 "Moderator (communications)," *http://en.wikipedia.org/wiki/Moderator_%28communications%29.*

37 Posted November 1, 2001, *http://meta.wikimedia.org/wiki/Is_Wikipedia_an_experiment_in_ anarchy.*

38 "Slashdot effect," *http://en.wikipedia.org/wiki/Slashdot_effect.*

39 "Wikipedia:The Wikipedia Militia," *http://en.wikipedia.org/w/index.php?title=Wikipedia:The_ Wikipedia_Militia&oldid=290128.*

group of people who were constantly on the watch for anything that smacked the least bit of authoritarianism; consequently, the name, and various aspects of how the proposal was presented, was vigorously debated.[40] Eventually, we switched to "The Wikipedia Welcoming Committee" and finally, the "Volunteer Fire Department"[41]—which eventually, it seems, fell into disuse.

The governance challenge

After the September Slashdotting, I composed a page originally called "Our Replies to Our Critics"[42] (and now called "Replies to Common Objections"[43]), in which I addressed the problem that "cranks and partisans" might abuse the system:

> Moreover—and this is something that you might not be able to understand very well if you haven't actually experienced it—there is a fair bit of (mostly friendly) peer pressure, and community standards are constantly being reinforced. The cranks and partisans, etc., are not simply outgunned. They also receive considerable opprobrium if they abuse the system.

This reflects the conception I had in September 2001 of Wikipedia's culture; the reply in the previous paragraph was as much hopeful and prescriptive as descriptive. But it turned out to be only partly true. As difficult users began to have more of a "run of the place," in late 2001 and 2002, opprobrium was in fact meted out only piecemeal and inconsistently. It seemed that participation in the community was becoming increasingly a struggle over principles, rather than a shared effort toward shared goals. Any attempt to enforce what should have been set policy—neutrality, no original research, and no wholesale deletion without explanation—was frequently if not usually met with resistance. It was difficult to claim the moral high ground in a dispute, because the basic project principles were constantly coming under attack. Consequently, Wikipedia's environment was not cooperative but instead competitive, and the competition often concerned what sort of community Wikipedia should be: radically anarchical and uncontrolled, or instead more single-mindedly devoted to building an encyclopedia. Sadly, few among those who would love to work on Wikipedia could thrive in such a protean environment.

It is one thing to lack any equivalent to "police" and "courts" that can quickly and effectively eliminate abuse; such enforcement systems were rarely entertained in Wikipedia's early years, because according to the wiki ideal, users can effectively

---

40 "Wikipedia talk:The Wikipedia Militia," *http://en.wikipedia.org/wiki/Wikipedia_talk:The_ Wikipedia_Militia.*
41 "Wikipedia:Volunteer Fire Department," *http://en.wikipedia.org/wiki/Wikipedia:Volunteer_Fire_ Department.*
42 "Wikipedia/Our Replies to Our Critics," *http://web.archive.org/web/20011112085441/www. wikipedia.com/wiki/Wikipedia/Our_Replies_to_Our_Critics.*
43 "Wikipedia:Replies to common objections," *http://en.wikipedia.org/wiki/Wikipedia:Replies_to_ common_objections.*

police each other. It is another thing altogether to lack a community ethos that is unified in its commitment to its basic ideals so that the community's champions *could* claim a moral high ground. So, why was there no such unified community ethos and no uncontroversial "moral high ground"? I think it was a simple consequence of the fact that the community was to be largely self-organizing and to set its own policy by consensus. Any loud minority, even a persistent minority of one person, can remove the appearance of consensus. In fact, I recall that (in October 2002, after I resigned) I felt compelled by ongoing controversies to request[44] that Jimmy declare that certain policies *were* in fact nonnegotiable, which he did.[45] Unfortunately, this declaration was too little, too late.

By late 2001, I had gained both friends and detractors. I think I had become, within the project, a symbol of opposition to anarchism, of the enforcement of standards, and consequently of the exercise of authority in a radically open project. But I was still trying to manage the project as I always had—by force of personality and "moral" authority. So, when people arrived who clearly and openly disrespected established policy, I was, in my frustration, very short with them; and when the project continued to try to establish new policies, my role in articulating those policies and actually establishing them (attempting to express a "consensus") was challenged. This undermined what remaining moral authority I had. I felt my job was on the line, and the project continued in turmoil day in and day out. From my point of view, fires were spreading everywhere, and as I had become a somewhat controversial figure, I did not have enough allies to help me put them out. Consequently, I was too peremptory and short with some users. This, however, exacerbated the problem, because the attitude could not be backed up by punishment; harsh words from a leader are empty threats if unenforceable. I thereby handed my antiauthoritarian "wiki-anarchist" opponents an advantage, because— ironically—they were able to portray me as dictatorial, when I was anything but. I came to the view, finally and belatedly, that it would be better to ignore the trolls. However, this is particularly hard to do on a wiki. Unlike on an email list, trollish contributions do not just disappear into the archives; they sit out in the open, as available as the first day they appeared and festering. Attempts to delete or radically edit such contributions were often met by reposting the earlier, problem version: the ability to do that is a necessary feature of collaboration. Persistent trolls could be a serious problem, particularly if they were able to draw a sympathetic audience. And there was often an audience of sympathizers: contributors who philosophically were opposed to nearly any exercise of authority, but who were not trolls themselves.

---

44 "What we need," *http://mail.wikipedia.org/pipermail/wikien-l/2002-November/000047.html.*
45 "Re: What we need," *http://mail.wikipedia.org/pipermail/wikien-l/2002-November/000086.html.*

It is ironic that it was I who initially supported the lack of any enforceable rules in the community. Some legal theorists would maintain that a community that lacks enforceable rules lacks any law at all. In retrospect, it is clear that there was a fundamental problem with my role in the system: to have real authority, I needed to be able to enforce the rules, and for both fairness and the perception of fairness, there needed to be clear rules from the beginning. But, by my own design, I had very early on rejected the label "editor in chief" and much real enforcement authority; a year into the game, it would have been difficult if not impossible to claim enforcement authority over active but problem users. Moreover, I was the author of the "ignore all rules" rule. My early rejection of any enforcement authority, my attempt to portray myself and behave as just another user who happened to have some special moral authority in the project, and my rejection of rules—these were all clearly mistakes on my part. They did, I think, help the project get off the ground; but I really needed a subtler and more forward-looking understanding of how an extremely open, decentralized project might work.

In retrospect, I wish I had taken Teddy Roosevelt's advice: "Speak softly and carry a big stick." Since my "stick" was very small, I suppose I felt compelled to "speak loudly," which I regret. As it turns out, it was Jimmy who spoke softly and carried the big stick; he first exercised "enforcement authority." Since he was relatively silent throughout these controversies, he was the "good cop," and I was the "bad cop": that, in fact, is precisely how he (privately) described our relationship. Eventually, I tired of this arrangement. Because Jimmy had kept a low profile in the early days of the project and showed that he *was* willing to exercise enforcement authority upon occasion, he was never as ripe for attack as I was.

Perhaps the root cause of the governance problem was that we did not realize well enough that a community would form, nor did we think carefully about what this entailed. For months I denied that Wikipedia was a community, claiming that it was, instead, only an encyclopedia project, and that there should not be any serious governance problems if people would simply stick to the task of making an encyclopedia. This was wishful thinking. In fact, Wikipedia was from the beginning both a community and an encyclopedia project. And for a community attempting to achieve something, to be serious, effective, and fair, a charter seems necessary. In short, a collaborative community would do well to think of itself as a polity with everything that that entails: a representative legislative, a competent and fair judiciary, and an effective executive, all defined in advance by a charter. There are special requirements of nearly every serious community, however, best served by relevant experts; and so I think a prominent role for the relevant experts should be written into the charter. I would recommend all of this to anyone launching a serious online community. But indeed, in January 2001, we were in both "uncharted" and "unchartered" territory. The world, I think, will be able to benefit from this and our other initial mistakes.

In fairness to ourselves, it was a good idea to allow the community to decide by experience and consensus what *article content* rules to endorse. This allowed us to generate a very sensible set of article content rules. Yet it was a mistake to apply the same thinking to the organization of the community itself. We should have acknowledged that a community would form, that it would have certain persistent and difficult issues that would need to be solved, and that a lack of any effective founding community charter might result in chaos.

My Resignation and Final Few Months with the Project

Throughout the governance controversy, I was preparing for my wedding, which took place December 1, 2001. A few days after I arrived back from my honeymoon, I was informed that I should probably start looking for another job, because Bomis had to lay off most of its workers. Bomis had 10 to 12 workers at the end of 2000, and by the beginning of 2002 it was back to its original 4 to 5. My salary was reduced in December and then halved in January. This seemed inevitable because Wikipedia was not bringing in any money at all for Bomis, even if Wikipedia was becoming even more of a publicly recognized, if still modest success. Our first anniversary came just before we announced having 20,000 articles, and I was invited to talk about the project at Stanford[46] on January 16.

I was officially laid off at the beginning of February, which I announced a few weeks later.[47] I had continued on as a volunteer; Wikipedia and Nupedia were, after all, volunteer projects. But I was laboring in the aftermath of the governance controversies of the previous fall and winter, which promised to make the job of a *volunteer* project leader even more difficult. Moreover, I had to look for a real job. So, throughout the month of February, I considered resigning altogether.

Jimmy had told me the previous December that Bomis would start trying to sell ads on Wikipedia to pay for my job. Even in that horrible market for Internet advertising, there were already enough page views on Wikipedia that advertising proceeds might have provided me a very meager living. We knew that this would be extremely controversial, because so many of the people who are involved in open source and open content projects absolutely *hate* the idea of advertising on the web pages of free projects, even to support project organizers. In fact, when this advertising plan was

---

46 The presentation may be viewed at *http://www.stanford.edu/class/ee380/winter-schedule.html*. The text of the talk is located at *http://meta.wikimedia.org/wiki/Wikipedia_and_why_it_matters*.

   You might notice that I was still plugging the notion of using Nupedia to vet Wikipedia articles, as an answer to the objection that Wikipedia articles are unreliable.

47 "Announcement about my involvement in Wikipedia and Nupedia," *http://meta.wikimedia.org/wiki/Announcement_about_my_involvement_in_Wikipedia_and_Nupedia—Larry_Sanger*.

announced, in late February of 2002, the Spanish Wikipedia[48] was forked[49] (something I urged them not to do[50]).

Bomis was not successful in selling any ads for Wikipedia *anyway*—early 2002 was the very bottom of the market for Internet advertising. I also had some hope that we might, *finally,* set up the project's managing nonprofit, which we had discussed doing for a long time (and which eventually did come into being: Wikimedia[51]). The job of setting up the nonprofit was left to me, but ongoing controversies seemed to eat up any time I had for Wikipedia, and frankly I had no idea where to begin. So, after a month without pay, I announced my general resignation;[52] I completely stayed away from the project for a few months.

Wikipedia's offshoot projects—a dictionary, a textbook project, a quotation project, a public domain book repository, etc.—were all started in 2002 or later, and I cannot claim any credit for them.

In the spring, a controversy erupted. Caring as I did—and as I still do—about the future of free encyclopedias, I felt compelled to get involved. The controversy featured a troll who was putting up huge numbers of screeds on the meta-wiki and on Wikipedia as well. The controversy began with a discussion of what to do about, and how to react to, this particular troll. I maintained that one should not "feed the troll," and that the troll should be "outed" (it was an anonymous user, but it was not hard to use Google to determine the identity of the troll) and shamed.

There resulted a broader controversy about how to treat problem users generally. There were, as I recall, two main schools of thought. One, to which I adhered and still adhere, was that bona fide trolls should be "named and shamed" and, if they were unresponsive to shaming, they should be removed from the project (by a fair process) sooner rather than later. We held that a collaborative project requires commitment to ethical standards which are—as all ethical standards ultimately are—socially established by pointing out violations of those standards. Hence naming and shaming. A second school of thought held that *all* Wikipedia contributors, even the most difficult, should be treated respectfully and with so-called *WikiLove*.[53] Hence trolls were not to be identified as such (since "troll" is a term of abuse), and were to be removed from the project only after a long (and painful) public discussion. I felt at the time that the prevalence of the second school entailed rejection of both objective standards and rules-

---

48 Located at *http://es.wikipedia.org/wiki/Portada.*

49 The fork is called *Enciclopedia Libre Universal en Español, http://enciclopedia.us.es/index.php/Enciclopedia_Libre_Universal_en_Espa%F1ol.*

50 "Wikipedia:Statement by Larry Sanger about the Spanish wiki encyclopedia fork," *http://es.wikipedia.org/wiki/Wikipedia:Statement_by_Larry_Sanger_about_the_Spanish_wiki_encyclopedia_fork.*

51 The Wikimedia Foundation's home page: *http://www.wikimedia.org/.*

52 "My resignation," *http://meta.wikimedia.org/wiki/My_resignation*—Larry_Sanger.

53 "Wikipedia:WikiLove," *http://en.wikipedia.org/wiki/Wikipedia:WikiLove.*

based authority. It is *impossible* to explain why one is removing some partisan screeds from the wiki without, in some way, identifying it as a partisan screed, and pointing out that such productions are inconsistent with the neutrality policy. This will necessarily be received as less than respectful and "loving," especially if one must engage the troll himself in a long, drawn-out dispute. In a very long dispute with any trollish type, it is only a matter of time before some epithet gets bandied about. More generally, the very application of rules, or laws, entails a moral judgment, or what for its effectiveness must have the force of a moral judgment. I suppose I agree with those legal theorists who say that there is necessarily, in its core, a *moral* component to the law. Consequently, the new policy of "WikiLove" handed trolls and other difficult users a very effective weapon for purposes of combating those who attempted to enforce rules. After all, any forthright declaration that a user is doing something that is clearly against established conventions—posting screeds, falsehoods, nonsense, personal opinion, etc. —is nearly always going to appear disrespectful, because such a declaration involves a *moral* accusation. The result is that, on pain of becoming *persona non grata* in the community, one had to treat brazen, self-conscious violators of basic policy with *particular* respect. It was a perfect coup for the resident wiki anarchists. I again left the project for several months.

In fall of 2002, I had started teaching at a local community college, and with some extra time on my hands, I started editing Wikipedia a little and engaging in mailing list discussions. I think my first new post to Wikipedia-L, from September 1, 2002, was "Why the free encyclopedia movement needs to be more like the free software movement."[54] In it I argued that the free software movement is led and dominated by highly qualified programmers, and that the "free encyclopedia movement"—that is, Wikipedia, Nupedia, and other newer projects—needs to move in that direction. I suggested that Nupedia be redesigned to release "approved" versions of Wikipedia articles; Wikipedia itself was not to be touched. This proposal met with a very cool reception. After a few months of discussion, Jimmy himself was "intending to revive Nupedia in the near future"[55] and "thinking very much along the lines of what is being discussed here." Unfortunately, this never happened.

By December, I proposed, and Magnus Manske very helpfully coded, an expert-controlled approval process for Wikipedia that was in fact to be independent of both Nupedia and Wikipedia.[56] It would not have affected the Wikipedia editorial process. It would have lived in a separate namespace or domain, as an independent add-on project for Wikipedia. Without explaining the details, expert reviewers, the recruitment of which I would organize, would examine Wikipedia articles and approve or disapprove of particular versions of those articles. We set up a mailing

---

54 *http://mail.wikipedia.org/pipermail/wikipedia-l/2002-September/022164.html.*
55 "Wikipedia subset proposal," *http://mail.wikipedia.org/pipermail/wikipedia-l/2002-November/024677.html*
56 *http://mail.wikipedia.org/pipermail/wikipedia-l/2002-November/024684.html.*

list, Sifter-L (archives no longer online, apparently), which for several weeks discussed policy issues.

There was not a great deal of support for the proposal on Wikipedia-L. There was little or no excitement that the new project might bring into Wikipedia a fresh crop of subject area specialists. But that was fine as far as I was concerned, since the project was to operate independently of Wikipedia. Still, I had the very distinct sense that any specialists arriving on the scene would not necessarily be met with open arms—particularly if before approving an article they wished to make whatever changes to articles that they felt necessary. There were even a few Wikipedians who made it clear that experts should not expect to be treated any differently than anyone else, even when writing about their areas of expertise.

I then considered whether the interaction between Wikipedians and the new reviewers might be a problem after all. Surely, I thought, most specialists would want to edit even very good articles before approving them (in the independent system). This would require that the reviewers interact with Wikipedians. Wikipedia's culture had become such that disrespect of expertise was tolerated, and, again, trolls were merely warned, but very politely (in keeping with the policy of WikiLove), that they please ought to stop their inflammatory behavior. Trolls would certainly find ripe targets in expert reviewers, I thought. I recalled that patient, well-educated Wikipedians like J. Hoffmann Kemp and Michael Tinkler had been driven off the project not only by trolls but also by some of the more abrasive and disrespectful regulars. I then considered: could I in good conscience really ask academics, who are very busy, to engage in this activity that would probably annoy most of them and do nothing to contribute to their academic careers? Recruiting for Nupedia had been easy by comparison and caused me no such pangs of conscience.

I believe it was this problem that finally prompted me in January of 2003 to inform Jimmy by private email that I was breaking with the project *altogether;* the only way he could prevent this, I told him, was that he personally crack down on problem users, and make the project more officially welcoming to experts. I also told him that I did not expect this information to change his mind, and that I did not mean to issue an ultimatum. And in fact our exchange did not change his mind. I concluded that we had a fundamental philosophical disagreement about how the project should be run. I respected and still respect his view. That is where matters ended, and it was then that I broke with Wikipedia altogether.

Final Attempts to Save Nupedia

Nevertheless, I was interested in pursuing Nupedia's development. It still seemed salvageable to me.

I recall two incidents in which I tried to have Nupedia revived. First, I approached Jimmy with the offer to try to find a buyer/managing organization for Nupedia. I suggested that since Bomis did not have enough money to support it, and since Jimmy did not appear to have any specific intentions with the project, I might be able to find a university or other organization that would take on the responsibility. In the end, we did not pursue this possibility. Later, I offered to buy Nupedia myself—that is, the domain name, the membership list, and whatever other proprietary material Bomis might have controlled. I wanted to start it up again as a simpler, more streamlined, but still fully peer-reviewed project. I thought, moreover, that if I owned it, I might be able to give it to a suitable sponsoring educational or nonprofit institution. Jimmy seemed cool to the idea, and did not ask for any specific offers.

Nupedia, then, didn't die just from the inefficiency of its system. To some extent it was also allowed to die, even after it was clear that its former editor in chief expressed an interest in continuing the project under an entirely different system. The result was that, without a leader or organization that could support its mission, Nupedia died a slow death. The server it lived on had some trouble in 2003, and as a result, the web site went offline. For whatever reason, the web site was never brought up again after that.

Perhaps there was a concern that Nupedia would essentially fork Wikipedia. I feel that such a concern would not have justified letting Nupedia wither untended. The projects, Wikipedia and Nupedia, were naturally complementary parts of a single, symbiotic whole. That at least is how I always regarded them. From the founding of Wikipedia, I always thought Wikipedia without Nupedia would have been unreliable, and that Nupedia without Wikipedia would have been unproductive. Together they were to be an "unstoppable high-quality article-creation juggernaut."[57]

It is still disappointing to me that we made plans and promises to thousands of Nupedians, including hundreds of extremely well-qualified people, some of them leaders in their fields. We spent many thousands of hours, all told, on the project. I apologize to those people, and I can only hope that they will find some future open content encyclopedia project worthy of their participation, one that will show the world the potential that Nupedia had.

Conclusions

I have some advice for anyone who would like to start new projects on the model of Wikipedia.

You can learn from Wikipedia's success; so, first and most importantly, note the principles I've articulated about why Wikipedia works.

---

57 "Britannica or Nupedia?" op. cit.

But you can also learn from our mistakes. Governance issues are, in my opinion, the primary failing of Wikipedia. Bear in mind, also, that these are only *rough* guidelines, for those who are starting projects that have enough resemblance to Wikipedia. These are not perfectly general rules:

- If you intend to create a very large, complex project, establish early on that there will be some nonnegotiable policy. Wikis and collaborative projects necessarily build communities, and once a community becomes large enough, it absolutely must have rules to keep order and to keep people at work on the mission of the project. "Force of personality" might be enough to make a small group of people hang together; for better or worse, however, clearly enunciated rules are needed to make larger groups of people hang together.

- There is some policy that, with forethought, can be easily predicted will be necessary. Articulate this policy as soon as possible. Indeed, consider making a project charter to make it clear from the beginning what the basic principles governing the project will be. This will help the community to run more smoothly and allow participants to self-select correctly.

- Establish any necessary authority early and clearly. Managers should not be afraid to enforce the project charter, even by removing people from the project. As soon as it becomes necessary, it should be done. Standards that are not enforced in any way do not exist in any robust sense. Do *not* tolerate deliberate disruption from those who oppose your aims. Tell them to start their own project; there's a potentially infinite amount of cyberspace.

- As any disagreements among project managers are apt to be publicly visible in a collaborative project, and as this is apt to undermine the moral authority of at least one manager, make sure management is on the same page from the beginning—preferably before launch. This requires a great deal of thinking through issues together.

- In knowledge-creation projects, and perhaps many other kinds of projects, make special roles for experts from the very beginning. Do not attempt to add those roles later, as an afterthought. Specialists are one of your most important resources, and it is irrational not to use them as much as you can. Preferably, design the charter so that they are included and encouraged. Moreover, make the volunteer project management a meritocracy, and not based on longevity but based on the ability to lead and contribute to the project. That is the only condition under which very many of the best-qualified people will want to participate.

Another point needs more in-depth development.

Radical and untried new ideas require constant refinement and adaptation to succeed. The first proposal is very rarely the best, and project designers must learn from their mistakes and constantly redesign better projects. Nupedia's Advisory Board failed to admit to inherent flaws in its system, and its delay in admission shut the window of opportunity on its improvement. The Wikipedia community fell into a mistake by thinking that just a few—the wiki feature and the neutrality policy and a few other things—explained Wikipedia's success and that those features can thus be applied with no significant changes to new projects. But there is no substitute for constant creativity and problem solving—nor for honesty about what problems need solving. The honesty to recognize problems and creativity in solving them is, after all, what made Wikipedia succeed in the first place.

This is a crucial point: if you use a tool or model from another project, think through very carefully how that tool or model should be adapted. Do not assume that you need to use every feature or every aspect of the surrounding culture, that you are borrowing. Wikipedia borrowed rather *too* much from (1) the culture of wikis, (2) unmoderated online discussions, and (3) freewheeling online culture generally. To be sure, Wikipedia is also a product of those cultures, and works as well as it does largely because of what it borrowed from those cultures. But it also shares some of its more serious current flaws[58] with such cultures. Those planning new projects, or wanting to overhaul old ones, might well bear in mind that a certain cultural context, including the context that has grown up around a tool, just might not be right for that project. Let me elaborate:

- Consider first the culture of wikis. On the one hand, I said we wanted to determine the best rules, and experience would help us determine that; so we had no rules to begin with. On the other hand, one might add that *another reason* we began without rules was that we were partaking in the *extremely* uncontrolled, freewheeling nature of "traditional" wikis. I think that's right. But there is an excellent reason why an encyclopedia project should *not* partake in that extremely uncontrolled nature of wiki culture, and why it *should* adopt actually enforceable rules. Unlike traditional wikis, encyclopedia projects have a very specific aim, with very specific constraints, and efficient work toward that aim, within those constraints, practically requires the adoption of enforceable rules. The mere fact that most wikis, when Wikipedia was created, did not have enforceable rules hardly meant that one could not innovate further, and create one that *did* have rules.

- Moreover, Jimmy and I and most of the first participants on Wikipedia were veterans of unmoderated Internet discussion groups, and hence, naturally, we could appreciate the advantages of letting a virtual community develop in the absence of any real authority. In unmoderated forums, there is often found a sense, among some participants, that any attempt to oust a particularly troublesome user amounts to unjustifiable censorship. The result is that the existence of many

---

58 See "Why Wikipedia Must Jettison Its Anti-Elitism," op. cit.

unmoderated forums online has created a small army of people militantly opposed to the slightest restriction on speech, who feel that they do and should have a right to say whatever they like, wherever they like, online. Any attempt to create and enforce rules for Internet projects, when that small army is ready to cry "censorship," will seem daring or even outrageous in many contexts online. But there is an excellent reason why such anarchy is inappropriate for many projects, including encyclopedia projects, even one that is self-policing like a wiki. There simply *must* be a way to *enforce* rules for rules to be effective. Given that encyclopedia project development happens almost entirely using words, nearly any rules will also be restrictions on speech. Anyone who advocates many enforceable rules on a collaborative project, in the cultural context of an Internet filled with so many unmoderated discussion groups, can be made to seem reactionary. But this is only a result of that cultural context; in any other context, the existence of rules would be perfectly natural and unobjectionable.

- Finally, and generally speaking, the Internet is a great leveler. Since social interaction *can* proceed among complete strangers who cannot so much as see each other, things that seem to matter in many "meatspace" discussions, such as age, social status, and level of education, are often dismissed as unimportant online. Many Internet forums, chatrooms, and blogs are populated by people who are identified by only a "handle," and any suggestion that communication should be restricted or in any way altered in accordance with "expertise" or "authority" is likely to be met with outrage in most forums. But there are several excellent and obvious reasons why expertise *does* need special consideration in an encyclopedia project, and in other collaborative projects. First, there are many subjects that dilettantes cannot write about credibly; I, for example, could not write very credibly about astronomy or speleology, but I have a passing interest in both. If I am working only with other dilettantes, our articles are apt to remain amateurish at best; we can fill in the gaps in each other's knowledge, and do research, but the results will remain problematic until someone with more knowledge of the subject contributes. Second, there are very many specialized subjects about which no one but experts have any significant knowledge at all. Third, it is only the opinions of experts that will be trusted by most of the public as authoritative in determining whether an article is generally reliable. Moreover, the standards of public credibility are not likely to be changed by the widespread use of Wikipedia or by online debate about the reliability of Wikipedia. Like them or hate them, those are the facts. But if one points out these facts online, culturally "leveled" as it is, particularly in forums or projects like Wikipedia which go out of their way to ignore individual differences among people, one finds a frosty reception at best.

Consider, if you will, that it was *because* Wikipedia was started in the context of the ingrained cultures of wikis, of unmoderated discussion forums, and of the leveling, anti-elitist influence of the Internet at large, that it was very difficult for us to exercise the maximal amount of creativity that a maximally successful project would require. In establishing a *new* cultural context, we were deeply constrained by the old. Now, to be sure, Wikipedia did not have to adopt the particular conjunction of policies that it did. But it is not *surprising* that it did adopt its particular conjunction of policies, considering the conjunction of influences on its development. It would have required much more explanation, persuasion, and struggle to have persuaded potential participants that some persons, *even* in a wiki environment, should have special standing. Constantly reinforced cultural habits die very hard indeed, and place strong constraints upon what can be imagined, and what bare possibilities seem worth consideration.

It was our willingness to exercise our creativity and follow our imagination and create what is a *new* kind of culture, that led to Wikipedia's success. For the overall project of creating open content encyclopedias—and indeed, for the fantastic collaborative Internet that has yet to be created—to reach its full potential, the processes of identifying mistakes honestly, and creatively seeking solutions, must be ramped up and continued unabated.

CHAPTER 21

Sonali K. Shah

Open Beyond Software

Teams of employees at firms innovate. Scientists and engineers at universities and research institutions innovate. Inventors at private labs innovate. Regular people consume. Wrong! Regular people innovate too. Users have been the source of many large and small innovations across a wide range of product classes, industries, and even scientific disciplines.

We are accustomed to thinking of firms as the primary engine of innovative activity and industrial progress. The research and development activities of most firms are based on a proprietary model; exclusive property rights provide the basis for capturing value from innovative investments, and managerial control is the basic tool for directing and coordinating innovative efforts. The proprietary model does not, however, stand alone.

The "community-based" model has generated many of the innovations we use on a daily basis. The social structure created by this model has cultivated many entrepreneurial ventures and even seeded new industries and product categories. In stark contrast to the proprietary model, the community-based model relies neither on exclusive property rights nor on hierarchical managerial control. The model is based upon the open, voluntary, and collaborative efforts of users—a term that describes enthusiasts, tinkerers, amateurs, everyday people, and even firms that derive benefit from a product or service by using it.

Open source software development is perhaps the most prominent example of the community-based model. Although often viewed as an anomaly unique to software production, the community-based model extends well beyond the domain of software. Innovative communities have been influential in product categories as diverse as automobiles, sports equipment, and personal computers.

In this chapter, I describe and discuss three elements of the community-based model. First, users and manufacturers generate different sets of information. This allows users to develop innovations distinct from those typically developed within firms. Specifically, innovations embodying novel product functionality tend to be developed by users. Second, users may choose to share their innovations within user communities. The structures of these communities vary, but those observed to date are built on the principles of open product design and open communication. Third, innovations developed by users and freely shared within user communities have provided the basis for successful commercial ventures. Data drawn from the windsurfing, skateboarding, and snowboarding industries illustrates these processes. Four additional examples of the community-based model—spanning fields and centuries—are then presented. I conclude by reframing my view of the innovation process as driven by the activities of firms and research institutions and discussing implications for firms and policy.

Sports Equipment Innovation by Users and Their Communities

Both users and manufacturers contributed to the development of equipment innovations in the windsurfing, skateboarding, and snowboarding industries. *Users* are defined as individuals or firms that expect to directly benefit from a product or service by using it (von Hippel 1988). In contrast, *manufacturers* are those who expect to benefit from manufacturing and selling a product, service, or related knowledge; thus, firms, entrepreneurs, and inventors seeking to sell ideas, products, or services are all examples of manufacturers. To illustrate, snowboarders are users of snowboards. Firms such as Burton and Gnu are manufacturers of snowboards. An inventor who hears that there is a market for improved snowboard bindings and develops a new type of binding with the intent of patenting and licensing it is categorized as a manufacturer.

The User Innovation Process in Three Sports

This section describes the process by which users and their communities develop innovations. I begin with an example that illustrates this process. The following passage describes how Larry Stanley and the community of windsurfing enthusiasts around him innovated in the sport of windsurfing.

Mike Horgan and Larry Stanley began jumping and attempting aerial tricks and turns with their windsurfing boards in 1974. The problem was that they flew off in midair because there was no way to keep the board with them. As a result, they hurt their feet and legs, damaged the board, and soon lost interest. In 1978, Jurgen Honscheid, of West Germany, came to participate in the first Hawaiian World Cup and was introduced to jumping. A renewed enthusiasm for jumping arose and soon a group of windsurfers were all trying to outdo each other. Then Larry Stanley remembered the Chip—a small experimental board that he had equipped with footstraps a year earlier for the purpose of controlling the board at high speeds—and thought:[1]

> It's dumb not to use this for jumping.

> I could go so much faster than I ever thought and when you hit a wave it was like a motorcycle rider hitting a ramp—you just flew into the air. We had been doing that, but had been falling off in midair because you couldn't keep the board under you. All of a sudden, not only could you fly into the air, but you could land the thing. And not only that, you could [also] change direction in the air!

> The whole sport of high-performance windsurfing really started from that. As soon as I did it, there were about 10 of us who sailed all the time together and within one or two days there were various boards out there that had footstraps of various kinds on them and we were all going fast and jumping waves and stuff. It just kind of snowballed from there.

News of the innovation spread quickly and instructions for how to make and attach footstraps to a windsurf board were shared freely. Later, Larry Stanley, Mike Horgan, and a small set of windsurfing friends would begin the commercial production and sale of footstraps (and other innovations). Today the footstrap is considered a standard feature on windsurf boards.

This example illustrates three key components of innovation development by users. First, the act of use itself creates new needs and desires among users that lead to the creation of new equipment and techniques. Second, user cooperation in communities is critical to prototyping, improving, and diffusing solutions to those needs. Working jointly allows rapid development and simultaneous experimentation, however working jointly also requires that users openly reveal their ideas and prototypes to others. Third, user innovations—even after they have been freely revealed—are sometimes commercialized. Each of these three key components is discussed in detail in the following subsections..

Discovery through use

Users generate and accumulate information based on product use in extreme or novel contexts, the creation of new (unintended) uses for the product or service, and accidental discovery—in addition to intended product use. In contrast, marketing teams at firms generally focus on understanding and improving the *intended* use(s) of a product. For example, until the handles of childrens' scooters accidentally fell off

---

1   Quotes from Larry Stanley in this chapter come from an interview of Stanley conducted by the author.

and children experimented with the resulting toy, it is unlikely that manufacturers would have identified skateboarding as a fun activity. These differences in usage and search patterns create an information asymmetry between users and manufacturers. Because users and manufacturers hold different stocks of information, they will tend to develop different types of innovations.

Two complementary sets of information are required for product development activity. The first is information regarding need and the use context. As discussed in the previous paragraph, this information tends to be generated by users.[2] The second is solution information. This information may be held by both manufacturers who specialize in a particular solution type and individuals with expertise in specific areas. It can be a challenge to bring these sets of information together. Both need and solution information can be difficult to communicate between individuals and can be difficult to transfer from the site where it is generated to other sites—in other words, information is both tacit and sticky (Polanyi 1958; von Hippel 1994; Nonaka and Takeuchi 1995). These difficulties in transferring information, combined with the potential idiosyncratic nature of the request and communication costs, can make it difficult for manufacturers and users to work together.

If information cannot be transferred, users and manufacturers will continue to hold different sets of innovation-related information. Not surprisingly, innovators will develop innovations based upon the information they possess. As a result, users and manufacturers will tend to develop *different types* of innovations. *Functionally novel innovations* will tend to be developed by users. These types of innovations allow users to do qualitatively different things that could not be done previously, that is, they create a new functional capability—e.g., adding footstraps to a windsurfing board so that "jumping" is possible. The development of such innovations requires a great deal of information regarding user needs and use context—information that is held by the user; it makes little sense for manufacturers to "guess" what novel functions users might want. *Dimension-of-merit innovations* may be developed by manufacturers or users. Dimension-of-merit innovations improve known product performance parameters—e.g., making a snowboard less expensive, faster, or lighter. Manufacturers, with their dedicated engineering and design staffs, can draw from their specialized expertise to improve dimensions of merit known to be of value to customers to maximize sales and market share. Users can also draw from what they know to make dimension-of-merit innovations.

---

2   Technique is as important as equipment when it comes to actual use activity. I will focus on innovations in equipment in this chapter, but innovations in technique are equally important—e.g., a surgeon with a new tool must devise a new surgical technique before using the tool. The example at the beginning of this section that describes the development of footstraps provides a particularly vivid illustration of the interplay between equipment and technique innovations.

Individual users hold limited stocks of information from which to draw when innovating.[3] Even a user who knows exactly what functionality she desires may be unable to independently create a solution that achieves that functionality, let alone create an efficient or elegant solution. Users frequently overcome this barrier by working together.

Communities: cooperation among users

Working together provides users with significant benefits. Working with others allows users to access resources to develop their innovations. Working with others also allows more rapid development due to simultaneous experimentation. To illustrate, consider the following description given by windsurfing innovator Larry Stanley:

> …we were all helping each other and giving each other ideas, and we'd brainstorm and go out and do this and the next day the [other] guy would do it a little better, you know, that's how all these things came about…I would say a lot of it stemmed from Mike Horgan because, if something didn't work, he would just rush home and change it or he'd whip the saw out and cut it right there at the beach.

Cooperation among users can take many forms. Informal one-to-one cooperation between users is frequent. Semistructured one-to-many interactions have also been documented (e.g., through publications in newsletters, magazines, and web sites). More structured cooperation within "innovation communities" is also widespread. Innovation communities provide social structures and, occasionally, tools that facilitate communication and interaction among users and the creation and diffusion of innovations. Open source software development communities are a good example of this.

Innovation communities are composed of loosely affiliated users with common interests. They are characterized by voluntary participation, the relatively free flow of information, and far less hierarchical control and coordination than seen in firms. These characteristics allow for rich feedback and the potential to match problems with individuals who possess the ideas and means to solve them. Due to the varied needs and skills of the individuals involved, user communities are often well equipped to identify and solve a wide range of design problems.

Innovation communities may be organized specifically around the development of a particular product or may be organized around a particular activity, with innovation being only one of the community's stated or emergent functions. The term *community*—rather than *network*, for example—is used, because these groups often call themselves communities and possess distinct social structures. User innovation communities develop norms and rules, methods for attracting new members, and methods for maintaining their structure and integrity.

---

3   Extending the information asymmetry argument one step further, we see that individual users and manufacturers will create and hold different stocks of information. As a result, different users (or manufacturers) will develop different solutions and some users (or manufacturers) will be able to more cheaply develop a solution or develop a better solution than others.

Two unique facets of innovation communities are their dedication to open product design and open communication. Open product design means that users are able to modify—"tinker with"—the product or service. Product design can be closed technologically (e.g., by distributing software code only in binary format) or via institutional and contractual mechanisms (e.g., warranties, intellectual property protection, government law and regulation, licensing, or usage agreements). For example, proprietary software by its very nature prevents user innovation: the code is closed both institutionally, through copyright protection, and technologically, through distribution in the form of binary code. In contrast, open source software not only allows but also encourages user innovation. This has two consequences: (a) user innovation will only flourish in open source, and (b) users inclined to innovate will gravitate toward open source. More generally speaking, open design is a prerequisite for facilitating user innovation and the formation of innovation communities.

In addition to open design, communities working with complex products or sets of information may choose to adopt modular project architectures. Modular design involves building complex products from smaller subsystems that can be designed independently yet function together as a whole. When a product or process is "modularized," the elements of its design are split up and assigned to modules according to a formal architecture or plan. Modularization makes complexity manageable; enables multiple individuals to work simultaneously and later integrate their work products; and makes it possible to accommodate unforeseen changes to the system, so long as the design rules are obeyed (Baldwin and Clark 2000).

Innovation communities embrace open communication. By making information and innovations accessible to as many interested users as possible in a timely manner, innovation communities increase the diversity of expertise that can be brought to bear on a problem and allow the results of trial-and-error experimentation by multiple parties to be exchanged. Both factors are likely to increase the likelihood that an effective solution will be created and will reduce the time required to create such a solution.

User communities utilize a number of communication channels. Today the Internet is one of the most common—and is being used for much more than open source software development. For example, kite-surfing enthusiasts have created an online community where they share innovation-related information on board and sail design. Mailing lists and web sites are well-suited communication platforms for communities. They allow many users to be reached very quickly and allow users to both share and record information; they are relatively inexpensive, widely accessible, and easily scalable. However, free and open diffusion of ideas and innovations occurred even before the advent of the Internet. Users have historically shared and continue to share ideas through word of mouth; at club meetings, conferences, and competitions; and in newsletters and magazines. For example, Newman Darby, who is credited with the invention of the windsurfer, published blueprints and instructions for making a windsurfer in *Popular Science* magazine.

The open revelation of information and innovations is a necessary input into cooperative work. Communities provide several innovation-related benefits that might lead an innovator to develop an innovation within or share a completed innovation with the community. First, community members work with innovators and provide innovation-related ideas and assistance (Franke and Shah 2003; Harhoff, Henkel et al. 2003). To get assistance, one must reveal the problem and possible solutions. Given that user-innovators are also enthusiasts who enjoy practicing their activity, much of the "reward" for innovation is in future improvements and continued use. It thus makes sense to reveal the innovation (unless the innovator believes the design is ideal), since revealing opens the door to getting feedback and improvement ideas from others. Interviews with innovators indicate that a desire to advance the technology motivates collaborative work[4]:

> We knew that we were just scratching the surface... The more we worked together, the sooner we'd go faster or do new things.

Second, innovators may share simply because they enjoy the innovation development process and working with others. This pattern emerged in this study, and in research examining the activities and motives of software, radio, and automobile enthusiasts (Weizenbaum 1976; Gelernter 1998; Torvalds 1998; Haring 2002):

> If you did not share... [others] would not be able to keep up with you. To do or experience something new and fantastic or go another step faster isn't much fun when you shout "Wow! Did you see that!" and nobody is there to hear you.

Third, user-innovators willing to share their work with others generally want to prevent third parties from appropriating that work. Third-party appropriation would prevent users from further modifying, improving, and producing the innovation. Communities take a variety of precautions to protect their work and make sure that it will remain available for others to use and modify. For example, public exhibition and documentation act to prevent appropriation by the manufacturer and encourage development by others. Protecting the innovation via available intellectual property protection mechanisms and then allowing others to use and modify it freely can have a similar effect. The sports enthusiasts described here engage in such practices, as do communities of open source software developers (O'Mahony 2003).

Finally, a generally unintended consequence of sharing the innovation in the community is the potential development of a market for the innovative product or product feature—and the opportunity to build a business to satisfy and further grow this market. Sharing the innovation with others can result in both improvement and widespread adoption of the innovation. While some adopters will be willing to construct the innovation for themselves, others will prefer to purchase the innovation, thereby paving the way for firm entry. The process by which user innovations were commercialized in the windsurfing, skateboarding, and snowboarding industries is described in the next section.

---

4   From interview with the author; interviewee unnamed for reasons of confidentiality.

Commercialization

Conventional wisdom argues that the open revelation of innovations and the commercialization of those same innovations for profit are antithetical. Yet a number of innovating users both freely revealed their innovations and started firms that produced those innovations for sale to others. The actions of snowboarding innovator Dimitrije Milovich show how a user-innovator can both profit from an innovation and contribute to community development and market growth. Milovich, granted a patent for his snowboard design in 1971, made it known that he would not enforce his patent against users and other firms in the industry. His actions encouraged experimentation by users and the founding of new firms; both of which are likely to have contributed to market development and growth. He also started his own snowboard manufacturing firm, called Winterstick. Many other user-innovators in these sports did not patent their innovations—purposefully or because they did not recognize the potential commercial value of the innovations—but later started companies that produced the innovations for sale to others.[5]

Not only can free revealing and commercial activity coexist, but "free revealing" can actually set the stage for profitable commercial production. As the innovation diffuses through the community, the reactions of community members to the innovation can be observed. Information regarding improvement ideas, usefulness, and new uses is openly communicated and discussed, making the community a rich source of information for innovating users, users, entrepreneurs, and existing firms seeking to make investment decisions. This is especially true in the context of new or emerging product categories where price and quantity information is not available and where it is difficult or impossible to engage in market research; recall that at this stage, many users are building their own products, distribution chains do not exist, and overall awareness of the product has not penetrated to the mainstream.

As user-innovators observed interest in their innovations, many chose to commercialize the product. This process is straightforward in some cases, and highly emergent in others. Some user-innovators did not think to produce their innovation for sale to others until after receiving a series of requests from enthusiasts—who had heard of the

---

5 A small handful of user-innovators responsible for key innovations patented their innovations. Their experiences suggest that the enforcement of intellectual property rights—i.e., the decisions of courts in upholding patents which have been granted—is worthy of further examination. In the few cases where the windsurfing, skateboarding, and snowboarding innovations studied were patented and then challenged in court by firms wishing to profit from the manufacture of the innovation without paying licensing fees to the innovator, courts tended to overturn the patents. It was argued that these patents did not meet the "nonobviousness" criteria required to be granted a patent: if a layperson could develop the innovation, how could it be nonobvious? In contrast, firms tended not to challenge patents granted to users who were also professionally trained engineers. The legal system is reliant on the knowledge held by society and is influenced by society's assumptions, norms, and biases. It is possible that user-innovators will not be afforded the same rights as inventors, formally trained scientists and engineers, and firms until the importance of innovation by users is more widely recognized.

equipment from other enthusiasts or in newsletters and magazines—interested in purchasing a copy of the innovation. Handmade copies of the equipment were initially constructed for free or at cost. Eventually, some user-innovators realized that they could sell the equipment at a profit and began to manufacture and market the product.

Firms founded by users in these industries functioned as lifestyle firms for many years. By lifestyle firm, I mean a firm with 10 or fewer employees that generates modest revenues for innovating users while continuing to innovate and advance their skills in a sport. These firms were initially operated out of garages or spare rooms. In their early years, these firms generally had no capital equipment beyond portable power tools and produced products one by one or in small lots. User-innovators who founded firms typically worked full time at other jobs and often had low opportunity costs for their time.

The activities of users who founded firms highlights the multiplicity of motives at play, and cautions us to not think of entrepreneurial motivation in purely material terms. First, the innovative activity observed does not appear to be driven by pecuniary motives as is commonly thought; rather, it was driven by motives such as use, enjoyment, challenge, and a desire to build the sport. Second, for many user-innovators, the benefits of starting a firm were not merely financial. Starting a firm also allowed them to spend more time practicing and building the sport they enjoyed, and as the business became more profitable, they could afford to give up other forms of employment and focus fully on the sport.

Over time, many of these firms became leaders in their fields and many were regarded as makers of exceptionally high-quality equipment. Several continue to operate independently, while the brands established by others have been acquired by larger manufacturers. Many of today's well-known brands in the windsurfing, skateboarding, and snowboarding industries—including Windsurfing Hawaii, Gnu, Winterstick, and Dogtown Skates—were created by innovative enthusiasts who later became entrepreneurs.

How Important Is Community-Based Innovation in These Sports?

In 2000, I conducted a longitudinal study of the development and commercialization histories of 57 key equipment innovations in the windsurfing, skateboarding, and snowboarding industries (Shah 2000).[6] The aim of the study was to understand the extent to which users did or did not contribute to innovative and commercial

---

6 The innovations were identified with the assistance of multiple experts in each industry. Detailed information on each innovation was gathered through one-on-one interviews with a variety of actors—innovators, designers, early manufacturers, current manufacturers, magazine editors, book authors, friends and acquaintances of the innovator who were involved in the innovation process, and occasionally professional competitors in the sport. Whenever possible, the innovator was interviewed to get a better understanding for the local information employed and the specific circumstances, needs, and problem-solving methods surrounding the innovative activity. *Innovator* is defined as the individual or set of individuals who first develops a working prototype of an equipment innovation.

activity in these sports. The study found that users and their communities were critical to the emergence and development of these sports.

Sports equipment users developed the first-of-type innovation in each of the three sports studied, that is, users developed the first skateboard, the first snowboard, and the first windsurfer. Users also developed 57% of all major improvement innovations in the sample, and manufacturers developed 27% of the major improvement innovations. The remaining 16% were developed by other functional sources of innovation, such as joint user-manufacturer teams or professional athletes.[7]

Product origins: first-of-type innovations

In each of the three sports studied, users developed the initial first-of-type innovation. In each instance, the innovator(s) engaged in the process of bricolage, using the skills and materials at hand to create the innovation.

For example, skateboarding began in the early 1900s. At that time, children played and rode on wooden scooters, often homemade, consisting of a board with roller skate wheels and a handle attached for control. Over the next five decades, adventurous users removed or did without the handle (it often broke off), thereby creating the first skateboards.

In the case of snowboards, people have been trying to stand up on their sleds for ages. Experts agree, however, that the "formal" history of the snowboard began with Sherman Poppen's Snurfer (Howe 1998; Stevens 1998). In 1965, Poppen noticed his daughter and a friend standing up on their sleds as they slid down a hill. He went to his workshop and used the materials available to create the first prototype—two skis bound together with a string attached at the nose for stability—of what would later become known as the Snurfer (a name created by combining the words *snow* and *surfer*).[8]

In the case of windsurfing, an individual user, Newman Darby, was the initial innovator. In 1964, Darby, a Pennsylvania sailboat enthusiast and amateur boat builder, created the first windsurfer by fixing a universal joint to the base of a mast on a floating platform. The universal joint—a fundamental feature of the windsurfer—allowed the board and mast to move relative to one another. This in turn meant that the sailor could directly manage the direction of sail by standing up and holding the boom and tipping the mast. Darby recollects his experience:

> I first designed the universal joint back in 1948 to use, but I was afraid it would be too dangerous...But [with designs lacking the universal joint] every time the

---

7   In the study, users and professional athletes are treated as distinct. Users benefit directly through product use. In contrast, professional athletes derive financial and career-related benefits from activities such as winning or placing well at competitions and being awarded advertising contracts.

8   Whether Poppen was a user is not clear, however his activities were first inspired and appreciated by a group of users important to him—his daughter and her friend!

wind blew too strong, it blew the sail out of the socket. So I decided, "Well I'm going to have to use the universal joint." I was a little afraid it would break your legs if you went over. Then I started developing one using rubber hoses…I even tried a metal universal joint, and I finally devised one using ropes (Darby 1997).

Major improvement innovations

Manufacturers developed 27% (n=12) of the major improvement innovations in the sample; users developed 57% (n=26).[9] Major improvement innovations are an important subset of overall innovative activity in the sport. They are those equipment innovations identified by multiple experts as being most critical to the development of the sport.

An existing manufacturer developed two major improvement innovations in the sample. Existing manufacturers might (theoretically) be of two types: those in closely related porduct categories (e.g., sailing, skiing, surfing) and those with production or design capabilities useful in mass-producing the product. The existing manufacturer observed in this study—NHS—was a small, Northern California firm founded by three surfing buddies to design and build surfboards. A surplus of fiberglass and a deficit of customers led the trio to begin designing skateboards. NHS ultimately developed two key skateboarding innovations: the use of precision ball bearings and skateboard truck modifications that allowed each wheel to move independently of the others.

Manufacturers organized specifically to produce for the sport in question developed three major improvement innovations in the sample. For example, F2, which was initially organized to distribute and manufacture windsurfers for the European market, is believed to have pioneered the use of polyester film as a sail material.

Existing sports equipment component suppliers developed seven major improvement innovations in the sample. These innovations generally involved transferring specific technology and know-how from an existing sport to the novel one. For example, a maker of fins for surfboards was asked to design a fin to solve some windsurfer-specific problems. Similarly, a producer of sailboat sails worked to improve the design of windsurfing sails and made several innovations. In most cases, the innovative components suppliers were small craft shops run by their founder-owners.

Users and *user-manufacturers* developed 58% of all improvement innovations in the sports studied. The term *user-manufacturer* describes innovative users who founded firms *after* prototyping and beginning to refine an innovation(s)—and, in most cases, also after sharing the innovation(s) with others.[10] These individuals benefited from

---

9   Percentage calculations throughout the paper exclude nine innovations for which the innovator is not known.

10   The first innovation produced by user-manufacturers was made prior to the creation of a firm. Subsequent innovations made by user-manufacturers with 10 or fewer employees are included in this category. Innovations developed by user-founded firms that grew beyond 10 employees are classified as manufacturer innovations to conservatively estimate innovative activity by users.

their innovation(s) both through use and financially. As discussed earlier, the firms they founded are generally best characterized as small, lifestyle firms rather than mass market producers.

Community-Based Innovation and Development: An Even Broader Phenomenon

We've seen how users and their communities shaped the windsurfing, skateboarding, and snowboarding industries and we observe that open source software communities have and continue to shape the software industry. Are these unique cases or are they representative of a broader phenomenon? It appears that users will innovate whenever they have the means and interest to do so. The following four examples show that community-based innovation has been influential in shaping product classes, industries, and even scientific disciplines for hundreds of years.

The Automobile

Franz (1999) describes innovations in automotive accessories made by middle-class American leisure travelers during the early 1900s. She reports that users built and added such features as radiator hoods, safety devices, interior heaters, automobile tops, trunks, reclining seats, and electric ignitions to their cars. Some even replaced the standard body altogether. "The rewards of tinkering lay… in the cultural space of leisure where amateurs produced their own narratives of ingenuity and claimed knowledge of the new machine" (Franz 1999, p.149).

Many of these innovators shouldered the cost of disseminating news of their innovations to other automobile enthusiasts. In the early 1900s, a high number of journals for automobile enthusiasts—"written by and for devotees of the new 'sport'" (Franz 1999, p.198)—published innovator-written "how-to" articles. Existing manufacturers often learned of innovations via the innovators themselves, through requests for repairs, phone calls suggesting that the manufacturer adopt the innovations, and articles in the hobbyist journals (one of the journals was sponsored by Ford). Despite these avenues for information transfer and the fact that many innovating users did not patent their innovations, substantial time lags existed between the time an innovation was made and communicated to other users and when manufacturers incorporated it into commercial products.

The Personal Computer

As is well known, the personal computer revolution was not instigated by R&D scientists and engineers toiling in well-equipped labs. The personal computer was initially developed by hobbyists working after hours in garages, warehouses, basements, and bedrooms (Freiberger and Swaine 2000). These individuals triggered a revolution through their own fascination with technology and willingness to openly share hard-won technical insights with fellow enthusiasts through local computer clubs (such as the Homebrew Computer Club) and hobbyist electronics magazines such as *Popular Electronics* and *Radio Electronics*. Over time, many hobbyists started companies to sell copies of their work to those unwilling or unable to construct their own. In fact, many well-known names in the computer and software industry today, including Bill Gates, Paul Allen, and Steve Wozniak, were active hobbyists before they became entrepreneurs.

User Firms in the 18th Century Iron Industry

We've seen many examples of individual users working together, but there are also examples of *user firms* working together. All firms use products that they do not sell to consumers—e.g., the information technology activities of investment banks. Allen's (1983) study of the 18th century iron industry found that firms cooperated and shared information pertaining to the design and construction of blast furnaces. Improved blast furnace design increased the temperature of the blast and significantly reduced fuel consumption. According to Allen, the science behind blast furnace technology was not well understood. No one could predict how design changes would affect furnace performance, so development took the form of trial-and-error learning. Firms were limited in their ability to independently experiment as construction costs were high. By sharing experiences with different designs, firms could multiply the number of experiments from which to learn and collectively improve the technology.

Amateur Astronomy

Users also contributed innovations and discoveries to the scientific disciplines. For example, amateurs played a significant role in the development of astronomy equipment (Lankford 1981). They pioneered the use of reflecting telescopes and applied photographic techniques to the study of the stars. Amateurs published papers in journals alongside professionals, received the same awards, and attended the same meetings. The activities of professionals and amateurs were similar, but because amateurs were allowed to take greater risks than professionals (who were concerned about their careers), the two groups often came into conflict. By the early 1900s, amateurs were unable to compete with the activities of trained astrophysicists, largely because only those with specialized training were allowed to access the increasingly sophisticated and expensive technologies housed within universities and research institutions. By restricting access to tools and technology, professionals effectively limited the ability of amateurs to contribute to and challenge the field.

Today thousands of amateurs are once again making meaningful contributions to the field of astronomy. A revolution triggered by three new and inexpensive technologies has reignited amateur astronomy over the past two decades (Ferris 2002). First, there was the creation of the Dobsonian, a powerful telescope built from inexpensive materials:

In the early 1950s, John Dobson spied a 12-inch piece of porthole glass on a friend's table and realized that it could be polished with sand into a reflecting telescope mirror. As an ascetic monk with no money, he was forced to scrounge for materials, cobbling the mount from such humble objects as a plywood box, the cardboard cores of garden hose reels, and roof shingles. Then he pointed his homemade contraption at the moon—and was astonished by how much detail he could see. Craters, mountains, crags leapt to life. "It was like I was coming in for a landing," he says. His eventual design for an affordable Newtonian reflecting telescope would later be named the Dobsonian (Campbell 2004).

Dobson actively reached out to other enthusiasts and provided them with instructions for building the telescopes. Enthusiasts willing to forego shortcuts can build a Dobsonian for about $20; for a few hundred dollars they can assemble one using materials available at most hardware stores or from a kit. Then came the creation of the CCD, a highly light-sensitive chip able to record very faint starlight with far greater accuracy than a photograph. Finally, the Internet multiplied the power of individual efforts by enabling rapid collaborative work.

Armed with Dobsonian telescopes and CCD sensors, thousands of amateurs are exploring space and recording events that might otherwise go unnoticed by professionals. This community of globally linked amateurs share their observations and expertise within minutes via email, community web sites, and mailing lists as they race to document, understand, and corroborate their findings. They also meet from time to time at meetings and conferences, and keep abreast of developments through magazines.

In these examples, we see the importance of use *and* community. Use drives the emergence and recognition of heterogeneous needs and desires. Community allows rapid experimentation and allows individuals with differing expertise to bring their skills and knowledge to bear on a particular problem. Users in a wide variety of fields work within communities where the open exchange of ideas, prototypes, and resources is commonplace.

Although communities are rarely created for the express purpose of encouraging and supporting innovation, many communities fulfill this function. The social structure provided by communities facilitates the development of user innovations by making resources—ideas, expertise, skills, and physical resources—more easily accessible and by creating incentives that support the sharing of resources and the creation and diffusion of innovation.

Reframing: Where Does Innovation Come From?

Why have we overlooked the fact that so much creative and innovative activity stems from the everyday behavior of regular people? Three factors are likely to have played a role: Schumpeter's legacy, the low visibility of user-innovators outside their own community, and the deliberate creation of a consumer culture.

Firms and entrepreneurs are generally recognized as the primary agents of product change and economic progress (Schumpeter 1934; Nelson and Winter 1977; Dosi 1982). Firms are motivated by profits and invest in research and development to create new products for consumers. As the instigators of change, it is incumbent upon firms to either educate the consumer to want what they produce or to identify and satisfy consumer needs. The consumer's role is a passive one: producers, not consumers, innovate and consumer preferences do not change without producer influence. The consumer merely chooses to make or not make a purchase based on price and comparison with other products and services. In broad and oversimplified terms, this is what is taught to students in management, marketing, economics, and engineering. There is no simple term by which to refer to the "everyday" person who also innovates. Enthusiast, hobbyist, tinkerer, and developer are all possibilities; but they all carry distinct connotations. The term *user-innovator* is better, but is neither perfect nor widely used.

The relatively low visibility of user-innovators may have also prevented us from noticing their activities or viewing them as more than mere anomalies: while firms are likely to heavily promote their innovations to the mass market, consumer innovations are more likely to be diffused through word of mouth or be written up in small, specialist newsletters, journals, or, more recently, web sites. Although it appears that users have always innovated, the advent of the Internet made their activities more visible to those outside of innovation communities and the success of some open source software development provided an extreme example of the power and effectiveness of user communities.

Nobel (1977) argues that the rise of the corporation and the engineer in the 1900s led to "the deliberate creation of a consumer culture, through advertising, to absorb and diffuse potential revolutionary energies." Institutions, namely corporations, sought to identify themselves with innovation, and relegate the consumer to a passive role (recall that historically individuals were anything but passive, producing much of what they used and consumed themselves). Corporations worked to inhibit innovation by consumers through a variety of means, including advertising and creating closed designs (i.e., product designs that made it difficult for a consumer to alter or tinker with the product).

As a result, two characters dominate the landscape of managerial, economic, and sociological thought in the area of innovation: firms and consumers. Firms produce. Consumers consume. As we have seen, however, users have played and continue to play a dramatic role in the development, diffusion, and commercialization of innovations. What does this mean for government policy and firm strategy?

Building and Preserving the Intellectual Commons

The commons are a crucial resource for fostering innovation. Keeping a resource in the commons both allows others to draw upon the resource and mitigates the number of strategic games played by those seeking to influence the innovative and commercial activities of competitors and potential competitors (Lessig 2001a, p. 72. For additional data and analysis regarding the strategic uses of patents and copyrights, see Parr and Sullivan 1996; Hall and Ziedonis 2001; and Shapiro 2001). Government policy plays an important role in developing and maintaining these commons.

The goal of intellectual property policy is to promote technological and cultural progress for the benefit of society. One of the underlying assumptions of these policies is that investment in innovative and creative activities is highly contingent on the ability to derive pecuniary profits from that investment. To that end, government policy in much of the world seeks to strike a balance between granting temporary control rights over innovative and creative work to originators of the work, and allowing others to access and build upon that work. These temporary control rights take the form of patent and copyright protection; patents generally offer protection for 14–20 years, copyrights for 95 years.

From the perspective of community-based innovation, however, benefit is derived primarily through use rather than pecuniary profit. As the examples in this chapter illustrate, users working within communities actively choose to partake of the benefits derived from allowing others to freely use their work rather than pursue benefits derived from control. Thus, protecting the ability of users to tinker and share their work is critical for fostering community-based innovation; the provision and exercise of exclusionary control rights, in contrast, might do little more than act to deter community-based innovation.

Both patent and copyright laws affect the users' "right" to tinker. Here, I will focus on some issues around fair use to show how these laws might influence community-based innovation. Fair use makes copyrighted work available to the public as raw material without the need for permission or clearance, so long as such use promotes progress. What activities do and do not constitute fair use? The answer to this question is unclear in many instances, providing users with little guidance regarding the legality of their actions. Law in this area is complicated and continuously evolving through legislative and judicial action. These decisions, however, do not move in lock-step. From the perspective of protecting fair use, the Digital Millenium Copyright Act (DMCA) is a setback and Sony v. Connectix (2000) is a victory.

Many are concerned that the DMCA has gone too far in restricting fair use in the digital domain (see, for example: Samuelson 1999; Nimmer 2000). The DMCA was intended to prevent consumers from illegally making *copies* of protected works. Unfortunately, the DMCA can also have a number of unintended side effects, one of which is preventing users from *modifying* the products that they purchase. Specifically, the DMCA outlaws technologies designed to circumvent technologies that protect copyrighted material. "The trouble, however, is that technologies that protect copyrighted material are never as subtle as the law of copyright. Copyright law permits fair use of copyrighted material; technologies that protect copyrighted material need not. Copyright law protects for a limited time; technologies have no such limit.

Thus, when the DMCA protects technology that in turn protects copyrighted material, it often protects much more broadly than copyright law does. It makes criminal what copyright law would forgive" (Lessig 2001b).

The judgment of the Ninth Circuit Court of Appeals in *Sony v. Connectix* upheld and extended the limits of fair use. In the case, Sony alleged that Connectix illegally reverse engineered the Sony BIOS to develop its Virtual Game Station, which played Sony PlayStation games on Windows. The court concluded that "Connectix's reverse engineering of the Sony BIOS extracted from a Sony PlayStation console purchased by Connectix's engineers is protected as a fair use. Other intermediate copies of the Sony BIOS made by Connectix, if they infringed Sony's copyright, do not justify injunctive relief." The court determined that it was acceptable for Connectix to not just copy and study Sony's code, but to actively use that code in the process of developing a noninfringing product and make multiple copies of the code. The judgment established new precedents in fair use law, opening up some areas for fair use that were previously risky from a legal perspective.

Restricting the ability of others to build upon ideas may slow the overall rate of innovation; the modification of existing ideas, products, and artistic work is the source of much creative and innovative production by firms, researchers, and users. Evidence of this can be found in many areas. Consider, for example, the development of Linux versus Minix (DiBona, Ockman et al. 1999, Appendix A: The Tanenbaum-Torvalds Debate). Software developers were free to tinker with Linux and adapt it to suit their own needs and desires. They were also able to share what they had learned with one another and build upon each other's efforts. In contrast, enhancements were generally not accepted to Minix to preserve its integrity as a teaching tool. As a result, disgruntled Minix users chose to adopt—and work to improve—Linux. Also consider the "anticommons" effect. The anticommons effect is a side effect of patent protection in fields where innovation is cumulative. A commons is a resource that everyone has the right to use. In contrast, an "anticommons" is a resource which many have the right to prevent others from using (Heller 1998; Buchanan and Yoon 2000). In such a context, innovation may be stifled as innovators become reluctant to innovate because too many others have the right to prevent or raise the costs of use and commercialization (see Heller and Eisenberg 1998 for evidence from biomedical research). Finally, recall the importance of tinkering and bricolage in the examples of community-based innovation presented in this chapter.

The impact of intellectual property policy on the activities of innovation communities deserves careful consideration. As a society, there are important decisions to be made regarding intellectual property protection that will influence not only the rate of technological progress, but also control over its direction, our own ability to "tinker" with and adapt those products to suit our own desires, and the variety of commercial products that are available to us.

Firm Strategy

Not all firms are choosing to enclose their intellectual property inside hermetically sealed black boxes. Some firms—ranging from video-game makers to manufacturers of airplane kits to Lego—have found that it is in their self-interest to permit and even encourage innovation by user communities. Contributions by user communities can complement a firm's own R&D and marketing efforts, extend a product's life, and cater to market niches not targeted by the firm's marketing department. As discussed, user communities often generate a variety of functionally novel and incremental, dimension-of-merit innovations; firms can observe which of these innovations are adopted by community members. Firms, with their specialized engineering, design, manufacturing, and marketing departments, can then streamline, promote, and produce these innovations for the many consumers who are unable or unwilling to construct the product or service themselves. Firms may choose to incorporate these innovations into the core product or service, sell these features as optional modules, or allow a third party to freely distribute or sell the modules.

User groups often form and operate independently of firms. Many groups, however, are open to participation by firms so long as firms support the general goals of the community and abide by the community's rules, norms, and practices. Businesses seeking to encourage user activity around their products have found it useful to open all or part of their product design and establish or support forums where users can congregate and share information (see von Hippel and Katz 2002; Jeppeson 2005, in press).

Building a business around freely revealed user innovations is more straightforward when the product is physical rather than virtual. In the case of physical products, a fraction of users will build their own, but many will prefer the convenience of purchasing a copy. In other words, even if product development by users displaces that of manufacturers, manufacturers can still profit from manufacturing activities and product innovation. Manufacturers may compete against each other for customers based on complementary assets such as brand name, and distribution and production capabilities. Firms may also choose to provide services that go with the product—e.g., in the case of sports equipment, lessons, facilities, or equipment maintenance.

The case of virtual products is more complicated for manufacturers, because many more users will be able to access and deploy the product themselves. One option is to sell services that support the product. A second option is to build and sell proprietary platforms on which users can develop and build their own products.

There are two general approaches to platforms—the "walled garden" and the "open range." Walled gardens place limits around the ability of others to build on and use the platform. For example, this may mean that outside vendors are restricted in their ability to offer commercial products based on the platform or that the platform owner controls the content available to users. While users and outside vendors may have considerable latitude within the walled garden, the platform owner often retains ultimate control rights and establishes both the boundaries and rules of the garden.

Open ranges, in contrast, allow users and other firms to build on and use the platform in limitless ways. The platform owner typically retains few, if any, control rights. NTT DoCoMo explicitly created a walled garden within a larger open range, with respect to content, when setting up its i-mode wireless Internet service. "Official" content partners—subject to strong editorial and usability rules—populate the walled garden, however users are also allowed open Internet access to "unofficial" sites. There is a long-standing debate between proponents of the walled garden and open range approaches. However, from the perspective of the platform owner, it is not yet clear which of these approaches will yield greater profits.

Conclusion

Community-based innovation has contributed to technological and industrial advances in many fields. Users are at the center of this model: they discover new needs and desires, cooperate with other users within innovation communities, and sometimes even commercialize their innovations. The community-based innovation model is pervasive across time and context, contributing to the development of physical and virtual products and shaping products, industries, and scientific disciplines. Yet, for a number of reasons, communities of users often go unnoticed by firms, policymakers, and society at large. Some firms have, however, recognized the contributions of users and their communities and actively work alongside them, providing consumers with novel and improved products and services and creating a revenue stream that contributes to the firms' profits. As intellectual property policy evolves, policymakers ought to consider the impact of proposed policy changes on the ability of users to innovate. Preserving the ability of users to collectively tinker and modify is necessary for continued innovation of the type that has provided us with many of the products and tools, and a substantial amount of the knowledge and know-how that we rely upon and enjoy on a daily basis. In short, the principle that Richard Stallman succinctly defined in the GNU General Public License—that people must be free to use, modify, and distribute—applies to creative and innovative activity in many fields, not just software.[11]

References

R. C. Allen, "Collective Invention." *Journal of Economic Behavior & Organization* 4, 1983: 1–24.

C. Baldwin and K. Clark, *Design Rules*. (Cambridge, MA: HBS Press, 2000).

J. M. Buchanan and Y. J. Yoon, "Symmetric Tragedies: Commons and Anti-Commons." *Journal of Law and Economics* 43, April 2000: 1–13.

---

11 Sincere thanks to Carliss Baldwin, Glenn, Hoether. Mark Stone, and Rosemarie Ziedonis for their feedback and enthusiasm.

B. Campbell, "The Father of Street-Corner Stargazing." *The New York Times*. (September 1, 2004): D10.

N. Darby, "Naomi & Newman Darby: The Interview." *American Windsurfer* 5, no. 1, 1997: 38–52, 94.

C. DiBona, S. Ockman, et al. (eds.), *Open Sources: Voices from the Open Source Revolution* (Sebastopol, CA: O'Reilly, 1997).

G. Dosi, "Technological Paradigms and Technological Trajectories: A Suggested Interpretation of the Determinants and Directions of Technical Change." *Research Policy* 11, no.3, 1982: 147–162.

T. Ferris, *Seeing in the Dark: How Backyard Stargazers Are Probing Deep Space and Guarding Earth from Interplanetary Peril* (New York, NY: Simon & Schuster, 2002).

N. Franke and S. Shah, "How Communities Support Innovative Activities: An Exploration of Assistance and Sharing among End-Users." *Research Policy* 32: 157–178.

K. Franz, "Narrating Automobility: Travelers, Tinkerers, and Technological Authority in the Twentieth Century" (Unpublished Doctoral Dissertation). (Providence, R.I.: Brown University, 1999).

P. Freiberger and M. Swaine, *Fire in the Valley* (New York, NY: McGraw-Hill, 2000).

D. Gelernter, *Machine Beauty* (New York, NY: Basic Books, 1998).

B. H. Hall and R. H. Ziedonis, "The Patent Paradox Revisited: An Empirical Study of Patenting in the US Semiconductor Industry, 1979–95." *Rand Journal of Economics* 32, no. 1, 2001: 101–128.

D. Harhoff, J. Henkel, et al., "Profiting from Voluntary Information Spillovers: How Users Benefit by Freely Revealing Their Innovations." *Research Policy* 32, no. 10, 2003: 1753–1769.

K. Haring, "Technical Identity in the Age of Electronics" (Unpublished Doctoral Dissertation). (Cambridge, MA: Harvard University, 2002).

M. A. Heller, "The Tragedy of the Anticommons: Property in the Transition from Marx to Markets." *Harvard Law Review* 111, 1998: 621.

M. A. Heller and R. S. Eisenberg, "Can Patents Deter Innovation? The Anticommons in Biomedical Research." *Science* 280. no. 5364, 1998: 698–701.

S. Howe, *(Sick) A Cultural History of Snowboarding* (New York, NY: St. Martin's Griffin, 1998).

L. B. Jeppeson, "User Toolkits for Innovation: Consumers Support Each Other." *Journal of Product Innovation Management* (2005, in press).

J. Lankford, "Amateurs and Astrophysics: A Neglected Aspect in the Development of a Scientific Specialty." *Social Studies of Science* 11, no. 3, 1981: 275–303.

L. Lessig, (2001a) *The Future of Ideas* (New York, NY: Random House, 2001).

L. Lessig, (2001b) "Jail Time in the Digital Age." *The New York Times*, July 30, 2001: A 17.

R. R. Nelson and S. G. Winter, "In Search of Useful Theory of Innovation." *Research Policy* 6, no. 1, 1977: 36–76.

D. Nimmer, "A Riff on Fair Use in the Digital Millennium Copyright Act." *University of Pennsylvania Law Review* 148, 2000: 673–742.

D. Nobel, *America by Design: Science, Technology, and the Rise of Corporate Capitalism* (New York, NY: Alfred A. Knopf, 1977).

I. Nonaka and H. Takeuchi, *The Knowledge-Creating Company* (New York, NY: Oxford University Press, 1995).

S. O'Mahony, "Guarding the Commons: How Community Managed Software Projects Protect Their Work." *Research Policy* 32, no. 7, 2003: 1179–1198.

R. L. Parr and P. H. Sullivan, *Technology Licensing: Corporate Strategies for Maximizing Value* (New York, NY: John Wiley & Sons, 1996).

M. Polanyi, *Personal Knowledge: Towards a Post-Critical Philosophy* (New York, NY: Harper Torchbooks, 1958).

P. Samuelson, "Intellectual Property and the Digital Economy: Why the Anti-Circumvention Regulations Need to Be Revised." *Berkeley Technology Law Journal* (14, 1999).

J. Schumpeter, *The Theory of Economic Development* (Cambridge, MA: Harvard University Press, 1934).

S. Shah, "Sources and Patterns of Innovation in a Consumer Products Field: Innovations in Sporting Equipment." MIT Sloan School Working Paper #4105. Cambridge, MA: 2000.

C. Shapiro, *Navigating the Patent Thicket: Cross Licenses, Patent Pools, and Standard-Setting. Innovation Policy & the Economy.* A. Jaffe, J. Lerner and S. Stern. (Cambridge, MA: MIT Press, 2001).

B. Stevens, *Ultimate Snowboarding* (New York, NY: Contemporary Books, 1998).

L. Torvalds, "First Monday Interview with Linus Torvalds: What Motivates Free-Software Developers?" *First Monday* 3, no. 3, 1998.

E. von Hippel, *The Sources of Innovation* (New York, NY: Oxford University Press, 1988).

E. von Hippel, "'Sticky Information' and the Locus of Problem Solving: Implications for Innovation." *Management Science* 40, no. 4, 1994: 429–439.

E. von Hippel and R. Katz, "Shifting Innovation to Users Via Toolkits." *Management Science* 48, no. 7, 2002: 821–833.

J. Weizenbaum, *Computer Power and Human Reason: From Judgment to Calculation* (San Francisco, CA: W. H. Freeman, 1976).

Steven Weber

CHAPTER 22

Patterns of Governance
in Open Source

The hardest problem facing a political community is how to increase the probability that the whole will be greater than, not less than, the sum of its parts. People join together voluntarily to solve problems because they believe that the group can do things that an individual cannot. They also believe (in some abstract sense and often implicitly) that the costs of organizing the group and holding it together will be smaller than the benefits the group gains. If you put aside for the moment the affective and emotional needs that individuals satisfy in groups and focus instead on the part of politics that is about problem solving, the bet that people make when they enter a political community is simply that "none of us is as smart as all of us"—maybe not on any particular issue or at any particular moment, but on the vast set of problems that human beings confront and try to manage over time.

It doesn't have to work out that way. Everyone has been part of a community or a company where the whole is less smart than the individuals who comprise it. Political systems often seem to suffocate under their own organizational costs—not just national governments, but smaller systems like city councils and co-op boards. And even if a community does create net benefits for at least some segment of the group, the distribution of those benefits can be so grossly unequal that most of the community members would be better off on their own. Get the balance wrong, and you can easily create situations where no one is as dumb as all of us.

These are very old problems confronting political thinkers. The rise of the Internet adds a small but significant twist by making it much easier to discover potential collaborators and pull together "affinity groups," networks of subcontractors, outsourced component makers for production systems, and the like—all of which are political communities that aim to solve some kind of problem. The core idea is *joint production at a distance*, the opening up of a universe of collaborative projects in which physically separated individuals contribute to the creation and refinement of a solution. Because the opportunities for creating collaborative communities have been expanded greatly by Internet technology, the boundaries and borders of existing communities are open for redefinition, and the possibility for new communities seems vast.

Which means that the stakes are high for getting it "right," or at least getting it right enough. This, I believe, is where some of the most important lessons of open source collaboration are likely to emerge. This chapter poses the question this way: if patterns of collaboration within open source communities were to become surprisingly pervasive, or pervasive in surprising places, what would this suggest about institutional design for communities of knowledge and practice in politics, outside of the realm of software or even technology per se?

To answer that question takes at least four steps. I first bound the question by limiting the argument to a class of problems most likely susceptible to open source-style principles. I then describe more precisely some of the theoretical issues at stake in group problem solving. The third section of the chapter lays out seven design issues that follow from the experience of what works (and does not work) within open source communities. The final section suggests some actionable implications. If we view the politics of problem solving through this kind of prism, what might or should we do differently?

The Empirical Problem Set: What Are We Aiming At?

Consider this proposition: some significant subset of social problems that communities confront are (or can be) structured as knowledge creation and/or problem-solving domains similar to the "problems" that the open source software community has found innovative ways to "solve." It would follow that the tools and governance principles of the open source software community, in some modified form, could yield new approaches to community organization and problem solving that build on but go beyond what is currently known about traditional institutions of formal government as well as the more informal notions of "civil society" and "communities of practice."

I think the proposition is defensible at least for a class of complex social problems that have three characteristics. The problems we are thinking about should be multidimensional in the sense that they call on several different realms of expertise. They should be large in scope, in the sense that they require some kind of division of labor to make progress. And they should be complex in their essence, not just in their implementation. I mean here problems that are substantively and inherently difficult to solve, not difficult only because of the failure of well-understood social or political processes to yield optimal outcomes. An example: if you want to build a new

100-story office building in Manhattan, you will need to pull together many different realms of expertise and organize a rather sophisticated division of labor. Some of the problems you confront will be idiosyncratic social and political issues—the metalworkers will strike because they know you really need them today, the neighbors will complain about the noise, and deliveries of certain materials will get "held up unexpectedly" somewhere until you pay a friendly fee to the person who can "fix" that problem. But even if you had the magic solution to all these issues, it would still be hard to create this building simply because it is a difficult engineering task to put together a mountain of steel, concrete, and glass that will hang together and stand up to wind, rain, gravity, and use over all the years that it will be there. It would still cost more and take longer than expected.

The analogy to software development should be obvious. Complex software is hard to build because it is multidimensional, because it demands a division of labor, and because the problems it is trying to solve are inherently hard. One of the earliest and still best analyses of complex software development projects is Frederick Brooks' *The Mythical Man-Month*.[1] Brooks' Law states one of the fundamental conclusions from Brooks' assessment: "Programming work performed increases with direct proportion to the number of programmers (N), but the complexity of a project increases by the square of the number of programmers (N2)." Brooks' Law, even if it is not precisely verifiable, is a powerful statement about the software engineering manifestation of a repeated observation on this point. It is hard to build complex systems in considerable measure because it is so hard for people to explain to each other what they are trying to do. Brooks' argument boils down to this simple but profound claim: human communication about complex, often tacit goals and objectives is imperfect. And it gets more imperfect, and at an increasing rate, as it travels between larger numbers of people. So, how do we ever get a functioning division of labor at a large scale to do things like build a New York skyscraper, or write a program with a million lines of code?

One way to manage this dilemma is to enclose the production process within a formal organization—for example, a proprietary software company. The ideal-type principles of organization here are command and control authority, hierarchical structure for decision making, and tight governance of principal-agent problems. Sustaining that kind of organization depends on maintaining control over the essential resources in the production process. In the software world, that means keeping source code secret. The open source community, by releasing source code, undermines the possibility of setting up the production system in the same way and energizes a quite different organizational model.

No one should bet on anything like a wholesale transfer of the organizational model(s) from the open source community to the nonsoftware world; that is too simplistic. What I think we should focus on instead is the means by which the open source community processes, collates, upgrades, corrects, distributes, and implements problem-solving

---

1   Frederick P. Brooks, *The Mythical Man-Month: Essays on Software Engineering*, 20th Anniversary Edition (Addison Wesley, 1995).

information. In other words, think of open source as a particular kind of information processing algorithm. (It then makes sense to treat the related issues of intellectual property rights and organizational structures that are typically seen as core to the open source community as instrumental, not foundational.) What is foundational to transfer is the information processing "system" that is enacted in this community, and how the results of that process are incorporated into real solutions to practical problems.

It may seem quixotic to think about complex social problem solving in political communities as an information processing challenge. After all, we know that innovation in this setting traditionally is slow, constrained, inefficient, and frustrating. And we know, from the work of Max Weber and Joseph Schumpeter and extending into modern public choice theory in political science and management theory in business, some of the reasons why that is the case, in particular the organizational disincentives and cultural impediments to change that are inherent parts of bureaucratic culture and institutions.[2]

Clearly there are a lot of things going on in political communities besides poor information processing. But any experiment, even a thought experiment, has to start somewhere. The proposition here is that information processing is a significant impediment to problem solving in some important political situations—and that, if we can define a set of problem domains that fit this description, we can do something interesting by attacking the nature of the information processing problem first and then thinking about the organizational structure and political problem secondary to that. In other words, design the governance institutions in ways that facilitate information paths that we think will work, rather than the other way around. This is worth experimenting with in part precisely because it is the reverse of many conventional ways of thinking; and in part because we know more about the trade-offs associated with governance institutions than we do about the information processing issues.

In sum: think of the target as a set of problem-solving practices which necessarily include an information processing algorithm and the associated institutional structures and incentives that make that algorithm function in real-world settings. These practices will tap into distributed knowledge that in some cases may be present in geographically dispersed individuals or communities; in some cases may be present in separate pieces that have not been integrated into a single, useful whole; and in some cases may be implicit in relatively undefined or tacit practices that "belong" to individuals' experiences—but are for that very reason not available for use, testing, and refinement by larger groups. Primary care medicine is a good example. My doctor in Berkeley is often solving the same problems of diagnosis and treatment that a primary care doctor in Manhattan solved yesterday, but she has to re-create complex, tacit, and multifaceted knowledge that already exists elsewhere, because there is no structure within which that knowledge can be effectively shared. The bet you

---

2   Some classic readings are in Max Weber, *Economy and Society* (University of California Press, 1978); *Essays in Sociology* (Routledge & Kegan Paul, 1958); and Joseph Schumpeter, *Capitalism, Socialism, and Democracy* (Allen and Unwin, 1958).

need to make to stay with me in this chapter is simply that an important subset of social and political problems fits in this category and might be attacked in this way.

The Theoretical Problem: How Is Knowledge Distributed?

Contemporary literature on "communities of practice" takes off from a very similar bet.[3] This literature offers a set of relatively obvious but useful design principles that appear to contribute to success. None of these principles really is well enough specified to be operational, but they are clearly worth keeping in mind as a checklist against which any system design can be compared. Roughly, they are:

- Design for evolution (allow the community to change).

- Open a dialog between inside and outside perspectives (tightly insulated communities tend to corrode).

- Allow for different and bursty levels of participation (different people will participate at different levels, and any single person will participate at different levels over time).

- Preserve both public and private community spaces (not all community interactions are public; backchannels should be available).

- Focus on the value that is created for the people in the community.

- Mix the familiar and the new.

- Facilitate the creation of a rhythm (pure bursty-ness and unpredictability tend to corrode commitment).

These design principles actually presuppose quite a lot about the nature of the knowledge that the community of practice is trying to generate, organize, and share. I want to parse out some of the assumptions about that knowledge and some of the different ways it may be embedded in communities to illustrate this point.

Consider again the common saying "none of us is as smart as all of us." The operative assumption is that each one of us has bits and pieces of "good" (useful) knowledge and "bad" (wrong, irrelevant, or mistaken) knowledge about a problem. If Frederick Brooks was even partially right about the social dynamics of complex reasoning (and I think he was right), the demonstrated success of the open source process cannot simply depend on getting more people or even the "right" people to contribute to a project. It depends, crucially, on how those people communicate information to each other. Put differently, depending upon how the community selects, recombines, and iteratively moves information forward over time, the collectivity will become either very smart or very stupid.

---

3   See, for example, Etienne Wenger, *Communities of Practice: Learning, Meaning, and Identity* (Cambridge University Press, 1999).

I am just saying explicitly here what Eric Raymond implied about the open source process. It is not simply that "with more eyeballs all bugs become shallow." It depends directly on how those eyeballs are organized. And since I am treating organization as an outcome of what kind of information processing algorithm the community needs, to get to operational design principles means understanding better at least these two aspects: how knowledge is distributed in the community, and what the error correction mechanisms you can apply to that knowledge. In simpler language, who knows what, and how do you fix the mistakes?

We know from both intuition and experience that much of what a group needs to "know" to do something is in fact coded in the experiences, tacit knowledge, implicit theories, and data that is accessible to individuals. The problem for the group is that these individuals often don't know how to, aren't incentivized to, or haven't thought of sharing it with others in a mutually beneficial way. We know also that there is noise in the signal. At best, the pieces of distributed knowledge that (if they could be brought together effectively) make up a solution to a problem, are floating around in a sea of irrelevant or incorrect "knowledge."

In a changing and uncertain environment, with strategic players who sometimes have economic incentives to mislead others, and a relatively low tolerance for cascading failures that hurt human lives, the law of large numbers won't solve this problem for us. That is a complicated way of saying that we can't afford to wait for evolutionary selection. Most of evolution is wasted resources. It is extremely inefficient and slow, destroys enormous amounts of information (and protoplasm), and can't backtrack effectively. No one wants this for human systems and it's not clear that we should tolerate it. We need an engineered system.

We also know that this is a very tall hurdle to get over. Large firms commit huge resources to knowledge management, and with very few exceptions (Xerox's Eureka project is notable here) these investments underperform. These systems fail in a number of distinct ways. The most common (and probably the most frustrating) is simply that nobody uses the system, or not enough people use it to generate sufficient interest. More troubling is the failure mode in which the "wrong" people use the system—people with good intentions who happen to have bad information, or people who might be trying to game the system or intentionally insert bad information to advantage themselves over others in a manner that is either cynical or strategic, depending on how you look at it.

There are other potential failure modes, but the point is to recognize that there is no inherent ratchet-up mechanism for knowledge management. The system could deteriorate over time in several ways. People could share mistakes with each other and scale them up. People could reuse past experiences which are seen as successful in the short term or by particular individuals, but actually are failures overall from the long-term perspective of the community. You could attract the wrong "experts" into your network, or perhaps more likely use experts for the wrong purpose. And you could populate a database with garbage and produce multiplying wastes of effort and cascading failures of behavior. All of us have worked in organizations or communities that have suffered from knowledge management failures of at least one of these types.

But put the community in the background for a moment, and consider the problem from a microperspective by imagining that you are a person searching for a solution to a problem within that community. Now, how knowledge is distributed directly affects the search problem that you face. There are at least three possibilities here.

Case 1 is where you have a question, some other individual has the answer, and the problem for you is whether you can find that person and whether that person is interested in sharing with you what she knows. Case 2 is where no single person has the answer to your question; instead, pieces of the answer are known by or embedded in many people's experiences. The relevant bits of information float in a sea of irrelevant information; your problem is to separate out the bits of signal from the noise and recombine them into an answer. Case 3 is a search and discovery problem. Some of the knowledge that you need is floating around in disaggregated pieces (as in Case 2) but not all of it; you need to find and combine the pieces of what is known and then synthesize answers or add to that new knowledge from outside the community itself.

Here's where your dilemma gets deeper. You don't know to start if you are facing Case 1, 2, or 3. And it matters for what kind of search algorithm you want the system to provide for you. For example, should you use a snowball method (go to the first node in the network and ask that node where to go next)? Or some kind of rational analysis rule? Or a random walk? Or maybe you should just talk to the people you trust.

And now consider the dilemma from the perspective of the person trying to design the system to help you. She doesn't know if you are an expert or a novice; or how entrepreneurial or creative you are; or what your tolerance will be for signal-to-noise ratios; or whether you can more easily tolerate false positives or false negatives.

The history of the open source community as it navigates some of these dilemmas, some of the time, suggests a big lesson: it's impossible to "get it right" and it's not sensible to try. What is more sensible is to try to parse the uncertainties more precisely so that we can design systems to be robust. More ambitiously, to design systems that can diagnose to some degree and adapt to uncertainties as the system interacts with the community over time. A second big lesson of open source is the high value of being both explicit and transparent about the choices embedded in design principles. The next section incorporates both of these lessons into a set of seven design principles for a referee function, inspired by patterns of collaboration within open source communities, that just might make sense for a community of knowledge and practice in politics.

Design Principles for a Referee Function

Voluntarism is an important force in human affairs, and the open source software process would not work without it. But harnessing the efforts of volunteers is not enough to build a piece of software or, for that matter, anything else that is even moderately complex. As I've said elsewhere, the reason there is almost no collective poetry in the world is not because it is hard to get people to contribute words. Rather, it is because the voluntary contributions of words would not work together as a poem. They'd just be a jumble of words, the whole less than the sum of its parts.[4]

In my view, this implies that the bulk of social science research that tries to parse the motivations of open source developers, while interesting, basically aims at the wrong target. Noneconomic motivations (or at least motivations that are not narrowly defined by money in a direct sense) are a principal source of lots of human behavior, not a bizarre puzzle that requires some major theoretical innovation in social science. The harder and more interesting question is governance. Who organizes the contributions and according to what principles? Which "patches" get into the codebase and which do not? What choices are available to the people whose contributions are rejected?

The real puzzles lie in what I'll call the "referee function," the set of rules that govern how voluntary contributions work together over time.

In other words, what makes the open source process so interesting and important is not that it taps into voluntarist motivations per se, but rather, that it is evolving referee functions that channel those motivations, with considerable success, into a joint product and that it does so without relying on traditional forms of authority. No referee function is perfect, and among the variety of open source projects, we can see people experimenting with different permutations of rules. I believe I can generalize from that set of experiments to suggest seven discrete design issues that any referee system will have to grapple with. Certainly this is not a comprehensive list, and the seven principles I suggest are not sharply exclusive of each other. Each incorporates a tradeoff between different and sometimes competing values. And I am not proposing at this point where to find the "sweet spot" for any particular community or any particular problem-solving challenge; my goal is much more modest than that. The point here simply is to lay out more systematically what the relevant tradeoffs are so that experiments can explore the underlying issues that might cause groups to move or want to move the "levers" of these seven principles in one direction or another over time.

Weighting of Contributions

No problem-solving community is homogeneous (in fact, that's why it makes sense for individuals to combine forces). Not everyone is equally knowledgeable about a particular problem. Different people know different things. And they know them with different levels of accuracy or confidence. A referee system needs a means for weighting contribu-

---

4   Steven Weber, *The Success of Open Source* (Harvard University Press, 2004).

tions and it should reflect these differences so that when information conflicts with other information, a more finely grained judgment can be made about how to resolve the conflict. Mass politics teaches us a great deal about bad ways to weight contributions (for example, by giving more credence to information coming from someone who is tall, or rich, or loud). One of the interesting insights from the open source process is the way in which relatively thin-bandwidth communication—such as email lists—facilitates removal of some of the social contextual factors in weighting which are ultimately dysfunctional. Tall, handsome men have a significant advantage in televised political debates, but not on an email list. Collaborative problem solving at a distance probably leans toward egalitarianism to start. But egalitarianism does not automatically resolve to meritocracy. The transparency of any algorithm is both desirable and risky—desirable because it makes visible whose contributions carry weight and why; and risky because, well, for exactly the same reasons.

Evaluating the Contributor Versus Evaluating the Contribution

A piece of information can in principle be evaluated on its own terms, regardless of its source. But in practice it is often easier to (partially) prequalify information based on the reputation of the person who contributes the information. Take this to an extreme—trusted people get a free ride and anything they say, goes—and you risk creating a winner-takes-all dynamic that is open to abuse. But ignore it entirely and you give up a lot of potential efficiency—after all, there is almost certainly some relevant metadata about the quality of a piece of knowledge in both what we can know about the contribution *and* what we can know about the contributor. eBay strongly substitutes the reputation of the person (seller or buyer) for information about what is at stake in a particular transaction. I suspect that software patches submitted to Linux from well-known developers with excellent reputations are scrutinized somewhat less closely than patches from unknown contributors, but that's only a hypothesis or a hunch at this point. We don't really have a good measure of how large, open source projects actually deal with this issue, and it would be a very useful thing to know, if someone could develop a reasonable set of measurements.

Status Quo Versus Change Bias

The notion of a refereed repository, whether it is made up of software code or social rules or knowledge about how to solve particular problems, is inherently conservative. That is, once a piece of information has passed successfully through the referee function, it gains status that other information does not have. Yet we know that in much of human knowledge (individual and collective), the process of learning is in large part really a process of forgetting—in other words, leaving behind what we thought was correct, getting rid of information that had attained special status at one time. The design issue here is just how conservative a referee function should be, how protective of existing knowledge. There are at least two distinct parameters that bear on that: the nature of the community that produces the knowledge, and the nature of the environment in which that community is operating. Consider, for example, a traditional community that is culturally biased toward the status quo, perhaps because of an ingrained respect for authority. This community might benefit from a referee function that compensates with a bias toward change. If the community is living in a rapidly shifting environment, the case for a change bias is stronger still. The parameters could point in the other direction as well. Too much churn in a repository would rapidly reduce its practical usefulness, particularly in a problem environment that is relatively stable.

Timing

Separate from the issue of status quo versus change bias is the question of timing. How urgently should information be tested, refereed, and updated? The clear analogy in democratic electoral systems is to the question of how frequently to hold elections—which is obviously a separable question from whether incumbents have a significant electoral advantage. A major design consideration here follows from a sense of just how "bursty" input and contributions are likely to be. Will people contribute at a fairly regular rate, or will they tend to contribute in short, high-activity bursts followed by longer periods of quiet? We know from the open source process that contributors want to see their work incorporated in a timely fashion, but we also know that speeding up the clock makes increasing demands on the referee. This is probably one of the most difficult design tradeoffs because it is so closely tied to levels of human effort. And it's made harder by the possibility that there may be elements of reflexivity in it—that is, a more rapidly evolving system may elicit more frequent input, and vice versa.

Granularity of Knowledge

Modular design is a central part of open source software engineering. The question is where to draw the boundaries around a module. And that is almost certainly a more complicated question for social knowledge systems than it is for engineered software. No referee function can possibly be effective and efficient against many different configurations of claims of knowing things. And there is likely to be a significant tradeoff between the generality of information, the utility of information, and the ease and precision of evaluation. Put differently, rather general knowledge is often more difficult to evaluate precisely because it makes broader claims about a problem, but it is also extremely useful across a range of issues and for many people if it is in fact valid. Highly granular and specific knowledge is often easier to evaluate, but it is often less immediately useful to as many people in as many different settings precisely because it is specific and bounded in its applicability.

System Failure Mode

All systems, technical and political, will fail and should be expected to fail. In the early stages of design and experimental implementation, failures are likely to be frequent. At least some failures and probably most will present with a confusing mix of technical and social elements. How failures present themselves, to whom, and what the respective roles of systems designers, community members, and outsiders are at that moment, are critical design challenges. In *Exit, Voice, and Loyalty*, Albert Hirschman distinguished three categories of response to failure—you can leave for another community (exit), you can stick with it and remain loyal, or you can put in effort to reform the system (voice). One of the most striking features of the Linux experience is that this community, by empowering exit and more or less deriding loyalty, has had the effect of promoting the use of voice. It is precisely the outcome we want—a system that fails transparently in ways that incentivize voice rather than exit (which is often extremely costly in political systems) or loyalty (which is not a learning mode).

Security

How to design and implement security functions within a referee system depends sensitively on the assumptions we make about what the system needs to guard against. In other words, what level and style of opportunism or guile on the part of potential attackers or "gamers" do we believe we ought to plan for. This is simply a way of saying that no system can be made secure against all potential challenges. Security is always a tradeoff against other considerations, in particular ease of use, privacy, and openness. And security likely becomes a greater consideration as the value that the system provides rises over time. Hackers and crackers—whether benign or malicious in their intentions—are an important part of software ecologies precisely because they test the boundaries of security and force recognition of weaknesses. Can political communities be designed to tolerate (and benefit from) this kind of stress testing on a regular basis?

What Should We Do Differently?

Political ideas, like democratic experimentalism and distributed community problem solving, share some central characteristics with the open source software process.[5] In my view, the most interesting intersections lie in the configuration of referee functions—how any system decides that some "code" is better than others and should be incorporated into an interim package, how long it should stay there, how it can be removed or modified and by whom, how it ought to be configured to interface with other code, and what happens when the system breaks down. These are constitutional questions in a profound sense, in that they reflect on the constitutive elements that make up a community (even if they are not legally enshrined in something that people call a constitution).

Open source communities are tackling all of these problems, with varying degrees of self-consciousness. The patterns and practices of collaboration within open source communities are evolving rapidly, and that provides interesting experimental insights

---

5   See, for example, M.C. Dorf and Charles Sabel, "A Constitution of Democratic Experimentalism" (Columbia Law Review, 1998).

that can travel outside the software and technology worlds. The seven design principles I laid out earlier are not optimization functions. They are explications of tradeoffs; understanding them is a prerequisite to smart experimentation. So, the first thing we should do differently, or more than we do at present, is to instrument some of these experiments and tighten up the feedback loops so that we learn more quickly and more precisely about what happens when you slide the levers of these seven design principles into different configurations.

The second thing we can do differently is to get precise and transparent about the overall goal of communities of knowledge and practice in politics. Open source communities provide a very good template for a broad statement of goals. I propose that when designing a system, you ensure the following:

- The system has effective individual incentives, organizational structures, and information technology tools...

- To pull together distributed knowledge within communities that are trying to solve practical problems...

- By combining pieces of knowledge into something useful in a manner that...

- Ensures that error correction exceeds the rate of error introduction as the system "learns," while...

- Maintaining the process over time in a sustainable, nonexploitable, and expandable way.

The challenge and opportunity here are highly general across political communities. And they are going to get more important in the future, particularly as the economics and demographics of advanced industrial countries continue to drive many of the social welfare functions that have for some time been provided by the public sector, out of that sector. Some of these functions, of course, get moved into the private sector. And social scientists have learned a great deal in the last 20 years about the upsides and downsides of what is commonly called "privatization." We argue around the margins about how to engineer the transition and we argue about the overall efficacy and desirability of the outcomes, but at a high level we do understand a fair amount about sensible governance principles and the tradeoffs they engender in the private sector setting.

We know much less about how to set up systems for moving some welfare provision functions into the civil society space, which is neither public sector nor private sector per se. Open source-style collaboration is, in a real sense, a form of technologically enabled civil society. And so the third thing we can do differently is to mine the experience of open source for lessons about how to create pragmatic, workable alternatives to privatization that can be implemented and can evolve within a developing civil society space.

CHAPTER 23

Jeff Bates and Mark Stone

Communicating Many to Many

Typical users in the Windows community and typical users in the Linux/open source community have different tendencies. For real-time communications, Windows users tend to prefer an Instant Messaging (IM) client like AOL Instant Messenger (AIM) and Linux users tend to prefer an Internet Relay Chat (IRC) client like XChat. On the surface, these clients and protocols differ little: both support channel or chat room messaging, both support one-on-one messaging, and both allow for some degree of moderated discussion. Yet IM users tend to communicate one-on-one by default, resorting to chat room discussions only for specific purposes and even then rarely, and IRC users tend to communicate in group channels, resorting to one-on-one communication only occasionally.

That tendency toward group activity underlies much of the collaborative instinct in the open source community, and ultimately provides a key to open source's remarkable success. Yet any IRC veteran knows well the scaling problems group communication encounters. A channel with a dozen or so participants, a handful of whom are vocal, can be a very productive center of communication. A channel with 20 to 50 participants suffers a crippling signal-to-noise ratio, absent some form of moderation: too much noise, not enough signal.

This pattern is reflected in all forms of online group communication: discussion forums, email lists, and Usenet from the very early days. Breaking this pattern stands as one of the chief challenges to effective collaboration. How can communicating many to many scale? What conditions enable network effects to take hold so that more participants improves rather than diminishes the quality of communication, and hence the power of collaboration?

Slashdot stands as a striking counterexample to the usual pattern. Over the years, the site has evolved into a high-quality, moderated discussion forum, one where the more people participate, the more valuable the discussion is. Network effects have taken hold; many-to-many communication is enabling a unique form of collaboration.

The Origins of Slashdot

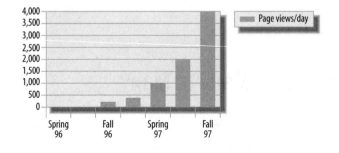

Like many web endeavors that evolved into dot-coms, Slashdot started as a hobby and a learning exercise. Rob Malda, Jeff Bates, Nate Oostendorp, and Kurt DeMaagd—the future "Blockstackers"—had been friends growing up in Holland, Michigan, and later at Hope College. At first, "online community" meant bulletin board systems (BBSs): email, discussion, and file sharing services available locally by dialing up and connecting directly to a BBS machine. College provided access to the Internet and the earliest web sites via one of the college's Unix or VAX machines.

Rob Malda's first home page at college (*http://cs.hope.edu/~malda*), dubbed "Chips and Dips," would be labeled by mainstream media today as a weblog, or blog. Superficially, Chips and Dips did resemble current blogs minus many of the interactive elements we take for granted in today's Internet. It was a personal page for Rob. It offered his opinions on everything from web design to science fiction book reviews. Yet to call it an early blog really misses the point.

Blogs are fundamentally inward facing. They share events from the blogger's life, together with opinions of the blogger on those events and the world at large. Chips and Dips had none of that intimate and voyeuristic sense of a diary. Instead, it had more in common with the early versions of Yahoo!: a hand-built directory of useful links, with guidance and commentary, aimed at like-minded people. Chips and Dips was, from the beginning, about that sense of community; it was outward facing.

Indeed, right from the start Chips and Dips offered more than just Rob's links and opinions. Friends submitted suggested entries for the directory listings, or reviews of movies that others had yet to see. The site was just flat HTML, no CGI or other dynamic elements. As a consequence, the only way to submit to the site or contribute to discussion was to email Rob, and then wait for Rob to post on the site. This created an implicit and autocratic moderation mechanism. If Rob did not have the time or interest to post something, it didn't get posted. As it was Rob's site, his decision, or even whim, was final.

In the fall semester of 1997, Rob Malda, Jeff Bates, and the other Blockstackers entered their junior year of college. Most had two years of computer science under their belts, and lots of hands-on experience from hobbies pursued and an assortment of student jobs. They also had a sense of the larger world of which Hope College was a part.

Netscape had completed its successful IPO. Graphical web browsers, and the Web itself, were becoming pervasive parts of popular culture. Microsoft had released Windows 95 and announced that the Internet was the future of the company. Linux was headed toward the 2.0 release of the kernel. Apache was, as it is today, running most public web sites. The dot-com boom was in full swing.

Registering a domain name had gone from an esoteric to a more commonplace activity, albeit an expensive one by student standards. While Rob had done a lot of computer programming, he was fundamentally a designer, and indeed one with a strong sense of the ironic. He approached problems visually, and in his spare time was as likely to be doodling cartoons as writing code.

The choice of the name "Slashdot" for a domain was a clever play on the line between the visual and the verbal. In the early web days, the idea of a URL and what it was still seemed alien to the mass media. Every ad for a web site began with the announcer carefully spelling out, "H-T-T-P colon slash slash..." Visually, "/" is simple and distinctive. In those days, verbally spelling out "H-T-T-P colon slash slash....dot org" was ridiculous. It appealed to Rob's sense of humor.

Moving from Chips and Dips to Slashdot brought several immediate changes, not all of them forseen.

Slashdot in the Early Days

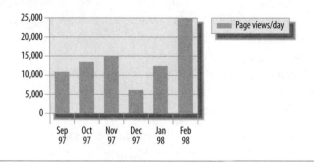

From fall of 1997 to spring of 1998, Rob Malda and the other Blockstackers went through a rapid education on emerging web technologies. Chips and Dips had been entirely static HTML. By spring of 1998, Slashdot had dynamic content through CGI, and then very quickly, given the performance limitations of CGI, dynamic content via Apache modules, specifically mod_perl. The quality of HTML improved, coming much closer to standards compliant.

The purpose at this stage was a learning exercise as much as anything else. New features of HTML or Perl were learned during "day jobs" at work, and then what was learned was applied to Slashdot. By the spring of 1998, this process had begun to reverse itself. New ideas were tested on Slashdot, often proving valuable at day jobs as well.

One of the important changes during this time was the creation of submit.pl, the Slashdot submissions bin. Making submissions an inherent part of the site rather than something that had to pass through the Inbox of Rob's email made the submissions process scalable, and also enabled others besides Rob to take a direct hand in the editorial process of reviewing submissions and approving them for posting to the site.

Another key technical change during this time was the move from a web site that was merely under an account on a university server to a standalone server running Slashdot. The hardware was a single DEC Multia that Rob had received as barter payment for drawing cartoons for a self-published business book. The server was hosted at Rob and Jeff's place of employment. Their employers needed an email server, and Rob suggested that they could use that box as an email server, as long as he could do some other hosting on the box as well. Initially this arrangement worked well, though the server was subject to sudden power outages and downtime if you weren't careful to avoid the power cord when putting you feet under Rob's desk. Of course, it was a server running Linux.

While Rob's interest in Linux dates back to Chips and Dips, running Slashdot and the technology behind it really increased that interest. Naturally, content on the site itself served this need, as Rob began to accumulate, and visitors continued to send in, an impressive list of Linux resources online.

This dynamic of needing Linux to run the site and using the site to learn about Linux created an unanticipated side effect, compounded by other changes happening at the same time. Chips and Dips, while it was Rob's personal home page, had always been a community site. Indeed, this community focus is one of the key differentiators between Chips and Dips and what today we would call a blog. However, that initial community consisted primarily of other geeks in and around Holland, Michigan, and specifically, others at Hope College (though even early on the larger Linux community had interest in some of Rob's graphics work).

A university home page and a top-level domain like "Slashdot.org" are, in some sense, equally public. Both could be accessed from any browser connected to the Web anywhere in the world. But Slashdot felt like a more prominent site than Chips and Dips, and was more likely to be bookmarked by others, or added to directories, which, at the time, were still largely compiled by hand. This alone drew a larger, more global audience to Slashdot.

In addition, in early 1998, there simply weren't that many thorough Linux resources online. As one of these few, Slashdot stood out to Linux enthusiasts everywhere, not just at Hope College. The site was still a community site. But somewhere along the way in winter of 1997–1998, the community had become global. This was still a community of like-minded peers; it was, as Slashdot has proclaimed from the early days, "News for Nerds." But it was becoming a much larger community. During this period, Slashdot passed 20,000 page views per day, a level of traffic that, especially circa 1998, signified a sizable, loyal audience beyond the confines of Holland, Michigan. By way of comparison, Holland has a population of a little over 30,000.

Some years later, then-Slashdot columnist Jon Katz speculated about why this important resource for the technically inclined had happened in rural Michigan rather than in a flourishing center of technology like Silicon Valley. Katz felt that it was a community born of necessity. In Silicon Valley, technical communities abounded, and face-to-face opportunities to meet and interact with like-minded peers were plentiful. Only in a more isolated place like Holland, Michigan, would it be necessary to actively pursue a community online to find the critical mass needed to form a genuine community of interest.

A community of like-minded peers is one thing. The arrival of the masses is quite another. The Slashdot community was about to change dramatically, starting with what seemed like a small incident.

The Slashdot Effect

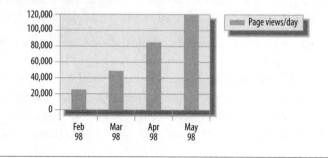

On January 12, 1998, Rob Malda posted a piece on Slashdot titled "Simple Solutions" that he had written, and that he described as "the first of hopefully many Slashdot.org Editorials" (see Appendix D). In this editorial, he challenged Netscape to open source its browser code as the best available alternative for a company losing both money and market share. On January 22, Netscape put out a press release with the headline "Netscape Announces Plans to Make Next-Generation Communicator Source Code Available Free on the Net." The release of Mozilla marks one of the signature events in open source development.

It's important to understand what did not happen. No one at either Slashdot or Netscape has ever claimed that Rob Malda's essay had any direct effect on Netscape's decision. Indeed, the complexity of the process and the proximity of the dates suggest that Netscape's decision must have been made well in advance of Rob's editorial. But the proximity of the two events—and a general lack of understanding about open source at the time—caused mainstream technology media to link the two.

Slashdot had been discovered by mass technology media.

Several changes began to take hold on Slashdot, and indeed, the effects of those changes are still playing out today. The mainstream technology media took a regular interest in Slashdot, and the mere fact that a story was covered on Slashdot became significant. Ironically, Slashdot seldom has been the originator of a news story, so this interest in Slashdot coverage was largely about watching what other people were watching, an obsession that seems to be a distinctive part of the Internet generation.

A peculiar side effect of this "watching the watchers" was that Slashdot became a source of journalistic research. Smart journalists looked for insightful comments, finding stories and ideas in those comments and their authors. Indeed, because Slashdot has always permitted anonymous posting, comments often had the insider's candor that journalists value.

An extension of this media interest was the involvement of media figures in Slashdot. Author Jon Katz was working at HotWired at the time, while also researching his book, *Geeks*. He contacted Rob and Jeff first as part of his research for the book, but more and more because of his genuine interest in Slashdot and the community for which it stood. The result was that Katz became a regular columnist on Slashdot for several years.

The reaction to Katz's presence was revealing. Many were impressed with his writings and insights into geek culture, and would quietly send their notes of appreciation to Rob, Jeff, or Jon privately. At the same time, a vocal minority of the audience objected strongly to Katz as an outsider, and posted their views bluntly as comments to any column he posted. When user accounts and customizations arrived on Slashdot, "filter out Jon Katz" was, for a time, the most frequently selected customization.

This response was, of course, out of all proportion to anything Katz said or did. Katz is a professional writer, a serious journalist, and had a genuine interest in the geek community. However, he became a symbol of the new crowd that had arrived at Slashdot, readers who were more interested in geek culture than geek technology. Katz became a lightening rod for all the resentment felt by the original core audience, some of whom felt the need to lash out at these "invaders."

Slashdot was reaching traffic levels that signified a vastly larger audience than the original Linux and open source enthusiasts and other geeks. On February 9, 1998, the site received its 1 millionth hit. Barely a month later, on March 18, it received its 2 millionth hit. Just a month after that, the site recorded more than 100,000 page views per day.

Yet if Slashdot was changing, it was also changing other sites as well. The appearance of a particularly noteworthy story at the top of Slashdot's home page would generate a flurry of discussion. Few stories posted left the front page with fewer than 200 comments, and many stories received in excess of 700 comments. Further, much of the Slashdot audience would descend, simultaneously and en masse, on the site from which the story originated. Many sites were unable to handle this sudden influx of new traffic, and would simply crash under the load.

This sequence of events—the posting of a story, the rush of traffic to the story's site, and the strain or failure of the site under the load—has become so notorious that it is now known as "the Slashdot Effect." Indeed the network characteristics of the Slashdot Effect have been worthy of academic study. Stephen Adler at Brookhaven National Laboratory had the first real study of the Slashdot Effect, available online at *http://ssadler.phy.bnl.gov/adler/SDE/SlashDotEffect.html*.

The term *Slashdot Effect* has entered popular culture; it has an entry in the Oxford English Dictionary's online edition, and Slashdot serves as the answer to a question in the '90s edition of Trivial Pursuit.

Trolls, Anonymous Cowards, and Insensitive Clods

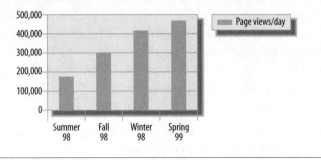

Many discussion forums, be they mailing lists, web-based discussions, or IRC channels, produce high-quality discussions among a few participants. The challenge is to scale this "few-to-few" communication all the way to "many to many." Adding a moderator helps to a point. Many moderated discussions, though, improve the "signal-to-noise ratio" at a cost: over time, the signal takes on more and more of the moderator's particular viewpoint. Other viewpoints, intentionally or not, are stifled with a hostile reaction. Rarely can a discussion forum have enough commonality in interest to draw a critical mass of audience, have enough variety in viewpoints to keep the discussion interesting, and scale to ever-larger audience sizes without losing the signal of the discussion in the noise of the chatter.

Discussion systems can avoid the tyranny of a single moderator by letting all users vote on moderation. However, letting every user vote equally on moderation assumes that all users are equally informed, equally concerned, and equally motivated about what they are moderating. These are unrealistic assumptions that lead to a tyranny of the masses as damning as the tyranny of a single moderator.

In the spring and summer of 1998, Slashdot faced all these challenges. What emerged was a mix of software engineering and social engineering to cope with the growth of the site. Not everything that was tried worked, and some solutions are still works in progress.

User accounts were created. Registered users could customize which *slashboxes* (sets of links to changing external content; a precursor to today's RSS) appeared on their Slashdot page, and where. Today we would call this *portalizing* the site. To the Slashdot team it was a way of minimizing complaints and encouraging visitors to register and log in. Users could filter content to see only stories on certain topics, or to exclude stories on certain topics. Having a low user account number also became a point of pride with the regular visitors.

An interesting and intended side effect of user accounts was to reduce the number of off-topic or deliberately inflammatory comments. People are, by nature, less inhibited when communicating anonymously. The simple step of encouraging people to identify themselves helped restore some order to discussions.

An issue continuously debated behind the scenes was whether anonymous posting should be allowed at all. In the end, free-speech considerations have trumped all other considerations in this debate. Anonymous posting allows an employee to speak out candidly about his employer without fear of retribution. Anonymous posting allows someone to express an opinion or political view without being chastised by their peers. Ultimately, anonymous posting contributed significantly to key events that have left an indelible mark on Slashdot.

One step that also occurred was the slowly diminishing value of anonymous comments—over time, their base score became set to 0, which was lower than the 1 threshold that the non-logged-in user reads at by default. This means that most of the people only read anonymous posts that have been moderated up.

As the audience grew, so did the number of submissions. It became impractical to get every worthy submission posted to the front page; stories simply would have scrolled by too fast. In May of 1998, sections were added to Slashdot, creating a separate front page for those interested in a specific topic. Today there are 14 sections on Slashdot, ranging from Book Reviews, to Science, to Politics.

The most controversial modification to Slashdot came in October 1998, with the introduction of moderation. Each comment is classified on two dimensions: one for the type of comment, and the other for the quality of the comment. Types of comments include: Overated, Underated, Troll, Insightful, Informative, Redundant, Offtopic, and Flamebait. Quality is numeric, from -1 to 5. All comments initially started at mod level 1, but could be moderated up or down for as long as the discussion on that story was open.

One of the important settings available to registered users is the moderation level. Users can select what level of comments they see by default, and can change that setting on a story-by-story basis. Thus, if the number of comments at a certain level is too many to read through, or the relevance drops off too much, users can filter out comments below a certain level. The default moderation level also enables a form of filtering without censoring. The most irrelevant or inflammatory comments routinely get moderated down to a level of 0 or -1. An unregistered user visiting the site has his default moderation level set to 1. Since 95% of all visitors to the site never change their moderation level, the vast majority of visitors never see the lowest moderated comments. Moderation also provided an additional motivation to get users to register. After a time, the policy was changed so that an anonymous user's posts generally started with a moderation level of 0, but a regsitered user's comments started with a moderation level of 1.

Many discussion forums have tried various forms of moderation. The challenge is to find a system that is scalable—Slashdot routinely generates tens of thousands of comments per day—and heterogeneous, namely reflecting more than a single point of view about how comments should be moderated.

Slashdot met these challenges by borrowing from the principles of collaboration in its open source roots. The audience is essentially self-moderating, and indeed the more people participate in the system, the better the moderation gets. This is the enormous differentiator for Slashdot. Where most discussion forums crumble under a deteriorating signal-to-noise ratio as their size increases, Slashdot actually benefits from network effects: more is better.

The key is that Slashdot tracks a wide range of information about its users: how many comments a user has posted, how many stories a user has submitted, how many submissions have been accepted, what moderation level a user's comments tend to settle on, what type of comments a user typically makes. All of this data is combined to produce a number that roughly quantifies the value of a user to the site. This number is referred to in the Slashdot system as *karma*.

Users with high karma are periodically selected to moderate. The system is automated, requiring little intervention from the Slashdot staff. Once selected, a user has his moderation authority turned on for a period of time, enabling him to moderate up or down, or classify comments he reads. After a period of time, moderation is turned off, and passes to another user. At any given time there are roughly 1,850 users moderating comments.[1]

---

[1]   If one thinks of the task of a moderator as similar to that of a copyeditor, it is possible to put an approximate monetary value on the work performed by users while moderating. If the typical moderator spends even an hour a day on moderation—and many spend much longer—this amounts to roughly $50,000 worth of work being done for the site for free each day. The key, though, is to see it in terms of value provided rather than money saved. Thanks to Slashdot's tiered moderation system, the site continues to scale and discussion continues to be valuable. There simple isn't anything one could spend $50,000 a day on to provide comparable value.

Originally, a user's karma number was viewable. This policy led to problems. Users viewed their karma rating as a score, and raising their karma as a game. Once people tried to deliberately game the system, the whole system no longer functioned as well.

The Slashdot staff was also inundated with email complaints about karma—for instance, "My latest comment was modded up to 5 but my karma went down; I think your system is broken." Of course, in this context, karma is just a technical term for the sum of a formula used in the Slashdot system; as such, it could not possibly be "broken." Furthermore, the code for Slashdot has always been available as open source, meaning that those who really wanted to understand their karma rating could have done so. Human nature being what it is, however, people quickly slipped into thinking that there was some real thing to which karma corresponded and which the system was trying to approximate. In the end, the only workable solution has been to keep karma ratings private and give users only a vague approximation of their karma ratings.

Slashdot also evolved to have a number of "social engineering" elements that did as much to channel users' behavior as any of the technical features. Some of these social engineering elements are blatant, like referring to not logged-in posters as "anonymous cowards." Similarly, wildly off-topic or deliberately inflammatory posters are referred to as "trolls."[2]

Some of these elements are subtler. Watch a first-time visitor try to find the "submit" link, for example. Visitors aren't actively encouraged to submit; you have to really want to get your submission in. Creating that barrier to entry means that the overall submission is of higher quality. Because some effort is required to learn how to submit, people who put some thought into their submissions are more likely to submit.

The site also has a distinct personality, one that does not take itself too seriously. This is apparent from the self-deprecating tag line ("News for nerds; stuff that matters") to the obvious humor in many of the weekly polls. Staff on the site are referred to by their nicknames, usually originating as nicks on an IRC network or handles on a BBS. These nicknames have deliberate cultural references meant to be understood by an audience with the right cultural background. "Cowboy Neal" derives from a character in Kerouac. "CmdrTaco" is a reference to a Dave Barry column. If you haven't watched "The Simpsons" or "South Park" regularly, much of the humor on Slashdot will pass you by.

In its own way, this too is part of the social engineering. The site personality and the cultural references are a subtle test for like-mindedness with the audience, a way of encouraging participation from those who "get it" and distancing those who don't.

---

2   The term *troll* does not originate with Slashdot, but in fact dates back to the early Usenet discussion forums on the Internet. Presumably it's a reference to the children's story "The Three Billy Goats Gruff," in which a troll lurks under the bridge waiting to ambush hapless passers by.

Slashdot grew up a lot as a site in the summer of 1998. That spring the site began running banner advertising as a means of generating revenue. Banner sales were originally outsourced to a third party, but out of frustration with the ad sales company's inability to manage sales or understand the Slashdot audience (who still vociferously complain at the use of Flash animations in ads), advertising sales were brought back in-house in July of 1998 and were managed by Jeff Bates.

With little fanfare, Rob Malda quit his job in August 1998 and became Slashdot's first full-time employee. While other staff continued to work on a part-time or volunteer basis, this was a significant milestone. Less than a year after the registration of the Slashdot domain name, the Blockstackers had gone from running a web site to running a business.

Columbine

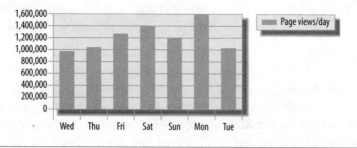

Slashdot has moved away from its technology roots only gradually, and only where a new topic clearly connects to its core audience. There is a "Book Review" section, but its focus is on computer books. Sections like "Your Rights Online" branch farther afield, but are still rooted in topics of concern to the technically inclined: the SCO lawsuit, the legality of file sharing and peer-to-peer networks, or the status of the Digital Millenium Copyright Act (DMCA).

Slashdot has never tried to be a general news site, and events of general interest typically are not covered. In this regard, what arrives in the Slashdot submission bin and what is published to the front page differ substantially.

On Wednesday, April 21, 1999, a number of Slashdot readers submitted reports of a shooting at Columbine High School in Littleton, Colorado. Ultimately, it would emerge that fifteen people died, including the two teenage boys who were the shooters. Twenty-three other people were injured before the two boys ended the massacre by committing suicide. These details were far from clear as the first submissions arrived at Slashdot that Wednesday.

A breaking story like Columbine brings out the best and worst in Internet news reporting. The Internet has an unprecedented capacity to cover events in real time, and to draw from a widely distributed network of sources. At the same time, standards for journalism and fact checking are largely undefined in this new medium.

Rumor is reported as fact, and the distributed nature of information flow makes correction in light of new information difficult.

When the Slashdot staff looked at the submissions regarding Columbine that Wednesday, it seemed clear that first of all, the story did not tie directly to the technology topics that are the core of Slashdot's coverage, and second, that the story was still an emerging one with important facts still unknown. Without much discussion, the staff decided not to post the story, and leave coverage of Columbine to the mainstream media.

Columnist Jon Katz, however, saw a different story emerging over the next two days. Katz had been hard at work on his book *Geeks*, and indeed it was this research into geek culture that had drawn him to Slashdot in the first place. An important part of that research concerned the isolation and alienation felt by geek teenagers simply because they were smart and different. Katz was appalled by the quick move of mainstream media to stereotype the Columbine shooters as a deranged byproduct of a violence-desensitizing subculture of hard rock, computers, and video games.

On Friday, April 23, Katz used his column on Slashdot to voice a response. He published a piece titled "Why Kids Kill." Katz's main point was to counter the stereotype. He argued that youth violence was dramatically on the decline, and that there was no research to establish a correlation between violence in movies, music, games, or television and violent behavior among youth. He also raised the question of why our expectations of who violent youth would be—urban, disadvantaged, and ethnic—do not match the reality of who violent youth actually are—suburban, middle class, and white. The suggestion, though subtly stated by Katz, was that stereotypes provided a convenient alternative to confronting the fact that today's parents do not understand today's youth. If technology was responsible for anything, it was for widening that gap in understanding.

What followed was unprecedented in the history of Slashdot. Most Slashdot readers view the site and comment from somewhere other than a home computer. Either they are at work, or they are students using a university computer. Consequently, traffic on the site, and number of comments, tend to decline on the weekend, with a big drop-off starting on Friday. Katz's column was posted at 11:00 A.M. on Friday. It received over 1,000 comments (see *http://slashdot.org/article.pl?sid=99/04/22/2136230&mode=thread*).

The comments included some from parents, teachers, and other adults. But the vast majority of comments were from teenagers. They spoke out not to defend the shootings at Columbine, but to express their own feelings of alienation. Much of this alienation was rooted in the struggle to grow up in a world of rapid, technology-driven change to which the adults in their life could not relate. The comments are heartfelt, surprisingly articulate, and seemingly countless (footnote: the "countless" part is, in some sense, true. Slashdot does not archive comments moderated below zero, so while we know that more than 1,000 comments moderated zero and higher, there is no record of the total number of comments posted).

It's worth quoting a representative example:

> When I was growing up, I wore a lot of black, I studied explosives and bomb-making, I learned how to shoot, and I memorized complete copies of _Jane's Infantry Weapons_ and various army and special forces survival manuals. It was a funky hobby that never really went anywhere. I've worn a black trenchcoat almost every day for ten years, I've played DOOM-like games since they first appeared, and I'm a big fan of John Woo films. To the best of my knowledge, I never went nuts and killed anyone.
>
> I also graduated at the top of my high school class and graduated with honors from an ivy-league college, and I'm now happily married and managing the support team for a successful tech startup. I give credit for all of my success to my parents, who took an active interest in what I was doing and why, without trying to control my life.

Katz and the others at Slashdot were stunned. While the site didn't crash, the volume of comments put the site under an unprecedented load. Katz's personal email was flooded with messages from young people contacting him directly to tell their own stories of alienation and ostracism. Everyone at Slashdot realized that they had tapped into a deep sentiment in urgent need of expression.

On Monday, Katz posted a new column, "Voices from the Hellmouth" (see _http://slashdot.org/articles/99/04/25/1438249.shtml_). Katz could have tried to assert himself and lead the discussion at this point. He chose not to. Instead, he recognized a still-pent-up need for the discussion to continue, and recognized that the most useful thing he could do was facilitate, rather than lead the discussion. His Monday column contained very little of his own words or opinion, and was instead his attempt to relay the most insightful or poignant stories he had received. More than 1,200 comments were posted in response to "Voices from the Hellmouth."[3]

Monday's comments continued the themes of Friday. Young people expressed how frustrated they were that parents and teachers disapproved of their interests, their community, and their culture simply because it was something adults did not understand. Young people expressed how isolated they felt when teased and persecuted by their peers for dressing different, acting different, and worst of all, being smart.

On Tuesday, sensing that the discussion had not yet run its course, Katz posted another column, titled "More Stories from the Hellmouth" (see _http://slashdot.org/features/99/04/27/0310247.shtml_). More than 500 comments were posted in response. It's interesting to note an update that Rob Malda inserted into the story late in the day (around 7:45 that evening): "Sharon Isaak from Dateline NBC wants to get in

---

3   The name _Hellmouth_ derives from the television series "Buffy the Vampire Slayer." The series is set in the fictional town of Sunnydale, mainly at Sunnydale High, which sits atop a nexus of power drawing everything evil toward it; this nexus is known as the Hellmouth.

touch with folks to do a story on this subject for this show. She's specifically seeking Jay of the Southeast, Anika78 of suburban Chicago, ZBird of New Jersey, Dan in Boise, Idaho, but she'd also like anyone who's been targeted as a result of this thing to contact her. Wonder if they make ya wear pancake makeup..."

The Columbine story was now a week old.

Two facts about Slashdot are easy to overlook in the course of more-routine day-to-day content that appears on the site. First, Slashdot is a discussion site, not a news site. Breaking stories are seldom reported on Slashdot, and while a great deal of news coverage is presented on the site, fundamentally the purpose of the news is to seed discussion. Second, the audience of Slashdot is not so much an audience as it is a community.

Before Columbine, even Slashdot's regulars may not have realized the extent to which they were a community. Rob Malda and the other Blockstackers had elevated themselves from an isolated community of geeks in Holland, Michigan, to a global community of like-minded peers. The "Hellmouth" series on Slashdot affirmed that it was not just the staff of the site that could make this transformation, but the audience as well. Over the course of that post-Columbine week, teenage geeks became the voice of Slashdot, speaking many to many. They recognized and celebrated that they were not alone, but were part of a larger community. The site, like Katz, receded into the background, and communication was direct between those who came to the site and posted. Nor was the discussion that week merely a collective pat on the back. Practical, meaningful advice was asked for, offered, and shared: crisis centers to contact, teachers to recommend who had been particularly understanding, programs and opportunities that catered to the aspiring geek.

That practical dynamic is an essential ingredient of community, and had been a characteristic of Slashdot for some time. Those who think of the site merely as a news site have overlooked a small but important section of the site called "Ask Slashdot." Debuting in May of 1998, this category of posts was not affiliated with any seeding news story, but instead was a direct plea for advice from the community.

At times the "Ask Slashdot" posts have looked suspiciously like questions that Rob or other staff members would like to have answered: how to set up a local wireless network, or wire a home theater system. This is part of the meaning of like-minded peers, however; questions of interest to one member of the community are usually of interest to many others as well.

Over the years, the archives of "Ask Slashdot" have grown to an impressive repository of advice and how-to information for those immersed in the geek lifestyle. More than any other section, "Ask Slashdot" exemplifies the community aspect of the site, with the audience speaking directly to each other, unfettered by any lead-in story.

The week following Columbine exemplified the finest characteristics of that community spirit.

Slashdot Grows Up

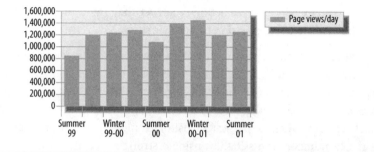

If the Netscape story had introduced Slashdot to mainstream technical media, Columbine introduced Slashdot to the mass media. The site's audience grew, and diversified. The attention focused on the site grew as well. With Columbine, the media recognized that Slashdot could be, on occasion, not just a discussion point for news, but the news story itself.

Comments on Slashdot have become increasingly sophisticated. View Slashdot comments at moderation level 5 and you see a number of lengthy and thoughtful commentaries on each story. You also see certain usernames recur as authors of particularly insightful comments. That combination of an easy online vehicle for expressing opinion and building reputation through regular posting of well-considered opinions has made Slashdot a precursor for the blogging movement of today.

Story submissions have also become more sophisticated. In the first couple of years, Slashdot accepted many submissions that were simply a link to an interesting bit of technology news and a brief description of the news item. Accepted submissions now typically have a main story link, one or more background links to other related stories, as well as links to past Slashdot discussions on the same topic, all contained within a paragraph or more of explanation. This richer form of submission is a result of one of the network effects behind the site. The site is now large enough that any important story will be submitted multiple times, enabling the site staff to pick the most complete and well-formed submission for actual publication to the front page of the site. This puts the regular submitters in tacit competition with each other to create the best submission for key stories.

The continued success and popularity of the site owe as much to its anonymous contributors as to its regular, registered users. Slashdot could not have been the singular "town hall" that it was in the wake of Columbine without allowing and supporting anonymous posting. That aspect of the site has become ever more complicated to manage and maintain, however.

There are legal threats to anonymous status. While Slashdot has yet to be asked to turn over its logfiles, legislation from the DMCA to the Patriot Act has the potential to force Slashdot's hand on this issue. Furthermore, the Motion Picture Association of America (MPAA) and the Recording Industry Association of America (RIAA) have made it clear that they will use civil action wherever they think they have a chance of uncovering the identity of supposed copyright violators.

One change at Slashdot in response to this new legal climate has been to handle log-files differently. Most sites keep logfiles of daily visitors to the site, and differentiate visitors by Internet Protocol (IP) number (the numerical address computers use to identify themselves to each other over the Internet). This IP number can potentially be used to backtrace where a particular visitor to a site came from. Slashdot now scrambles the IP numbers in its logfiles using a strong encryption scheme, and then discards the encrypting key. The result is a unique encrypted number associated with each unique IP number, but no way, even by the Slashdot staff, to derive the original IP number from the encrypted number stored in the logfiles.

As much of a problem as external challenges to anonymity is the mischievous behavior of a few anonymous posters. Given the high profile of Slashdot, there is prestige in the "troll" community associated with defacing or bringing down the site.

Many attacks on the site attempt some form of exploit on the comment system. These can include proxy flooding (repeated comment submissions dispersed via different sources using open proxies; similar to a "denial-of-service" attack); comment binging (attempting to overload the system with large comment submissions); and script attacks (using a script to generate a nonsense comment that can be submitted repeatedly at high speed).

Slashdot now has a maximum comment size, as well as a test for and block of open proxies. It has an internal definition of a well-formed comment, and roughly 500 separate regular expression tests, written in Perl, to which each comment submission is subjected. The attacks continue, but anonymous posting has never been disabled, and remains a bedrock principle of the site.

Sheer scaling issues have presented a different challenge. As of this writing, the site delivers 3.9 million page views per day to 400,000 unique visitors.[4] While Slashdot's initial moderation system worked well for a while, moderation has had to become more sophisticated to both handle the larger volume of submissions and

---

4   *Unique visitors* (known in the trade as "uniques") is not a firm number. The industry norm is to count the number of IP numbers on client machines of visitors. This approach can undercount when visitors are behind a certain kind of proxies that shows only one IP number for everyone behind the proxy, and can overcount because not every IP number is associated with an actual person at the other end. The Slashdot staff uses a different method. Looking at historical data, Slashdot has an idea of the ratio of registered to unregistered visitors, as well as the page views per visit typical of registered and unregistered users. From this data, the number of unique visitors can be extrapolated. By industry norms, the number of daily "uniques" on Slashdot would be roughly 750,000.

comments and take better advantage of the network effects inherent in an audience of this size. The most notable change has been the introduction of metamoderation. In metamoderation, select users are asked to moderate the moderators. The meta-moderators review both comments and moderation decisions about those com-ments, and respond with a simple "agree, disagree, or no comment" response. Those selected for metamoderation typically have about 20 moderation decisions to review when metamoderation is turned on, and then might not metamoderate again for sev-eral weeks or months. The results give the staff and the Slashdot system a more fine-grained picture of which regular users of the site are effective at moderation, and which are consistently contrarian.

One of the lessons of Columbine was that the site not only had to be restructured to meet regular, steady growth in traffic, but also had to be capable of responding to surges in traffic associated with an extrordinary news event. While the sequence of events around Columbine never brought the site down, the staff realized they had, in many ways, been fortunate. They had not covered Columbine the day of the event. Katz's first story had been posted on a Friday, a low-traffic day. They had a weekend to recognize the effects of his story and anticipate the follow-up. Overall they had a whole week to work through the process. In many ways, Columbine was the excep-tion; it is unusual for a news event to play out that gradually.

September 11

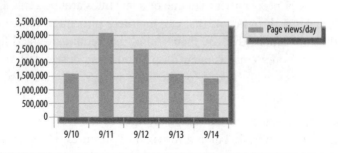

Department of Defense money originally funded the research that became the archi-tecture of the Internet. That research mandate was based on the perceived needs of a late-'60s Cold War nation. One of the design constraints was this: in the event of a selective, possibly nuclear attack on the United States, could a data network be designed that would continue to function despite outages to significant grid sections of the network? In other words, could the architecture of the network allow for graceful degradation and an opportunity to route around outages?

Two principles that underly the Internet architecture are decentralization and redun-dancy. Think, for example, about the way in which the Domain Name System (DNS) is implemented. This is the protocol by which a computer knows how to associate a human-readable address such as Slashdot.org with a computer-readable IP number. DNS has no single, canonical server to provide an authoritative list of these mappings. Such an approach would create a single point of failure that would serve as a bottleneck under high-traffic conditions and would bring down all DNS-dependent traffic in the event of a server failure.

Instead, DNS is more of a peer-to-peer network, with thousands of servers across the Internet functioning as DNS servers. Any server can update its records, and its updates will gradually be propagated to other DNS servers. In the course of a couple of days, any change to one DNS server can reach all others. Nor does one need "permission" to put up a DNS server. The requisite software is open source, and the Internet architecture is designed to automatically accept new servers or respond when encountering missing or offline servers. The system is highly distributed and redundant.

Even the basic network rules about how packets are routed from one destination to another follow these principles. Any computer sends a "test packet" first, attempting to establish a route to its destination. Once a route is established, actual data packets are sent. If at any point, the originating computer fails to get an authenticating response for that route, it explores for a new route and continues sending packets along the new route. There are no canonical, authoritative routes from Point A to Point B; each network route is a process of discovery based on current network conditions.

The principles are simple enough: avoid single points of failure by relying on a highly distributed network of peers rather than one or a few hubs around central, authoritative servers. The network protocols that employ these principles, however, are only as robust as the applications that use them. All of that redundancy and flexibility in routing does no good once email is queued up at an unresponsive destination server. If millions of requests are all headed for the same web server, that becomes the de facto center of an unresponsive hub.

In other words, to benefit from the design features of the Internet architecture, an application must be specifically tailored to use those features. In fact, relatively few applications do make use of this underlying structure. One application that does is IRC, the staple of online communication in the open source community.

A look at the network list in a default setup of XChat (a common IRC client) reveals dozens of IRC networks. Some are based around a common interest, like QuakeNet; some are based around a geographical location, like OzNet. Many, like Freenode, are general purpose. Within each of these networks will be dozens, or even hundreds of channels, each of which represents a particular community or topic of interest. The more popular networks easily have tens of thousands of users connected simultaneously at any given time.

IRC puts very slight demands on a server; all of the transmissions are short strings of text. Many universities and a large number of commercial sites volunteer server space to run an IRC server. All of the servers that are part of a given network work together to mirror the activity on the network. Typically servers in a network are partitioned into groups, with each group responsibile for mirroring a subset of the channels on that network. In the event that a given server goes offline, clients connected to channels for which that server had responsibility automatically reconnect to another server in that group. The view of a channel conversation that a particular user has, remains the same even through several reconnects to different servers.

As early as 1998, the Slashdot staff had set up an IRC network, called Slashnet. Initially this included a work channel for the staff to communicate with each other. This made sense since the staff was not always together in one place, but it was also just a natural form of communication for those with a Linux/open source background. A public channel was also added, for members of the Slashdot audience to communicate with the staff. The work channel quickly split into two channels, one for actual work communication, and another "water cooler" channel for idle conversation among staff members. Over time, other channels appeared, many from users treating Slashnet as just another IRC network, who were unaware that Slashnet and Slashdot were in any way affiliated.

By September of 2001, Slashnet had become an indispensable form of communication for the Slashdot staff. By this time, the staff was very distributed: Rob Malda and a core group of programmers remained in Holland, Michigan, but editors Timothy Lord and Robin Miller worked remotely; Timothy from various midwest locations, and Robin from Maryland. Jeff Bates had moved to Boston, working out of the offices of the parent company that had acquired Slashdot.[5]

Slashnet was, in many ways, the last refuge for Slashdot's original core audience. As the web site itself had become more mainstream, more about culture and less about technology, Slashnet represented a technical hard core of the site's open source roots. The barrier was not a very rigid one. While IRC channels can be moderated, and access can be password restricted, Slashnet, like most networks, was wide open for anyone to participate. In fact, though, the more mainstream online audience tended to gravitate to one-to-one IM systems like AIM or Yahoo! Instant Messenger, rather than the more text-based, more complex, and less user-friendly IRC.

Jeff Bates began the morning of September 11 at home in Boston before heading to the company office. He started with a call to Northwest Airlines, hoping to rearrange some business travel scheduled for later in the month. The call to Northwest was the first he knew that anything out of the ordinary was transpiring that day.

---

5   Originally this was online media company Andover.net. Andover was acquired by VA Linux Systems, and reformed as the wholly owned subsidiary OSDN, the Open Source Development Network. Since then, VA Linux Systems has changed its name to VA Software, and OSDN has changed its name to OSTG, the Open Source Technology Group.

The woman at customer service told Jeff that a plane had flown into the World Trade Center. She had not seen or heard a news report directly, but was instead repeating what she had heard from other customers calling in that morning. Still on the phone, Jeff turned on the television to watch events unfolding on CNN, all the while describing what he was seeing to the woman at Northwest.

After talking to Northwest, Jeff called a friend's cell phone in Manhattan to make sure he was OK. This call went through; many others, from many other people that day, would not. Jeff left for the office wondering, as many people did in those early hours, if this was some sort of freak accident or something more sinister.

Nine hundred miles away, in Holland, Michigan, Rob Malda was also beginning his workday. For Rob, this involved logging on to the Slashnet staff IRC channel, checking his email, and reviewing the Slashdot submissions bin. Rob's first word of the World Trade Center attacks came from monitoring discussions on IRC. With no radio or television at hand, he attempted to look at the CNN and MSNBC web sites, but both sites were already struggling under heavy load, and other than one small, grainy photo from the CNN web site, Rob was unable to get any information. Only the first plane had hit at this point, but the Slashdot submissions bin was already filling with related submissions, and Rob quickly realized this was not going to be an ordinary news day.

Slashdot reviews submissions 24 hours a day, 7 days a week. To provide this coverage, the staff rotates who is in charge of the submissions bin. While any staff online at a given time can review submissions and make suggestions, one person has to be the final authority: only one person at a time can wear the pants in this family. That's an official Slashdot job description: "Daddy Pants." On the morning of September 11, Rob was wearing Daddy Pants. He made the decision that they would depart from their normal coverage and focus exclusively on the World Trade Center story.

By the time Jeff arrived in the Boston office, he had heard on the radio that the second plane had hit, and everyone knew that some form of terrorist attack was underway. Rob's decision to focus the coverage was the right one. By 9:30 A.M.. EST, Slashdot was serving 30–40 pages per second off of its six mirrored web servers, roughly double the usual traffic load.

It's significant to note the range of communication media the Slashdot staff was involved with during the first 90 minutes of that morning: land line telephone, cellphone, television, radio, web sites, email, and IRC. In his 1991 book *Virtual Reality* Howard Reingold described cyberspace as where you are when you're on the phone. His point, in part, was that we live in a vast and evermore pervasive telecommunications network, and that oftentimes our location within that network is more significant than our geographical location.

All of the Slashdot staff described that working day as one of feeling intimately connected to others at work, despite the fact that they were operating from at least five different geographical locations. In fact, neither Rob nor Jeff can recall, and probably never knew, where Timothy was that day. It could have been Texas, but it could just as easily have been Tennessee. All that mattered was that he was there in channel on IRC to contribute and help out.

It's also significant to note what part of the telecommunications network suffered that day. Cellphone calls in and out of New York and Washington became increasingly difficult, though some calls went through under remarkable and tragic circumstances. Cellphone calls provide our most intimate historical record of what happened within the World Trade Center itself, as well as what happened on the doomed Flight 93 that crashed in Pennsylvania. Land-line long-distance calls would experience bottlenecks nationally throughout the day. Television and radio provided important early reports, but these became less effective later in the day as reporters had difficulty getting on the scene.

News web sites suffered the most, many completely unprepared for the deluge of traffic. Slashdot began a daylong battle to stay up and stay on top of events. The Columbine experience had alerted them to the need to overhaul the site infrastructure. Many changes had been made; now those changes would be put to the test.

Thirty page views per second was well above Slashdot's normal load, but also about the limit of what the site architecture was designed to handle. Around 10:00 that morning, the backend database crashed, and the site was temporarily down. In fact, Slashdot had a backup database server on hand, one with a more current version of the database software (MySQL). This server was not yet online only because the staff didn't want to take the site offline to make the switch. The database crash provided an opportunity to quickly make that switch.

The new database server performed well, and now the bottleneck shifted to the web servers themselves. The staff made some on-the-fly adjustments to caching limits on the servers, and for the moment everything was functioning. It was now noon EST, and the site was serving 50 pages per second. Rob took a short break, and for the first time saw the actual video footage of the two crashes that morning.

As the staff struggled to keep the Slashdot web site up, a parallel phenomenon was emerging. Traffic on Slashnet was swelling, as more and more people turned to IRC as a way to communicate. The 200,000 pages per hour Slashdot's web site was now serving was impressive enough. Yet at the same time, Slashnet had thousands, and perhaps as many as 20,000 simultaneous users sharing information even more rapidly. The Slashdot staff set up a moderated channel to bring some organization to the process, but unmoderated public channels were springing up as well. There was one channel for communicating with the staff, and another for just general discussion.

Many of the links Slashdot posted that day, and many of the inline comments and quotes that appeared in stories, were pulled directly out of IRC on Slashnet. At a time when major news networks had difficulty getting reporters to the scene, Slashdot had eyewitness accounts coming in over IRC. At a time when major news web sites could not keep up with the traffic or rapidly changing information, Slashdot persevered. Some it was information, but some of it was a matter of dispelling misinformation ("I heard a truck bomb went off outside the State Department".... "No, my dad works at State, I just spoke to him, and nothing like that's going on"). Some of the concerns were global ("Who's behind the attacks?"). Some of the concerns were terribly personal ("I have a friend/loved one/family member who works in midtown Manhattan...").

Behind the scenes, the Slashdot staff frantically stripped down functionality on the site to keep the bare minimum of processes running and the maximum number of pages flowing. Dynamic content was turned off. Reverse DNS lookup was turned off. Eventually the ad server was turned off. Logfiles were turned off. After the initial database failure early in the day, however, the site stayed up. At the peak, Slashdot was serving 70 pages per second. For the day, it served more than 3 million page views.[6]

As a nation, we've never faced the kind of global telecommunications breakdown that the Internet was architected to handle as gracefully as possible. We have, however, seen episodes like September 11 that put a sudden and unexpected strain on the telecommunications infrastructure, and where the graceful degradation for which the Internet allows, matters a great deal. Taking advantage of that inherent robustness, however, requires a communication medium that follows the same architectural design and a group of communicators comfortable using that medium. Slashdot's successful coverage that day would not have been possible without IRC, a protocol as distributed and robust as the Internet itself, and without Slashnet, a community of users who knew how to make the most of that medium.

The word *disintermediation* has been much abused in the online world. The idea is simple enough: where traditional news media "mediates" between audience and events, the directness of the Internet should make this traditional mediation unnecessary. In practice, disintermediation happens far less often and far less effectively than one might think. Several elements need to be in place. First, there must be a genuine community of like-minded communicators looking to interact directly. Second, the medium through which they interact must be, to the degree possible, both responsive and transparent. Finally, those enabling the medium must have the humility to do nothing more than facilitate.

With Columbine, the Slashdot staff learned very quickly that they were not presenting a story, but were instead in the midst of a story that was happening all around them. The most useful thing they could do was get out of the way and let the story

---

6   Rob Malda has nicely summarized the day's behind-the-scenes work in his piece, "Handling the Loads," at *http://slashdot.org/article.pl?sid=01/09/13/154222*.

happen, let the Slashdot community connect to each other. September 11 proved again the value of facilitating rather than mediating. By and large, the audience that day did not notice or care that it was Slashdot they were using as their medium. Any channel of communication that was real time, up-to-date, and available would have sufficed. September 11 revealed which communications channels were up to that challenge; which could be effective but informative; which could disintermediate when disintermediation was needed most.

Conclusion

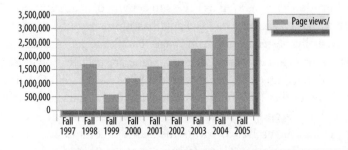

In his influential essay, "The Cathedral & the Bazaar," published in the original *Open Sources*, Eric Raymond offered an arresting metaphor to contrast the top-down approach of traditional software development with the more grass-roots nature of open source software development. Cathedral-style development happens in isolation from users and with rigid authority from the top. Bazaar-style development is more community driven, without a clear line between developer and user, and follows a more evolutionary design process.

People often mistake Raymond's metaphor, however. Too often the assumption is made that the open source development community is a legion of programmers with hundreds or thousands of contributions coming in to each project, as if somehow by sheer numbers open source will triumph against its proprietary competitors. The mistake in this view is to look at the open source community as a flat, homogeneous organization.

Organizations seldom have so simple a structure. Online communities are not accidental organizations, thrown together by geography or family ties. Online communities are a subset of intentional communities, groups formed by those with common interests seeking out like-minded peers and exploiting the low communication cost of the Internet to make those connections. These communities indeed have a structure and a hierarchy. They are not so much a bazaar as they are a tribe.

Slashdot today serves more than 3.5 million page views a day. It is tempting to think of its viewers as an audience, not a community. It is easy to think of its viewers as a bazaar-style, unorganized mass. However, facilitating successful many-to-many communication requires a more sophisticated view. Fundamentally, Slashdot is a community, with the complex hierarchy all online communities have.

Rob Malda and Jeff Bates have commented that they see themselves not as staff versus audience, but as one group with a continuum of privileges. At the base of this hierarchy are the "anonymous cowards," who can read stories, submit stories, and comment on stories, though their comments start from a lower ranking. Registered users start from a higher position, with comments initially moderated higher, and with their activity tracked and evaluated within the system. Moderators are selected from among registered users, as are metamoderators. The paid staff have access to the actual submissions bin, as well as access to activity data about users. Finally, the staff member currently wearing Daddy Pants has ultimate authority over what submissions are actually posted to the front page. Parallel to all of this is the Slashnet IRC network. Some channels are moderated. Some are private. Some are public and open. Some moderated channels are moderated by Slashdot staff. Others are moderated and cover topics that have nothing to do with Slashdot itself. Some participants on IRC see Slashnet as an important way to connect to the larger Slashdot community. Others think of themselves as part of the Slashdot community, but never participate on IRC. Yet as September 11 revealed, even those uninvolved in or unaware of the IRC part of the community nonetheless benefit from it.

The system is authoritarian. So too is a tribe. Every position in this community has its unique privileges, however. Staff members do not have the right to moderate or metamoderate. Only registered users can do that. Furthermore, the system, though authoritarian, relies essentially on a practice of "term limits." A moderator receives a small number of moderation points, and once these are used, moderation rotates to someone else. Metamoderators get about 20 or so moderations to evaluate, and then metamoderation rotates to someone else. Even among the paid staff, no one has absolute editorial authority; Daddy Pants rotates among all of them.

Slashdot and other online communities, like many traditional tribes, bestow authority and privilege based on actions and reputations. This form of many-to-many communication works because these communities are egalitarian, but not democratic: everyone gets a voice, but not everyone gets a vote.

Appendixes

No discussion of open source could be complete without laying out the basic documents that are foundational to the movement. The Open Source Definition and the GPL have much the same status within the open source community as the Declaration of Independence and the Bill of Rights. Those documents, and other key open source licenses, are reprinted here in the appendixes.

Additionally, several short columns encapsulate some pivotal events in the evolution of Slashdot discussed by Bates and Stone. While these columns are archived on Slashdot, they are fairly brief and it seemed expedient to simply reprint them here.

The Open Source Definition

The Open Source Definition is maintained and applied by the nonprofit Open Source Initiative (OSI; *http://opensource.org*). The OSI has a board of directors made up of key people from industry and the community involved in open source. Licenses may be submitted to the OSI for review, and if, in the judgment of the OSI, they meet the terms of the Open Source Definition, the term "open source license" may be applied to them as OSI approved.

The Open Source Definition, Version 1.9

Introduction

Open source doesn't just mean access to the source code. The distribution terms of open-source software must comply with the following criteria:

1. Free Redistribution

The license shall not restrict any party from selling or giving away the software as a component of an aggregate software distribution containing programs from several different sources. The license shall not require a royalty or other fee for such sale.

2. Source Code

The program must include source code, and must allow distribution in source code as well as compiled form. Where some form of a product is not distributed with source code, there must be a well-publicized means of obtaining the source code for no more than a reasonable reproduction cost—preferably, downloading via the Internet without charge. The source code must be the preferred form in which a programmer would modify the program. Deliberately obfuscated source code is not allowed. Intermediate forms such as the output of a preprocessor or translator are not allowed.

3. Derived Works

The license must allow modifications and derived works, and must allow them to be distributed under the same terms as the license of the original software.

4. Integrity of The Author's Source Code

The license may restrict source-code from being distributed in modified form only if the license allows the distribution of "patch files" with the source code for the purpose of modifying the program at build time. The license must explicitly permit distribution of software built from modified source code. The license may require derived works to carry a different name or version number from the original software.

5. No Discrimination Against Persons or Groups

The license must not discriminate against any person or group of persons.

6. No Discrimination Against Fields of Endeavor

The license must not restrict anyone from making use of the program in a specific field of endeavor. For example, it may not restrict the program from being used in a business, or from being used for genetic research.

7. Distribution of License

The rights attached to the program must apply to all to whom the program is redistributed without the need for execution of an additional license by those parties.

8. License Must Not Be Specific to a Product

The rights attached to the program must not depend on the program's being part of a particular software distribution. If the program is extracted from that distribution and used or distributed within the terms of the program's license, all parties to whom the program is redistributed should have the same rights as those that are granted in conjunction with the original software distribution.

9. License Must Not Restrict Other Software

The license must not place restrictions on other software that is distributed along with the licensed software. For example, the license must not insist that all other programs distributed on the same medium must be open-source software.

Referenced Open Source Licenses

The Open Source Initiative (OSI; *http://www.opensource.org*) is a nonprofit organization with responsibility for approving open source licenses as applied to software. Currently there are more than 50 OSI-approved licenses. Included here are the major licenses referenced by essays in this volume.

The BSD License

The following is a BSD license template. To generate your own license, change the values of OWNER, ORGANIZATION, and YEAR from their original values as given here, and substitute your own.

Note: the advertising clause in the license appearing on BSD Unix files was officially rescinded by the Director of the Office of Technology Licensing of the University of California on July 22, 1999. He states that clause 3 is "hereby deleted in its entirety."

Note the new BSD license is thus equivalent to the *MIT License*, except for the no-endorsement final clause.

\<OWNER\> = Regents of the University of California
\<ORGANIZATION\> = University of California, Berkeley
\<YEAR\> = 1998

In the original BSD license, both occurrences of the phrase "COPYRIGHT HOLDERS AND CONTRIBUTORS" in the disclaimer read "REGENTS AND CONTRIBUTORS."

Here is the license template:

Copyright (c) <YEAR>, <OWNER>
All rights reserved.

Redistribution and use in source and binary forms, with or without modification, are permitted provided that the following conditions are met:

- Redistributions of source code must retain the above copyright notice, this list of conditions and the following disclaimer.

- Redistributions in binary form must reproduce the above copyright notice, this list of conditions and the following disclaimer in the documentation and/or other materials provided with the distribution.

- Neither the name of the <ORGANIZATION> nor the names of its contributors may be used to endorse or promote products derived from this software without specific prior written permission.

THIS SOFTWARE IS PROVIDED BY THE COPYRIGHT HOLDERS AND CONTRIBUTORS "AS IS" AND ANY EXPRESS OR IMPLIED WARRANTIES, INCLUDING, BUT NOT LIMITED TO, THE IMPLIED WARRANTIES OF MERCHANTABILITY AND FITNESS FOR A PARTICULAR PURPOSE ARE DISCLAIMED. IN NO EVENT SHALL THE COPYRIGHT OWNER OR CONTRIBUTORS BE LIABLE FOR ANY DIRECT, INDIRECT, INCIDENTAL, SPECIAL, EXEMPLARY, OR CONSEQUENTIAL DAMAGES (INCLUDING, BUT NOT LIMITED TO, PROCUREMENT OF SUBSTITUTE GOODS OR SERVICES; LOSS OF USE, DATA, OR PROFITS; OR BUSINESS INTERRUPTION) HOWEVER CAUSED AND ON ANY THEORY OF LIABILITY, WHETHER IN CONTRACT, STRICT LIABILITY, OR TORT (INCLUDING NEGLIGENCE OR OTHERWISE) ARISING IN ANY WAY OUT OF THE USE OF THIS SOFTWARE, EVEN IF ADVISED OF THE POSSIBILITY OF SUCH DAMAGE.

The GNU General Public License (GPL)

Version 2, June 1991

Copyright (C) 1989, 1991 Free Software Foundation, Inc.
59 Temple Place, Suite 330, Boston, MA 02111-1307 USA

Everyone is permitted to copy and distribute verbatim copies of this license document, but changing it is not allowed.

Preamble

The licenses for most software are designed to take away your freedom to share and change it. By contrast, the GNU General Public License is intended to guarantee your freedom to share and change free software—to make sure the software is free for all its users. This General Public License applies to most of the Free Software Foundation's software and to any other program whose authors commit to using it. (Some other Free Software Foundation software is covered by the GNU Library General Public License instead.) You can apply it to your programs, too.

When we speak of free software, we are referring to freedom, not price. Our General Public Licenses are designed to make sure that you have the freedom to distribute copies of free software (and charge for this service if you wish), that you receive source code or can get it if you want it, that you can change the software or use pieces of it in new free programs; and that you know you can do these things.

To protect your rights, we need to make restrictions that forbid anyone to deny you these rights or to ask you to surrender the rights. These restrictions translate to certain responsibilities for you if you distribute copies of the software, or if you modify it.

For example, if you distribute copies of such a program, whether gratis or for a fee, you must give the recipients all the rights that you have. You must make sure that they, too, receive or can get the source code. And you must show them these terms so they know their rights.

We protect your rights with two steps: (1) copyright the software, and (2) offer you this license which gives you legal permission to copy, distribute and/or modify the software.

Also, for each author's protection and ours, we want to make certain that everyone understands that there is no warranty for this free software. If the software is modified by someone else and passed on, we want its recipients to know that what they have is not the original, so that any problems introduced by others will not reflect on the original authors' reputations.

Finally, any free program is threatened constantly by software patents. We wish to avoid the danger that redistributors of a free program will individually obtain patent licenses, in effect making the program proprietary. To prevent this, we have made it clear that any patent must be licensed for everyone's free use or not licensed at all.

The precise terms and conditions for copying, distribution, and modification follow.

Terms and Conditions for Copying, Distribution, and Modification

This License applies to any program or other work which contains a notice placed by the copyright holder saying it may be distributed under the terms of this General Public License. The "Program," below, refers to any such program or work, and a "work based on the Program" means either the Program or any derivative work under copyright law: that is to say, a work containing the Program or a portion of it, either verbatim or with modifications and/or translated into another language. (Hereinafter, translation is included without limitation in the term "modification.") Each licensee is addressed as "you."

Activities other than copying, distribution and modification are not covered by this License; they are outside its scope. The act of running the Program is not restricted, and the output from the Program is covered only if its contents constitute a work based on the Program (independent of having been made by running the Program). Whether that is true depends on what the Program does.

1. You may copy and distribute verbatim copies of the Program's source code as you receive it, in any medium, provided that you conspicuously and appropriately publish on each copy an appropriate copyright notice and disclaimer of warranty; keep intact all the notices that refer to this License and to the absence of any warranty; and give any other recipients of the Program a copy of this License along with the Program.

   You may charge a fee for the physical act of transferring a copy, and you may at your option offer warranty protection in exchange for a fee.

2. You may modify your copy or copies of the Program or any portion of it, thus forming a work based on the Program, and copy and distribute such modifications or work under the terms of Section 1 above, provided that you also meet all of these conditions:

   a. You must cause the modified files to carry prominent notices stating that you changed the files and the date of any change.

   b. You must cause any work that you distribute or publish, that in whole or in part contains or is derived from the Program or any part thereof, to be licensed as a whole at no charge to all third parties under the terms of this License.

   c. If the modified program normally reads commands interactively when run, you must cause it, when started running for such interactive use in the most ordinary way, to print or display an announcement including an appropriate copyright notice and a notice that there is no warranty (or else, saying that you provide a warranty) and that users may redistribute the program under these conditions, and telling the user how to view a copy of this License. (Exception: if the Program itself is interactive but does not normally print such an announcement, your work based on the Program is not required to print an announcement.)

These requirements apply to the modified work as a whole. If identifiable sections of that work are not derived from the Program, and can be reasonably considered independent and separate works in themselves, then this License, and its terms, do not apply to those sections when you distribute them as separate works. But when you distribute the same sections as part of a whole which is a work based on the Program, the distribution of the whole must be on the terms of this License, whose permissions for other licensees extend to the entire whole, and thus to each and every part regardless of who wrote it.

Thus, it is not the intent of this section to claim rights or contest your rights to work written entirely by you; rather, the intent is to exercise the right to control the distribution of derivative or collective works based on the Program.

In addition, mere aggregation of another work not based on the Program with the Program (or with a work based on the Program) on a volume of a storage or distribution medium does not bring the other work under the scope of this License.

3. You may copy and distribute the Program (or a work based on it, under Section 2) in object code or executable form under the terms of Sections 1 and 2 above provided that you also do one of the following:

   d. Accompany it with the complete corresponding machine-readable source code, which must be distributed under the terms of Sections 1 and 2 above on a medium customarily used for software interchange; or,

   e. Accompany it with a written offer, valid for at least three years, to give any third party, for a charge no more than your cost of physically performing source distribution, a complete machine-readable copy of the corresponding source code, to be distributed under the terms of Sections 1 and 2 above on a medium customarily used for software interchange; or,

   f. Accompany it with the information you received as to the offer to distribute corresponding source code. (This alternative is allowed only for noncommercial distribution and only if you received the program in object code or executable form with such an offer, in accord with Subsection b above.)

The source code for a work means the preferred form of the work for making modifications to it. For an executable work, complete source code means all the source code for all modules it contains, plus any associated interface definition files, plus the scripts used to control compilation and installation of the executable. However, as a special exception, the source code distributed need not include anything that is normally distributed (in either source or binary form) with the major components (compiler, kernel, and so on) of the operating system on which the executable runs, unless that component itself accompanies the executable.

If distribution of executable or object code is made by offering access to copy from a designated place, then offering equivalent access to copy the source code from the same place counts as distribution of the source code, even though third parties are not compelled to copy the source along with the object code.

4. You may not copy, modify, sublicense, or distribute the Program except as expressly provided under this License. Any attempt otherwise to copy, modify, sublicense or distribute the Program is void, and will automatically terminate your rights under this License. However, parties who have received copies, or rights, from you under this License will not have their licenses terminated so long as such parties remain in full compliance.

5. You are not required to accept this License, since you have not signed it. However, nothing else grants you permission to modify or distribute the Program or

its derivative works. These actions are prohibited by law if you do not accept this License. Therefore, by modifying or distributing the Program (or any work based on the Program), you indicate your acceptance of this License to do so, and all its terms and conditions for copying, distributing or modifying the Program or works based on it.

6. Each time you redistribute the Program (or any work based on the Program), the recipient automatically receives a license from the original licensor to copy, distribute or modify the Program subject to these terms and conditions. You may not impose any further restrictions on the recipients' exercise of the rights granted herein. You are not responsible for enforcing compliance by third parties to this License.

7. If, as a consequence of a court judgment or allegation of patent infringement or for any other reason (not limited to patent issues), conditions are imposed on you (whether by court order, agreement or otherwise) that contradict the conditions of this License, they do not excuse you from the conditions of this License. If you cannot distribute so as to satisfy simultaneously your obligations under this License and any other pertinent obligations, then as a consequence you may not distribute the Program at all. For example, if a patent license would not permit royalty-free redistribution of the Program by all those who receive copies directly or indirectly through you, then the only way you could satisfy both it and this License would be to refrain entirely from distribution of the Program.

If any portion of this section is held invalid or unenforceable under any particular circumstance, the balance of the section is intended to apply and the section as a whole is intended to apply in other circumstances.

It is not the purpose of this section to induce you to infringe any patents or other property right claims or to contest validity of any such claims; this section has the sole purpose of protecting the integrity of the free software distribution system, which is implemented by public license practices. Many people have made generous contributions to the wide range of software distributed through that system in reliance on consistent application of that system; it is up to the author/ donor to decide if he or she is willing to distribute software through any other system and a licensee cannot impose that choice.

This section is intended to make thoroughly clear what is believed to be a consequence of the rest of this License.

8. If the distribution and/or use of the Program is restricted in certain countries either by patents or by copyrighted interfaces, the original copyright holder who places the Program under this License may add an explicit geographical distribution limitation excluding those countries, so that distribution is permitted only in or among countries not thus excluded. In such case, this License incorporates the limitation as if written in the body of this License.

9. The Free Software Foundation may publish revised and/or new versions of the General Public License from time to time. Such new versions will be similar in spirit to the present version, but may differ in detail to address new problems or concerns.

   Each version is given a distinguishing version number. If the Program specifies a version number of this License which applies to it and "any later version," you have the option of following the terms and conditions either of that version or of any later version published by the Free Software Foundation. If the Program does not specify a version number of this License, you may choose any version ever published by the Free Software Foundation.

10. If you wish to incorporate parts of the Program into other free programs whose distribution conditions are different, write to the author to ask for permission. For software which is copyrighted by the Free Software Foundation, write to the Free Software Foundation; we sometimes make exceptions for this. Our decision will be guided by the two goals of preserving the free status of all derivatives of our free software and of promoting the sharing and reuse of software generally.

NO WARRANTY

1. BECAUSE THE PROGRAM IS LICENSED FREE OF CHARGE, THERE IS NO WARRANTY FOR THE PROGRAM, TO THE EXTENT PERMITTED BY APPLICABLE LAW. EXCEPT WHEN OTHERWISE STATED IN WRITING THE COPYRIGHT HOLDERS AND/OR OTHER PARTIES PROVIDE THE PROGRAM "AS IS" WITHOUT WARRANTY OF ANY KIND, EITHER EXPRESSED OR IMPLIED, INCLUDING, BUT NOT LIMITED TO, THE IMPLIED WARRANTIES OF MERCHANTABILITY AND FITNESS FOR A PARTICULAR PURPOSE. THE ENTIRE RISK AS TO THE QUALITY AND PERFORMANCE OF THE PROGRAM IS WITH YOU. SHOULD THE PROGRAM PROVE DEFECTIVE, YOU ASSUME THE COST OF ALL NECESSARY SERVICING, REPAIR OR CORRECTION.

2. IN NO EVENT UNLESS REQUIRED BY APPLICABLE LAW OR AGREED TO IN WRITING WILL ANY COPYRIGHT HOLDER, OR ANY OTHER PARTY WHO MAY MODIFY AND/OR REDISTRIBUTE THE PROGRAM AS PERMITTED ABOVE, BE LIABLE TO YOU FOR DAMAGES, INCLUDING ANY GENERAL, SPECIAL, INCIDENTAL OR CONSEQUENTIAL DAMAGES ARISING OUT OF THE USE OR INABILITY TO USE THE PROGRAM (INCLUDING BUT NOT LIMITED TO LOSS OF DATA OR DATA BEING RENDERED INACCURATE OR LOSSES SUSTAINED BY YOU OR THIRD PARTIES OR A FAILURE OF THE PROGRAM TO OPERATE WITH ANY OTHER PROGRAMS), EVEN IF SUCH HOLDER OR OTHER PARTY HAS BEEN ADVISED OF THE POSSIBILITY OF SUCH DAMAGES.

   END OF TERMS AND CONDITIONS

How to Apply These Terms to Your New Programs

If you develop a new program, and you want it to be of the greatest possible use to the public, the best way to achieve this is to make it free software which everyone can redistribute and change under these terms.

To do so, attach the following notices to the program. It is safest to attach them to the start of each source file to most effectively convey the exclusion of warranty; and each file should have at least the "copyright" line and a pointer to where the full notice is found.

One line to give the program's name and a brief idea of what it does.

> Copyright (C) <year> <name of author

> This program is free software; you can redistribute it and/or modify it under the terms of the GNU General Public License as published by the Free Software Foundation; either version 2 of the License, or (at your option) any later version.

> This program is distributed in the hope that it will be useful, but WITHOUT ANY WARRANTY; without even the implied warranty of MERCHANTABILITY or FITNESS FOR A PARTICULAR PURPOSE. See the GNU General Public License for more details.

> You should have received a copy of the GNU General Public License along with this program; if not, write to the Free Software Foundation, Inc., 59 Temple Place, Suite 330, Boston, MA 02111-1307 USA

Also add information on how to contact you by electronic and paper mail.

If the program is interactive, make it output a short notice like this when it starts in an interactive mode:

> Gnomovision version 69, Copyright (C) year name of author Gnomovision comes with ABSOLUTELY NO WARRANTY; for details type 'show w'. This is free software, and you are welcome to redistribute it under certain conditions; type 'show c' for details.

The hypothetical commands 'show w' and 'show c' should show the appropriate parts of the General Public License. Of course, the commands you use may be called something other than 'show w' and 'show c;' they could even be mouse-clicks or menu items—whatever suits your program.

You should also get your employer (if you work as a programmer) or your school, if any, to sign a "copyright disclaimer" for the program, if necessary. Here is a sample; alter the names:

> Yoyodyne, Inc., hereby disclaims all copyright interest in the program 'Gnomovision' (which makes passes at compilers) written by James Hacker.

> signature of Ty Coon, 1 April 1989

> Ty Coon, President of Vice

This General Public License does not permit incorporating your program into proprietary programs. If your program is a subroutine library, you may consider it more useful to permit linking proprietary applications with the library. If this is what you want to do, use the GNU Library General Public License instead of this License.

The Sleepycat License

Copyright (c) 1990-1999 Sleepycat Software. All rights reserved.

Redistribution and use in source and binary forms, with or without modification, are permitted provided that the following conditions are met:

- Redistributions of source code must retain the above copyright notice, this list of conditions and the following disclaimer.

- Redistributions in binary form must reproduce the above copyright notice, this list of conditions and the following disclaimer in the documentation and/or other materials provided with the distribution.

- Redistributions in any form must be accompanied by information on how to obtain complete source code for the DB software and any accompanying software that uses the DB software. The source code must either be included in the distribution or be available for no more than the cost of distribution plus a nominal fee, and must be freely redistributable under reasonable conditions. For an executable file, complete source code means the source code for all modules it contains. It does not include source code for modules or files that typically accompany the major components of the operating system on which the executable file runs.

THIS SOFTWARE IS PROVIDED BY SLEEPYCAT SOFTWARE "AS IS" AND ANY EXPRESS OR IMPLIED WARRANTIES, INCLUDING, BUT NOT LIMITED TO, THE IMPLIED WARRANTIES OF MERCHANTABILITY, FITNESS FOR A PARTICULAR PURPOSE, OR NON-INFRINGEMENT, ARE DISCLAIMED. IN NO EVENT SHALL SLEEPYCAT SOFTWARE BE LIABLE FOR ANY DIRECT, INDIRECT, INCIDENTAL, SPECIAL, EXEMPLARY, OR CONSEQUENTIAL DAMAGES (INCLUDING, BUT NOT LIMITED TO, PROCUREMENT OF SUBSTITUTE GOODS OR SERVICES; LOSS OF USE, DATA, OR PROFITS; OR BUSINESS INTERRUPTION) HOWEVER CAUSED AND ON ANY THEORY OF LIABILITY, WHETHER IN CONTRACT, STRICT LIABILITY, OR TORT (INCLUDING NEGLIGENCE OR OTHERWISE) ARISING IN ANY WAY OUT OF THE USE OF THIS SOFTWARE, EVEN IF ADVISED OF THE POSSIBILITY OF SUCH DAMAGE.

Copyright (c) 1990, 1993, 1994, 1995 The Regents of the University of California. All rights reserved.

Redistribution and use in source and binary forms, with or without modification, are permitted provided that the following conditions are met:

- Redistributions of source code must retain the above copyright notice, this list of conditions and the following disclaimer.

- Redistributions in binary form must reproduce the above copyright notice, this list of conditions and the following disclaimer in the documentation and/or other materials provided with the distribution.

- Neither the name of the University nor the names of its contributors may be used to endorse or promote products derived from this software without specific prior written permission.

THIS SOFTWARE IS PROVIDED BY THE REGENTS AND CONTRIBUTORS "AS IS" AND ANY EXPRESS OR IMPLIED WARRANTIES, INCLUDING, BUT NOT LIMITED TO, THE IMPLIED WARRANTIES OF MERCHANTABILITY AND FITNESS FOR A PARTICULAR PURPOSE ARE DISCLAIMED. IN NO EVENT SHALL THE REGENTS OR CONTRIBUTORS BE LIABLE FOR ANY DIRECT, INDIRECT, INCIDENTAL, SPECIAL, EXEMPLARY, OR CONSEQUENTIAL DAMAGES (INCLUDING, BUT NOT LIMITED TO, PROCUREMENT OF SUBSTITUTE GOODS OR SERVICES; LOSS OF USE, DATA, OR PROFITS; OR BUSINESS INTERRUPTION) HOWEVER CAUSED AND ON ANY THEORY OF LIABILITY, WHETHER IN CONTRACT, STRICT LIABILITY, OR TORT (INCLUDING NEGLIGENCE OR OTHERWISE) ARISING IN ANY WAY OUT OF THE USE OF THIS SOFTWARE, EVEN IF ADVISED OF THE POSSIBILITY OF SUCH DAMAGE.

Copyright (c) 1995, 1996 The President and Fellows of Harvard University. All rights reserved.

Redistribution and use in source and binary forms, with or without modification, are permitted provided that the following conditions are met:

- Redistributions of source code must retain the above copyright notice, this list of conditions and the following disclaimer.

- Redistributions in binary form must reproduce the above copyright notice, this list of conditions and the following disclaimer in the documentation and/or other materials provided with the distribution.

- Neither the name of the University nor the names of its contributors may be used to endorse or promote products derived from this software without specific prior written permission.

THIS SOFTWARE IS PROVIDED BY HARVARD AND ITS CONTRIBUTORS "AS IS" AND ANY EXPRESS OR IMPLIED WARRANTIES, INCLUDING, BUT NOT LIMITED TO, THE IMPLIED WARRANTIES OF MERCHANTABILITY AND FITNESS FOR A PARTICULAR PURPOSE ARE DISCLAIMED. IN NO EVENT SHALL

HARVARD OR ITS CONTRIBUTORS BE LIABLE FOR ANY DIRECT, INDIRECT, INCIDENTAL, SPECIAL, EXEMPLARY, OR CONSEQUENTIAL DAMAGES (INCLUDING, BUT NOT LIMITED TO, PROCUREMENT OF SUBSTITUTE GOODS OR SERVICES; LOSS OF USE, DATA, OR PROFITS; OR BUSINESS INTERRUPTION) HOWEVER CAUSED AND ON ANY THEORY OF LIABILITY, WHETHER IN CONTRACT, STRICT LIABILITY, OR TORT (INCLUDING NEGLI-GENCE OR OTHERWISE) ARISING IN ANY WAY OUT OF THE USE OF THIS SOFTWARE, EVEN IF ADVISED OF THE POSSIBILITY OF SUCH DAMAGE.

The Creative Commons License

Attribution-NonCommercial-NoDerivs 2.5

You are free:

* to copy, distribute, display, and perform the work

Under the following conditions:

* Attribution. You must attribute the work in the manner specified by the author or licensor.
* Noncommercial. You may not use this work for commercial purposes.
* No Derivative Works. You may not alter, transform, or build upon this work.

For any reuse or distribution, you must make clear to others the license terms of this work.

Any of these conditions can be waived if you get permission from the copyright holder.

Your fair use and other rights are in no way affected by the above.

Full Text of License Follows:

Attribution-NonCommercial-NoDerivs 2.5

CREATIVE COMMONS CORPORATION IS NOT A LAW FIRM AND DOES NOT PROVIDE LEGAL SERVICES. DISTRIBUTION OF THIS LICENSE DOES NOT CRE-ATE AN ATTORNEY-CLIENT RELATIONSHIP. CREATIVE COMMONS PROVIDES THIS INFORMATION ON AN "AS-IS" BASIS. CREATIVE COMMONS MAKES NO WARRANTIES REGARDING THE INFORMATION PROVIDED, AND DISCLAIMS LIABILITY FOR DAMAGES RESULTING FROM ITS USE.

License

THE WORK (AS DEFINED BELOW) IS PROVIDED UNDER THE TERMS OF THIS CREATIVE COMMONS PUBLIC LICENSE ("CCPL" OR "LICENSE"). THE WORK IS PROTECTED BY COPYRIGHT AND/OR OTHER APPLICABLE LAW. ANY USE OF THE WORK OTHER THAN AS AUTHORIZED UNDER THIS LICENSE OR COPY-RIGHT LAW IS PROHIBITED.

BY EXERCISING ANY RIGHTS TO THE WORK PROVIDED HERE, YOU ACCEPT AND AGREE TO BE BOUND BY THE TERMS OF THIS LICENSE. THE LICENSOR GRANTS YOU THE RIGHTS CONTAINED HERE IN CONSIDERATION OF YOUR ACCEPTANCE OF SUCH TERMS AND CONDITIONS.

1. Definitions

   a. "Collective Work" means a work, such as a periodical issue, anthology or encyclopedia, in which the Work in its entirety in unmodified form, along with a number of other contributions, constituting separate and independent works in themselves, are assembled into a collective whole. A work that constitutes a Collective Work will not be considered a Derivative Work (as defined below) for the purposes of this License.

   b. "Derivative Work" means a work based upon the Work or upon the Work and other pre-existing works, such as a translation, musical arrangement, dramatization, fictionalization, motion picture version, sound recording, art reproduction, abridgment, condensation, or any other form in which the Work may be recast, transformed, or adapted, except that a work that constitutes a Collective Work will not be considered a Derivative Work for the purpose of this License. For the avoidance of doubt, where the Work is a musical composition or sound recording, the synchronization of the Work in timed-relation with a moving image ("synching") will be considered a Derivative Work for the purpose of this License.

   c. "Licensor" means the individual or entity that offers the Work under the terms of this License.

   d. "Original Author" means the individual or entity who created the Work.

   e. "Work" means the copyrightable work of authorship offered under the terms of this License.

   f. "You" means an individual or entity exercising rights under this License who has not previously violated the terms of this License with respect to the Work, or who has received express permission from the Licensor to exercise rights under this License despite a previous violation.

2. Fair Use Rights. Nothing in this license is intended to reduce, limit, or restrict any rights arising from fair use, first sale or other limitations on the exclusive rights of the copyright owner under copyright law or other applicable laws.

3. License Grant. Subject to the terms and conditions of this License, Licensor hereby grants You a worldwide, royalty-free, non-exclusive, perpetual (for the duration of the applicable copyright) license to exercise the rights in the Work as stated below:

a. to reproduce the Work, to incorporate the Work into one or more Collective Works, and to reproduce the Work as incorporated in the Collective Works;

b. to distribute copies or phonorecords of, display publicly, perform publicly, and perform publicly by means of a digital audio transmission the Work including as incorporated in Collective Works;

The above rights may be exercised in all media and formats whether now known or hereafter devised. The above rights include the right to make such modifications as are technically necessary to exercise the rights in other media and formats, but otherwise you have no rights to make Derivative Works. All rights not expressly granted by Licensor are hereby reserved, including but not limited to the rights set forth in Sections 4(d) and 4(e).

4. Restrictions.The license granted in Section 3 above is expressly made subject to and limited by the following restrictions:

a. You may distribute, publicly display, publicly perform, or publicly digitally perform the Work only under the terms of this License, and You must include a copy of, or the Uniform Resource Identifier for, this License with every copy or phonorecord of the Work You distribute, publicly display, publicly perform, or publicly digitally perform. You may not offer or impose any terms on the Work that alter or restrict the terms of this License or the recipients' exercise of the rights granted hereunder. You may not sublicense the Work. You must keep intact all notices that refer to this License and to the disclaimer of warranties. You may not distribute, publicly display, publicly perform, or publicly digitally perform the Work with any technological measures that control access or use of the Work in a manner inconsistent with the terms of this License Agreement. The above applies to the Work as incorporated in a Collective Work, but this does not require the Collective Work apart from the Work itself to be made subject to the terms of this License. If You create a Collective Work, upon notice from any Licensor You must, to the extent practicable, remove from the Collective Work any credit as required by clause 4(c), as requested.

b. You may not exercise any of the rights granted to You in Section 3 above in any manner that is primarily intended for or directed toward commercial advantage or private monetary compensation. The exchange of the Work for other copyrighted works by means of digital file-sharing or otherwise shall not be considered to be intended for or directed toward commercial advantage or private monetary compensation, provided there is no payment of any monetary compensation in connection with the exchange of copyrighted works.

c. If you distribute, publicly display, publicly perform, or publicly digitally perform the Work, You must keep intact all copyright notices for the Work and provide, reasonable to the medium or means You are utilizing: (i) the name of the Original Author (or pseudonym, if applicable) if supplied, and/or (ii) if the Original Author and/or Licensor designate another party or parties (e.g. a sponsor institute, publishing entity, journal) for attribution in Licensor's copyright notice, terms of service or by other reasonable means, the name of such party or parties; the title of the Work if supplied; and to the extent reasonably practicable, the Uniform Resource Identifier, if any, that Licensor specifies to be associated with the Work, unless such URI does not refer to the copyright notice or licensing information for the Work. Such credit may be implemented in any reasonable manner; provided, however, that in the case of a Collective Work, at a minimum such credit will appear where any other comparable authorship credit appears and in a manner at least as prominent as such other comparable authorship credit.

d. For the avoidance of doubt, where the Work is a musical composition:

    i. Performance Royalties Under Blanket Licenses. Licensor reserves the exclusive right to collect, whether individually or via a performance rights society (e.g. ASCAP, BMI, SESAC), royalties for the public performance or public digital performance (e.g. webcast) of the Work if that performance is primarily intended for or directed toward commercial advantage or private monetary compensation.

    ii. Mechanical Rights and Statutory Royalties. Licensor reserves the exclusive right to collect, whether individually or via a music rights agency or designated agent (e.g. Harry Fox Agency), royalties for any phonorecord You create from the Work ("cover version") and distribute, subject to the compulsory license created by 17 USC Section 115 of the US Copyright Act (or the equivalent in other jurisdictions), if Your distribution of such cover version is primarily intended for or directed toward commercial advantage or private monetary compensation.

e. Webcasting Rights and Statutory Royalties. For the avoidance of doubt, where the Work is a sound recording, Licensor reserves the exclusive right to collect, whether individually or via a performance-rights society (e.g. SoundExchange), royalties for the public digital performance (e.g. webcast) of the Work, subject to the compulsory license created by 17 USC Section 114 of the US Copyright Act (or the equivalent in other jurisdictions), if Your public digital performance is primarily intended for or directed toward commercial advantage or private monetary compensation.

5. Representations, Warranties and Disclaimer

UNLESS OTHERWISE MUTUALLY AGREED BY THE PARTIES IN WRITING, LICENSOR OFFERS THE WORK AS-IS AND MAKES NO REPRESENTA-TIONS OR WARRANTIES OF ANY KIND CONCERNING THE WORK, EXPRESS, IMPLIED, STATUTORY OR OTHERWISE, INCLUDING, WITH-OUT LIMITATION, WARRANTIES OF TITLE, MERCHANTIBILITY, FITNESS FOR A PARTICULAR PURPOSE, NONINFRINGEMENT, OR THE ABSENCE OF LATENT OR OTHER DEFECTS, ACCURACY, OR THE PRESENCE OF ABSENCE OF ERRORS, WHETHER OR NOT DISCOVERABLE. SOME JURIS-DICTIONS DO NOT ALLOW THE EXCLUSION OF IMPLIED WARRANTIES, SO SUCH EXCLUSION MAY NOT APPLY TO YOU.

6. Limitation on Liability.

EXCEPT TO THE EXTENT REQUIRED BY APPLICABLE LAW, IN NO EVENT WILL LICENSOR BE LIABLE TO YOU ON ANY LEGAL THEORY FOR ANY SPECIAL, INCIDENTAL, CONSEQUENTIAL, PUNITIVE OR EXEMPLARY DAMAGES ARISING OUT OF THIS LICENSE OR THE USE OF THE WORK, EVEN IF LICENSOR HAS BEEN ADVISED OF THE POSSIBILITY OF SUCH DAMAGES.

7. Termination

a. This License and the rights granted hereunder will terminate automatically upon any breach by You of the terms of this License. Individuals or entities who have received Collective Works from You under this License, however, will not have their licenses terminated provided such individuals or entities remain in full compliance with those licenses. Sections 1, 2, 5, 6, 7, and 8 will survive any termination of this License.

b. Subject to the above terms and conditions, the license granted here is per-petual (for the duration of the applicable copyright in the Work). Notwith-standing the above, Licensor reserves the right to release the Work under different license terms or to stop distributing the Work at any time; pro-vided, however that any such election will not serve to withdraw this License (or any other license that has been, or is required to be, granted under the terms of this License), and this License will continue in full force and effect unless terminated as stated above.

8. Miscellaneous

a. Each time You distribute or publicly digitally perform the Work or a Collec-tive Work, the Licensor offers to the recipient a license to the Work on the same terms and conditions as the license granted to You under this License.

b. If any provision of this License is invalid or unenforceable under applicable law, it shall not affect the validity or enforceability of the remainder of the terms of this License, and without further action by the parties to this agreement, such provision shall be reformed to the minimum extent necessary to make such provision valid and enforceable.

c. No term or provision of this License shall be deemed waived and no breach consented to unless such waiver or consent shall be in writing and signed by the party to be charged with such waiver or consent.

d. This License constitutes the entire agreement between the parties with respect to the Work licensed here. There are no understandings, agreements or representations with respect to the Work not specified here. Licensor shall not be bound by any additional provisions that may appear in any communication from You. This License may not be modified without the mutual written agreement of the Licensor and You.

Creative Commons is not a party to this License, and makes no warranty whatsoever in connection with the Work. Creative Commons will not be liable to You or any party on any legal theory for any damages whatsoever, including without limitation any general, special, incidental or consequential damages arising in connection to this license. Notwithstanding the foregoing two (2) sentences, if Creative Commons has expressly identified itself as the Licensor hereunder, it shall have all rights and obligations of Licensor.

Except for the limited purpose of indicating to the public that the Work is licensed under the CCPL, neither party will use the trademark "Creative Commons" or any related trademark or logo of Creative Commons without the prior written consent of Creative Commons. Any permitted use will be in compliance with Creative Commons' then-current trademark usage guidelines, as may be published on its website or otherwise made available upon request from time to time.

Creative Commons may be contacted at http://creativecommons.org/.

This license may be found at: http://creativecommons.org/licenses/by-nc-nd/2.5/

Columns from Slashdot

The chapter written by Bates and Stone refers to several columns that appeared in Slashdot around the time of certain key events. Those columns are republished here.

Simple Solutions

Contributed by CmdrTaco on Mon Jan 12 at 8:50AM EST

From the editorials dept

This is the first of hopefully many Slashdot.org Editorials. In addition to just reporting the news, the Slashdot Team really wishes to try to put out new ideas, or share other information that our readers may find helpful, interesting, or entertaining.

We're standing at an amazing crossroad here. The Free Software Foundation, and especially the *Linux* OS have gained amazing ground. The mainstream press (e.g. the ZiffDavis marketing monopoly) actually now regularly acknowledge Linux along side MacOS and Windows as being a "Real" Operating system.

And then there is the browser world, where the race was once one horse, then hundreds, and now 2. Microsoft and Netscape have been battling it out for some time now, and Netscape's once unstoppable 70% market share has begun crumbling.

Meanwhile the Free Software world is facing a battle of its own. The Commercial browser world has been reduced to the big ones, but the free world is producing Mneumonic, Gzilla, and various other smaller projects. Many talented programmers slave away on these products, but each day, Microsoft gains ground.

Add the final piece of data to the mix:Netscape is losing money as well as browser market share. What's a company to do? Maybe the solution is simple:GPL Netscape's Source Code.

So now that you've stopped laughing, let's talk about this seriously for a moment. Let's look at why Netscape should seriously consider this:

Talented programmers from around the world would actively improve Netscape's browser. The Free Software Movement has proven that if some control is enforced at the center (eg Linus) programs can develop communally. Netscape would not have to pay most of the development cost of their software. Coordination, and key programmers would be essential, but minor once coders around the world join in.

Netscape needs browser dominance to fuel its server market, and to remain synonymous with the Internet. If current trends continue, MS will = the Internet in another year.

Netscape is losing money on the browser market. They need to release their browser for free to compete with Microsoft anyway.

Source code would allow compilation on other systems- say a Pentium optimized version, or whatever other optimizations become available for platform X.

Excellent Publicity generated by such an original move would earn Netscape respect from the Free Software junkies who often have somewhat negative feelings towards Netscape. These Free Software Junkies are gaining control of much of the world's IS departments, and Netscape's good name will get them places in these corporate worlds.

So that's all well and good for Netscape, but what about the rest of us. Netscape has taken a lot of heat for its gapping shortcomings. In particular its bloated size and slow performance. Why would the Free Software World want to take on this project?

GPL means we would have a state of the art free browser.

Netscape could be ported to GTK or Qt for faster performance and lower memory requirements than Motif.

Various web browser efforts could focus on a single project (which could have many faces) which already has so many of the features they need. Instead of these projects dividing the effort, they could unify.

The superior programming talents of the world's programmers would make Netscape the superior browser, which would win over converts back from Microsoft even on Wintel boxes where MS is gaining support.

New browsers derived from Netscape for more specific tasks could share things like an HTML rendering engine for commonality.

Now I realize that there are problems. Large parts of Netscape's code aren't really Netscape's to give away. The "about:" screen of Netscape Communicator lists 12 companies besides Netscape including Apple, Macromedia, Symantec and many others. Perhaps these modules are removed. Perhaps these modules could also GPLd. Netscape does need to maintain the primary code base, and finding someone with the charisma of Linus to steerhead the development of code from hundreds of people will me a challange. Then there are problems with large portions of the Free Software world disliking Netscape. I really hope this could change, especially if they were given the opportunity to maintain it.

I really think this could be the answer to a lot of problems. With the power of an Internet full of programmers, even Microsoft's Billions of R&D dollars would be threatened. And we would be guaranteed a real choice even if IE4 becomes the standard on Windows boxes.

What do you think?

—ROB "CMDRTACO" MALDA

Why Kids Kill

Posted by JonKatz on Fri Apr 23, '99 10:00 AM

from the hysteria-on-the-net dept.

Nightmarish high school massacres like the one in Littleton are now an almost ritualistic part of American life. And increasingly when they occur, journalists and educators blame new media like the Internet, computer games like Doom or violent movies. Why kids kill this way is an urgent and complicated question. But teenaged crime isn't rising, it's falling. And there's no evidence that the Net or other new media are the reason for massacres.

The images were familiar, yet surreal.

Media reports of books about "Doom," animated clips from the computer game, TV shots of websites with ugly images, ominous reports of heavy metal bands and film clips of "Natural Born Killers."

"What is known," said a CNN correspondent Wednesday night, "is that the members of the Trench Coat Mafia spent a lot of time playing computer games on the Internet." They had become obsessed with online killing, reported another TV reporter. They had delved into militia and hate-group websites, some papers said.

The fallout was, as always, nearly instantaneous.

In Vancouver, Washington, e-mailed Enzo Falzon, high school students were pulled aside as they came through the front door and told they weren't allowed to wear trenchcoats. In a Philadelphia suburb, e-mailed Tim, (who asked that his last name remain anonymous), kids who play Doom were offered counseling. In Maine, e-mailed Vektor, who's 14, his parents made him open his private computer files so they could look through and make sure he wasn't doing anything "anti-social."

By now, this schoolyard nightmare is as ritualistic as it is horrific.

We see televised scenes of kids running and sobbing, of SWAT teams creeping through schools and bloodied bodies carted out—followed by dark reports about hate on the Net, violence on TV and in movies. Everyone seems bewildered, uncomprehending.

Almost always, we are as confused as we are horrified, since young killers take their own lives or offer no coherent explanation, leaving us with questions but not answers. Since there are rarely trials, there is rarely any resolution, any understanding.

In June of 1988, writing for Hotwired, I wrote a column called "Why Kids Kill" after Kipland Kinkel of Springfield, Oregon, killed four people, including his parents, and wounded 22 more.

Not much has changed a year later, especially when it comes to knee-jerk, ignorant stereotypes from the media and from educators about kids, the Net, geeks and the violence allegedly inspired by the digital screen culture.

Federal agencies and academics studying this kind of episodic, uniquely American massacre, find little of any, connection between murders and media, digital or otherwise.

Kids being warned and counseled by fearful administrators and teachers ought to know that overall, teenage violence is way down in America, at its lowest levels since the Depression. In supposedly media-saturated, violent urban areas like New York City, Chicago and LA, schoolyard massacres are unknown. Nor has one ever occurred in Canada, even though Canadian kids watch almost the same media as American kids, and use the Net in even greater numbers.

What do we know about these horrible eruptions? Almost all of the killers have been white, teenaged males who are emotionally disturbed. Almost all lived in suburban or rural areas, the children of working or middle-class families. They've been generally described as well-parented.

And in almost single case, nobody really knows why they did what they did. They suffered various forms of social cruelty and exclusion, as so many of their peers also have, and they got their hands on especially lethal weaponry, particularly guns. Almost always, their friends and classmates and teachers are stunned and disbelieving. Some of the shooters have been avid media and computer users. Others weren't.

According to federal statistics, no school shootings occurred in 1994; in 1997, there were four incidents. In 1998, apart from the Springfield killings, an 11-year-old-old boy and his 13-year-old friend were charged with killing four students and a teacher and wounding 10 others in Jonesboro, Arkansas. A high-school senior shot and killed a student in a parking lot in Fayetteville, Tennessee. In Edinboro, Pennsylvania, a 14-year-old boy was accused of killing a teacher and wounding two students and another teacher at an eighth grade graduation. Two days later, a 15-year-old girl was shot in the leg in suburban Houston high-school classroom. In Washington, a 15-year-old boy got off his school bus carrying a gun, then went home and shot himself in the head. Now there is Littleton, Colorado, 1999's first school massacre, with at least fifteen dead.

Although experts, therapists and sociologists have crammed TV talk shows to offer various theories about the contagion of teenage violence, it is clear that no one yet understands why these incidents occur. Sociologists like Elaine Showalter of Princeton have written about media hysterias, contagions transmitted by the speed and power of media imagery in stories about the killings themselves. Some psychologists believe that when disturbed kids see the massive amount of media attention these shootings get, they begin fantasizing about this kind of attention being focused on their own, often unhappy, lives.

Other experts blame the availability of guns. Obviously, the ready availability of lethal weapons is significant in this kind of violence, but crime among teenagers has been plummeting for years now, even as the number of guns in the United States has risen.

And persistent efforts by journalists to link the massacres to hate-sites on the Net or to games like "Doom" and, before that, to "Dungeons & Dragons" don't hold up either. There are no consistent patterns of media behavior to link these killers, no single trait of movie-going, gaming or Net use.

Tens of millions of kids all over the world play computer games. The biggest users of new media recreational technologies are middle-class kids, since they have the money to afford the technology. Yet violence among this group, never very high, again has been plummeting even as online use has mushroomed.

Yet despite the confusion about the cause of these killings, all across America, newspapers and TV stations are warning parents about computer games, suggesting that their sons and daughters might be secretly turning into potential mass murderers online.

This is willful ignorance. There's no mystery about the greatest dangers to children. Every day, writes Don Tapscott in Growing Up Digital, three children in the United States are murdered or die as a result of injuries inflicted by their parents or caretakers. Of the annual three million reported cases of child abuse, 127,000 cases involve child abandonment. Each year, and throughout the 90's, the National Center for Missing and Exploited Children reports only a handful of child abuse cases related to the Internet. Of the 23 cases tracked from March 1996, to March, 1997, 10 involved the transfer of pornography, an adult soliciting sexual favors from minors, or sexual contact initiated over the Net. Of the remaining 13 cases, two involved police officers posing as children, and in two others the girls had previous histories as runaways. Nine others involved children over age 16 running away from home, allegedly to meet online acquaintances.

What these statistics indicate, Tapscott says, is that "children are 300,000 times more likely to be abused by their own relatives than by someone they have met over the Internet."

As horrific as massacres like Littleton are, they are also extraordinarily rare. Statistically, children are more likely to have an airplane fall out of the sky and kill them than they are to be shot in school, despite the staggering amount of media coverage.

Sissella Bok of Harvard, whose book Mayhem examined the effects of violence in media, writes that young people's lives are saturated with graphic violence in a way that's different and more dangerous than in previous generations.

"We have movie role models showing violence as fun, and video games where you kill, and get rewarded for killing, for hours and hours." It is, she wrote, a "very combustible mix, enraged young people with access to semiautomatic weapons, exposed to violence as entertainment, violence shown as exciting and thrilling."

There's no question that violent imagery is ubiquitous in screen culture, from gaming to TV. But these comparisons seem facile and unknowing. Gaming is intensely creative, in some contexts—Quake 3, Unreal, Ultima—almost approaching a new art form. The animation is rich and multi-dimensional, and violence is stylized, often presented more as a strategic challenge like chess than anything truly brutal or graphically violent. If the stylization of violence is a problem, it doesn't show up anywhere in crime or violence statistics involving computer users.

If Bok is right, it would. Why would there be a decline in youth violence even as "violent imagery" in the media has indeed increased, along with Web use, cable's share of audience, rap and hip-hop (also supposed to be inducing the young to violence), and movie attendance?

More relevant questions might be: Why are so many of these killers male and middle-class, rather than the poor or the underclass? Why do these assaults occur almost exclusively in rural or suburban areas? Why are these kids able to hide even severe emotional disturbance from the people closest to them?

Perhaps the most shocking thing about massacres like Littleton is that, for all of the massive amounts of coverage brought to bear on them, there really isn't anything approaching a consensus about why they occur. Since educators and authorities don't know what to do, what they tend to do is dumb.

Since the kids they're supposed to be protecting know quite well that wearing trench coats, going online or watching movies isn't dangerous in and of itself, mostly what educators and journalists end up demonstrating to kids is that they're clueless.

—JON KATZ

Index

biotechnology
    commercial, beginnings of, 283
    intellectual property and growing
        challenges, 285–287
    modern, the rise of, 282–284
    patents aggressively sought
        for, 284
Biotechnology Industry Organization
    (BIO), 284
bioweapon threats, 292
BitKeeper, 29
Blizzard and the bnetd project, 152
blockbuster drugs, 286
Blockstackers, 374–375, 383, 386
blog entry about open source
    security, 57–59
BluePoint (Chinese Linux
    distributor), 201
bnetd project (Blizzard), 152
Bok, Sissella, 422
Bomis.com, 310–312, 330
bootstrapping product
    complements, 132
Boston Consulting Group, surveys done
    by, 127
"both source" business model, 114
Boyer, Herbert, 282
branching in version control
    systems, 28
Brand, Stewart, 237, 242
Brazil
    developing software livre
        movement in, 214
    Java-based tools used in, 226
    livre vs. gratis (terms meaning
        free), 212
    market issues and software livre
        movement, 213
    Microsoft vs. FOSS
        movement, 216
Brent, Roger, 291
Broadcast Flag rule (FCC), 150, 158
Brooks, Frederick, 32, 363, 365
browser wars, 4
BSD License, 401
Buchanan, J. M., 355

bugs
    in open source and proprietary
        software, 32
    vs. security vulnerabilities, 60
bug-tracking system for Mozilla, 6
Bugzilla, 6
Building 20 at MIT (a Low Road
        building), 244
The Bunker, 57
Bureaucracy Bottleneck, roadblock to
        company growth, 108
Burton Matrix, 247, 248
Burton, Craig, 234, 237, 246
Burton, Michael, 302
Bush, Vannevar, 282
business and Freemacs, 139–141
business and politics, 71
business models
    ASP model, 117
    "both source" model, 114
    code-level service model, 118
    dual-license model, 116
    for Linux, 97–99
        consequences of future
            directions, 100–102
    for open source, 113–118
        in China, 202
    managed source model, 118
    marketplace view of, 127–129
    mixed source model, 114
    professional open source
        model, 115
    services model, 115
business tutorial, 143
byte range locks, setting up in
        POSIX, 41–43

C

CALIBRE project, 175, 187
Campbell, B., 352
Campbell-Kelly, Martin, 298, 305
Campeau, Liz, 313
canned libraries, problems with, 24
capabilities, 67
CAPerl project, 67

culture of secrecy in early computer industry, 298
culture of wikis, 315
culture of work in India, challenging to OSS adoption, 194
Cunningham, Ward, 315
Curry, Adam, 234
customer-centric solution networks, 132, 135
customizability of software
    software
        customizability of, 255, 265–270
Cutler, Dave, 46
CVE (Common Vulnerabilities and Exposures) project, 70
CVS (concurrent versioning system), 28

D

"Daddy Pants" (Slashdot job description), 392
DaimlerChrysler sued by SCO, 251
Danish strategies for libre software, 178
Darby, Newman, 344, 348
"The Darker Bioweapons Future", 292
DARPA (Defense Advanced Research Projects Agency), 58
David, Paul A., 164
Davis, Michael, 310
de jure vs. de facto standards, 123
de Mello, Ricardo "Gandhy", 226
de novo code, 291
Debian project, countries of origin of developers, 164
decommoditization, 94, 101
    of Linux platform, by Red Hat, 100
Decrem, Bart, 15
DELETE permission, 49
Dell Computer
    in China, 205
    making money from free software, 96
Dell, Michael, 254
DeMaagd, Kurt, 374
demand-side developments (Linux and open source), 232
Demarco, Tom, 30
denial-of-service (DoS) attacks suffered by Apache, 58, 62
Dent, Chris, 305
Department of Information Technology (D-IT), 193

dependencies and software distribution, 31
design principles, 372
    for referee function, 368–371
    of "communities of practice", 365–367
development community of libre software, 163–165
development styles, changing from proprietary to open source models, 6
DG-INFO (Directorate General on Information Society), 172
Diamond v. Chakrabarty, 283
DiBona, Chris, xv, 21–36, 355
Digital Computers Association (DCA), 299
Digital Millennium Copyright Act (DMCA), 35, 240, 354
    anticircumvention/antitools provisions of, 154–156
digital rights management (DRM), 156
digital video discs and content scrambling system, 155
dimension-of-merit innovations, 342
directive on software patents (EU), 181
Direto (email and collaboration tool), 226
discoveries of new uses for products, 341–343
discussion forums, communicating many to many, 373–396
disintermediation, 33, 394
disrupters of open source projects, dealing with, 276
disruptive technology, treating OSS as, in China, 207
disruptive vendors, 107–113
distributed computation, 263, 269
distributed development, 27–29
distribution strategies
    freeware vs. open source, 109–113
    vs. development strategies, 72
DKUUG (Danish Unix User Group), 167
DLLs and Win32 API, 47
DMCA (see Digital Millennium Copyright Act)
DNA constructs, manipulating, 282
DNA, synthetic, 281
    future trends for, 293
    risk of biological hacking, 292
DNS (Domain Name System), 389
Dobson, John, 352
Doctorow, Cory, 236

R

R&D processes and drug development, 285
Radio Userland, 274
Rai, Arti, 288
RAND (reasonable and non-discriminatory) terms
    Java technology and, 220
    making patents available on, 123
Rangaswami, J. P., 252
Raymond, Eric S., ix, 60, 125, 238, 256, 259, 270, 366, 395
READ_CONTROL permission, 49
Real Programmers, ix
Really Simple Syndication (RSS), 234
recalling drugs with dangerous side effects, 287
reciprocal licenses, 75, 83
reciprocity and dual licensing, 76
recombinant DNA technology, 282
Recording Industry Association of America (RIAA)
    challenging conference paper, 157
recovery effort at World Trade Center, 301–303
Red Flag (Chinese Linux distributor), 200
Red Hat, 99
    consequences of new business model, 100–102
    distributing Linux in China, 201–203
    Enterprise Linux, 99
Reed, D. P., 233
refactoring code, 24
referee function, design principles for, 368–371
reframing and origins of innovation, 352
The Register news site, 165
regulatory vs. voluntary standards, 123
Reimer, Neils, 282
Reingold, Howard, 392
Release Candidates, 12
repository of refereed information, 369
repository of source code
    controlling, 9
    having write access to, 10
reputation property vs. traditional intellectual property, 106
reputation, importance of, in hacker culture, 145, 235–236

research into encryption, threatened by DMCA, 156–158
research on libre software in Europe, 185–187
research process vs. development process, 285
reverse engineering and copyright law, 152
rewriting code, 25
RFCs (Requests for Comments), 266
Richey, Jason, 314
Richie, Dennis, 38
rigid hierarchy in work culture of India, 194
The Rise of the Stupid Network, 233
risk assessment ability, 65
Robles, Gregorio, xx, 161–188
Rooney, Paula, 121
RSS (Really Simple Syndication), 234
Ruffolo, Robert Jr., 286

S

Sabel, Charles, 371
Sali, Andrej, 288
SALOMÉ platform, 170
Saltzer, J. H., 233
Samba, 37–55
Samuelson, P., 354
Sanger, Larry, xx, 307–338
SAP
    in China, 206
    participating in OSS development, 131
Saville Row tailor, weblog of, 235
Saxena, Sunil, xxi, 197–210
scaling issues for Slashdot, 388
scaling of projects, 32
Scheider, Hendrik, 163
Schumpeter, E. F., 257
Schumpeter, Joseph, 353, 364
SCO and the FUD (Fear, Uncertainty, and Doubt) effects of its litigation, 251
SCO litigation and Groklaw, 273–280
search algorithms, choosing, 367
Searls, Doc, xxi, 231–252
secondary liability for copyright infringement, 153
secrecy necessary for legal research, 278
Secure Digital Music Initiative (SDMI), 156
security
    designing within referee systems, 371
    future of, 66–69

windsurfing industry
    commercialization of, 346–347
    developing equipment innovations
        in, 340
    importance of community-based
        innovation, 347–350
Wine project, 48, 54
winelib shared library, 54
Winer, Dave, 234
Winter, S. G., 353
Winterstick, 346
workstations being replaced by PCs, 94, 134
world languages, applications written in, 52
"World of Ends" by Searls &
        Weinberger, 233
World Trade Center recovery effort, 301–
        303
write access to source code repository, 10
WRITE_DAC permission, 49

X

XChat (IRC client), 373, 390
XPCOM (cross-platform component
        model), 7
Xteam (Chinese Linux distributor), 201
XUL (cross-platform XML-based UI
        language), 7

Y

Yankee Group, 111
YAPC::Europe, 166
Yeo, Boon-Lock, xxiii, 197–210
Yoon, Y. J., 355
Young, Bob, 257, 260

Z

Zawinski, Jamie, 5
Ziedonis, R. H., 354

Colophon

JAMIE PEPPARD was the production editor and proofreader for *Open Sources 2.0*. Audrey Doyle was the copyeditor. Adam Witwer and Claire Cloutier provided quality control. Judy Hoer wrote the index. Loranah Dimant, Jansen Fernald, and Lydia Onofrei provided production assistance.

MIKE KOHNKE designed the cover of this book. Karen Montgomery produced the cover layout in Adobe InDesign CS using Akzidenz Grotesk and Orator fonts.

MIKE KOHNKE designed the interior layout. This book was converted to FrameMaker 5.5.6 by Andrew Savikas. The text font is Adobe's Meridien; the heading font is ITC Bailey. The illustrations that appear in the book were produced by Robert Romano, Jessamyn Read, and Lesley Borash using Macromedia FreeHand MX and Adobe Photoshop CS and using the ORA hand font.

Better than e-books

Buy *Open Sources 2.0* and access the
digital edition FREE on Safari for 45 days.

Go to www.oreilly.com/go/safarienabled
and type in coupon code B7LN-WNJM-4J13-5IM3-XHSP

Search
thousands of
top tech books

Download
whole chapters

Cut and Paste
code examples

Find
answers fast

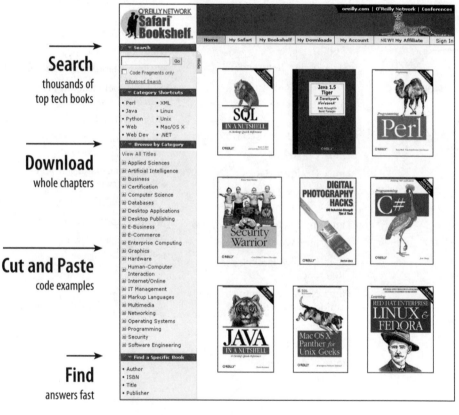

Search Safari! The premier electronic reference
library for programmers and IT professionals.

Related Titles from O'Reilly

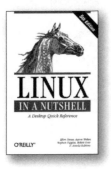

Linux

Building Embedded Linux Systems

Building Secure Servers
with Linux

The Complete FreeBSD,
4th Edition

Even Grues Get Full

Exploring the JDS Linux
Desktop

Extreme Programming Pocket
Guide

GDB Pocket Reference

Knoppix Hacks

Knoppix Pocket Guide

Learning Red Hat Enterprise
Linux and Fedora, *4th Edition*

Linux Cookbook

Linux Desktop Hacks

Linux Device Drivers,
3rd Edition

Linux in a Nutshell, *5th Edition*

Linux in a Windows World

Linux iptables Pocket
Reference

Linux Network Administrator's
Guide, *3rd Edition*

Linux Pocket Guide

Linux Security Cookbook

Linux Server Hacks

Linux Unwired

Linux Web Server CD
Bookshelf, *Version 2.0*

LPI Linux Certification in a
Nutshell

Managing RAID on Linux

More Linux Server Hacks

OpenOffice.org Writer

Programming with Qt,
2nd Edition

Root of all Evil

Running Linux, *5th Edition*

Samba Pocket Reference,
2nd Edition

Test Driving Linux

Understanding the Linux
Kernel, *3rd Edition*

Understanding Open Source &
Free Software Licensing

User Friendly

Using Samba, *2nd Edition*

Version Control with
Subversion

Keep in touch with O'Reilly

Download examples from our books

To find example files from a book, go to: *www.oreilly.com/catalog* select the book, and follow the "Examples" link.

Register your O'Reilly books

Register your book at *register.oreilly.com* Why register your books? Once you've registered your O'Reilly books you can:

- Win O'Reilly books, T-shirts or discount coupons in our monthly drawing.
- Get special offers available only to registered O'Reilly customers.
- Get catalogs announcing new books (US and UK only).
- Get email notification of new editions of the O'Reilly books you own.

Join our email lists

Sign up to get topic-specific email announcements of new books and conferences, special offers, and O'Reilly Network technology newsletters at:

elists.oreilly.com

It's easy to customize your free elists subscription so you'll get exactly the O'Reilly news you want.

Get the latest news, tips, and tools

www.oreilly.com

- "Top 100 Sites on the Web"—PC Magazine
- CIO Magazine's Web Business 50 Awards

Our web site contains a library of comprehensive product information (including book excerpts and tables of contents), downloadable software, background articles, interviews with technology leaders, links to relevant sites, book cover art, and more.

Work for O'Reilly

Check out our web site for current employment opportunities:

jobs.oreilly.com

Contact us

O'Reilly Media, Inc.
1005 Gravenstein Hwy North
Sebastopol, CA 95472 USA
Tel: 707-827-7000 or 800-998-9938
    (6am to 5pm PST)
Fax: 707-829-0104

Contact us by email

For answers to problems regarding your order or our products:
order@oreilly.com

To request a copy of our latest catalog:
catalog@oreilly.com

For book content technical questions or corrections: **booktech@oreilly.com**

For educational, library, government, and corporate sales: **corporate@oreilly.com**

To submit new book proposals to our editors and product managers:
proposals@oreilly.com

For information about our international distributors or translation queries:
international@oreilly.com

For information about academic use of O'Reilly books:
adoption@oreilly.com
or visit:
academic.oreilly.com

For a list of our distributors outside of North America check out:
international.oreilly.com/distributors.html

Order a book online

www.oreilly.com/order_new

Our books are available at most retail and online bookstores.
To order direct: 1-800-998-9938 • *order@oreilly.com* • *www.oreilly.com*
Online editions of most O'Reilly titles are available by subscription at *safari.oreilly.com*